IET ELECTROMAGNETIC WAVES SERIES 23

Series Editors: Professor P.J.B. Clarricoats
Professor E.D.R. Shearman
Professor J.R. Wait

Ferrites at Microwave Frequencies

Other volumes in this series:

Volume 1	**Geometrical theory of diffraction for electromagnetic waves, 3rd edition** G.L. James
Volume 10	**Aperture antennas and diffraction theory** E.V. Jull
Volume 11	**Adaptive array principles** J.E. Hudson
Volume 12	**Microstrip antenna theory and design** J.R. James, P.S. Hall and C. Wood
Volume 15	**The handbook of antenna design, volume 1** A.W. Rudge, K. Milne, A.D. Oliver and P. Knight (Editors)
Volume 16	**The handbook of antenna design, volume 2** A.W. Rudge, K. Milne, A.D. Oliver and P. Knight (Editors)
Volume 18	**Corrugated horns for microwave antennas** P.J.B. Clarricoats and A.D. Oliver
Volume 19	**Microwave antenna theory and design** S. Silver (Editor)
Volume 21	**Waveguide handbook** N. Marcuvitz
Volume 23	**Ferrites at microwave frequencies** A.J. Baden Fuller
Volume 24	**Propagation of short radio waves** D.E. Kerr (Editor)
Volume 25	**Principles of microwave circuits** C.G. Montgomery, R.H. Dicke and E.M. Purcell (Editors)
Volume 26	**Spherical near-field antenna measurements** J.E. Hansen (Editor)
Volume 28	**Handbook of microstrip antennas, 2 volumes** J.R. James and P.S. Hall (Editors)
Volume 31	**Ionospheric radio** K. Davies
Volume 32	**Electromagnetic waveguides: theory and applications** S.F. Mahmoud
Volume 33	**Radio direction finding and superresolution, 2nd edition** P.J.D. Gething
Volume 34	**Electrodynamic theory of superconductors** S.A. Zhou
Volume 35	**VHF and UHF antennas** R.A. Burberry
Volume 36	**Propagation, scattering and diffraction of electromagnetic waves** A.S. Ilyinski, G. Ya.Slepyan and A. Ya.Slepyan
Volume 37	**Geometrical theory of diffraction** V.A. Borovikov and B.Ye. Kinber
Volume 38	**Analysis of metallic antenna and scatterers** B.D. Popovic and B.M. Kolundzija
Volume 39	**Microwave horns and feeds** A.D. Olver, P.J.B. Clarricoats, A.A. Kishk and L. Shafai
Volume 41	**Approximate boundary conditions in electromagnetics** T.B.A. Senior and J.L. Volakis
Volume 42	**Spectral theory and excitation of open structures** V.P. Shestopalov and Y. Shestopalov
Volume 43	**Open electromagnetic waveguides** T. Rozzi and M. Mongiardo
Volume 44	**Theory of nonuniform waveguides: the cross-section method** B.Z. Katsenelenbaum, L. Mercader Del Rio, M. Pereyaslavets, M. Sorella Ayza and M.K.A. Thumm
Volume 45	**Parabolic equation methods for electromagnetic wave propagation** M. Levy
Volume 46	**Advanced electromagnetic analysis of passive and active planar structures** T. Rozzi and M. Farinai
Volume 47	**Electromagnetic mixing formulae and applications** A. Sihvola
Volume 48	**Theory and design of microwave filters** I.C. Hunter
Volume 49	**Handbook of ridge waveguides and passive components** J. Helszajn
Volume 50	**Channels, propagation and antennas for mobile communications** R. Vaughan and J. Bach-Anderson
Volume 51	**Asymptotic and hybrid methods in electromagnetics** F. Molinet, I. Andronov and D. Bouche
Volume 52	**Thermal microwave radiation: applications for remote sensing** C. Matzler (Editor)
Volume 502	**Propagation of radiowaves, 2nd edition** L.W. Barclay (Editor)

Ferrites at Microwave Frequencies

A.J. Baden Fuller

The Institution of Engineering and Technology

Published by The Institution of Engineering and Technology, London, United Kingdom

First published 1987
Reprinted 2005, 2008

The Institution of Engineering and Technology
Michael Faraday House
Six Hills Way, Stevenage
Herts, SG1 2AY, United Kingdom

www.theiet.org

British Library Cataloguing in Publication Data
Baden Fuller, A.J.
Ferrites at microwave frequencies –
(Electromagnetic waves series no. 23).
1. Ferrite devices 2. Microwave devices
I. Title II. Series
621.381'3 TK7872.F4

ISBN (10 digit) 0 86341 064 2
ISBN (13 digit) 978-0-86341-064-2

Printed in the UK at the University Press, Cambridge
Reprinted in the UK by Lightning Source UK Ltd, Milton Keynes

Contents

Preface

Between 1950 and 1965, there was a spate of intense activity to investigate the theory and applications of ferrite materials at microwave frequencies. In the early 1960s a number of workers wrote textbooks on the microwave applications of ferrites. Some books have been published subsequently which concentrate on particular microwave devices or specialise in one aspect of ferrites at microwave frequencies, but it is now 20 years since any comprehensive textbook on ferrites was published. The design of microwave ferrite devices has become a mature technology and wideband isolators and circulators are readily available for most microwave frequencies. However, new applications for ferrite materials in stripline and microstrip and for millimetre wave frequencies are also being investigated. There is the danger that each new generation of microwave engineers will not know the very wide range of applications of microwave ferrites that were investigated in the 1960s. A comprehensive book is needed to consolidate all the investigations of ferrites for microwave applications in the 1960s, to look back at earlier publications from the view point of a mature technology, and to bring the story up to date.

This book attempts to give *all* the structures and applications using ferrites at microwave frequencies that have been investigated or contemplated. In a new situation, an unlikely structure may be the one to use, for example, Faraday rotation circulators and isolators have been used at millimetre wave frequencies in the 1980s although in the late 1960s they appeared to be obsolescent and replaced by the much smaller Y-junction circulators and resonant isolators for use at normal microwave frequencies. Discussion is confined to electromagnetic wave interaction with ferrite material. There is no attempt to discuss magnetostatic wave or spinwave devices or effects.

The book can be divided into two parts. After an introductory chapter about the physics of magnetism and the chemistry of ferrites and garnets, the first part of the book describes low-power ferrite devices in wveguide; the second part looks at stripline devices and other microwave applications of ferrite material. For the mature science of ferrite in free space and in waveguide, the exact theory is given in Chapters 2, 3, 4, 6 and 12, and all likely applications of these waveguide structures to make useful devices are given in Chapters 5 and 7. Junction circulators and stripline devices have complicated shapes which have not yet been analysed completely, and so in the later chapters only a summary of the theory is given together with references to sources of further information.

After a quarter of a century of interest in microwave ferrite devices and their applications, it is difficult to thank all the many people from whom I have learned and I hope that lack of acknowledgment will not be taken to imply lack of gratitude. It is hoped that the references at the end of each chapter adequately represent the sources of information as well as providing suitable texts for further reading.

A.J. Baden Fuller

Chapter 1

Introduction

1.1 Introduction

Magnetism is an important part of electrical engineering, and iron and its alloys are the most abundantly used magnetic materials. However, eddy currents limit the usefulness of iron alloys at high frequencies, whereas ferrites are magnetic insulators which are able to provide magnetic cores for high frequency circuits. Ferrite materials are also transparent to radio frequency and microwave signals, and can be used to create a family of unique microwave circuit components that are nonreciprocal. Ferrite materials, with the exception of magnetite, do not occur naturally and have been created in the last 50 years. Those developed for use in radio and video circuits have a high loss at microwave frequencies, thus special low microwave loss ferrites have also been developed. The rest of this chapter discusses magnetism and the chemistry, structure and preparation of ferrite materials. The rest of this book gives a summary of the theory of the interaction between electromagnetic waves and low-loss insulating magnetic ferrites.

One typical nonreciprocal component is the circulator, whose circuit symbol is shown in Fig. 1.1. For the ideal circulator, a signal input to port 1 is connected to port 2 and nothing goes to port 3 and so on in the direction of the arrow. If a matched termination is connected to port 3, a signal input to port 1 is still transmitted to port 2 but a return signal to port 2 is absorbed in the matched termination and nothing is transmitted back to port 1. Such a device is an isolator. Alternatively, an isolator can be constructed in a simple two-port length of waveguide.

A typical use of a circulator is shown in Fig. 1.2, where the circulator enables the radar transmitter and receiver to use the same aerial. The circulator connects the transmitter to the aerial and then connects the return signal to the receiver. Practical radar systems are much more sophisticated than that shown in Fig. 1.2. For one thing, no practical circulator is ideal and additional protection is needed to shield the receiver from unwanted signals from the transmitter.

The microwave properties of a ferrite material are dependent on its magnetic state so an external d.c. magnetic field can be used to control microwave signals in amplitude or phase. Since 1955, a large number of different configurations of ferrite in waveguide or other transmission lines have been investigated to

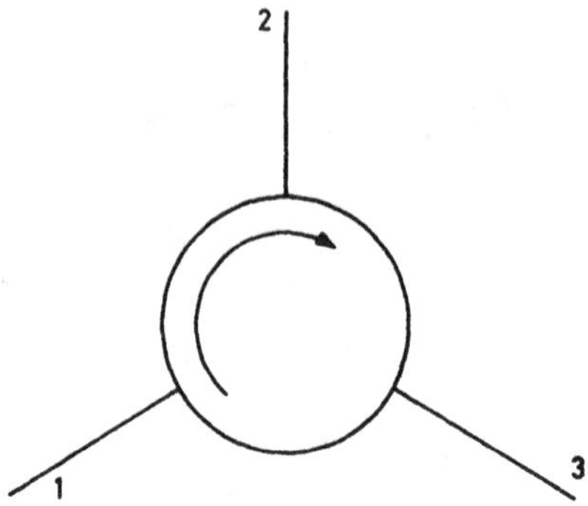

Fig. 1.1 *Circulator circuit symbol*

provide nonreciprocal or control functions. Most of these possible or contemplated applications of ferrite materials at microwave frequencies are discussed in this book.

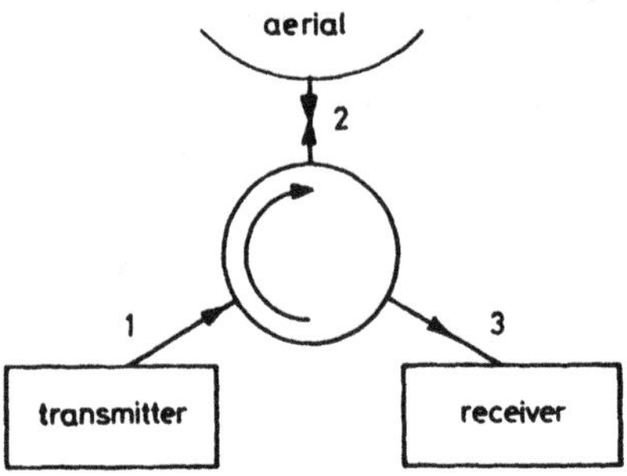

Fig. 1.2 *Simplified diagram of a radar system using a circulator*

1.2 Historical survey

Magnetism is probably that part of electrical science with the earliest practical applications. Lodestone, magnetite, the magnetic iron ore, Fe_3O_4, now known to be ferrous ferrite, was used at least 30 centuries ago. It is mentioned in the literature of ancient Greece by Thales of Miletus, in about 600BC, who noted its peculiar property of always pointing in a particular direction. The Chinese are thought to have used lodestone, floating in a bowl of water, as a compass. Peter

Peregrinus in 1269 gave the first detailed description of the floating compass, but he was obviously not describing a new invention. He also described a new pivoted compass furnished with a 360° scale marked with the cardinal points of the compass. In 1567 a compass factory was opened in London docks and the compass was used by most of the great explorers of the time.

Subsequently, it was found that iron and its alloys had a much stronger magnetic effect than lodestone, and electrical and magnetic research and applications concentrated on the magnetic metals. In the early 1900s it was appreciated that materials similar to magnetite might provide high resistance magnetic core materials, thus reducing eddy-current loss in inductors and transformers. Unfortunately, the materials produced that time still exhibited large losses and had only low permeability. The breakthrough really came with the work of Snoek and his co-workers after 1933 at the Philips Research Laboratories at Eindhoven in Holland. He first published the magnetic and electrical properties of the compounds $MO\ Fe_2O_3$ in 1936[1] where M can be Fe, Cu, Ni or Mg. Subsequently he made a systematic investigation of ferrites producing materials having low hysteresis, high resistivity and high permeability suitable for use as magnetic cores up to a few megahertz. Research continued during the war and culminated with the announcement of *ferroxcube* ferrite materials, suitable for use in radio circuits, in 1946.[2,3] A further book on the lower frequency properties of ferrites was published in the Philips' Technical Library by Smit and Wijn[4] in 1959.

Low-loss ferrites for the higher radio frequencies led to the investigation of their properties at microwave frequencies. In 1946, Griffiths[5] observed ferromagnetic resonance using thin films of iron, cobalt or nickel in a microwave resonant cavity. Polder[6] developed the theory of the interaction of a magnetically biased ferrite material with an electromagnetic wave. He gave the tensor form of the microwave permeability and showed how a plane wave propagating through an infinite ferrite medium would experience Faraday rotation or birefringence depending on the direction of the biasing magnetic field. Many people then began to investigate the microwave properties of ferrite materials. One very significant early paper was written by Hogan,[7] who used Faraday rotation to make a microwave gyrator and showed how it could make an isolator, circulator, switch, variable attenuator or modulator. Suhl and Walker[8] investigated the exact theoretical analysis of ferrite-filled circular waveguide, and Fox, Miller and Weiss[9] published a significant paper giving a very large number of possible configurations of ferrite in waveguide and their effect. Hogan edited a special issue of the Proceedings of the IRE[10] which was an excellent textbook on the theory and microwave applications of ferrites at the time. There were many outstanding papers in that issue and a number of them have been quoted as references later in this book. In the early 1960s a number of workers wrote textbooks on the microwave applications of ferrites,[11–16] that by Lax and Button[14] being the most comprehensive, but each having its strengths and weaknesses, as shown by the use of most of these books as references in different parts of this book. Subsequently, books were also written by von Aulock and Fay,[17] Helszajn[18–20] and Soohoo.[21]

Subsequent to the work by Snoek[1–3] in Holland, a large number of centres started investigating the manufacture of ferrite materials for different applications. A large number of different materials were found to be suitable for microwave applications. The information available up to 1964 is given by von Aulock.[22] Subsequent information is given by Nicolas[23] and Gilleo.[24]

This section has only been a short summary. A much fuller history of ferrite research and applications has been given by Lax and Button as Chapter 1 of their book[14] and more recently by Button.[25]

1.3 Magnetism

Magnetism is represented by the two field quantities $\boldsymbol{B}$ and $\boldsymbol{H}$. $\boldsymbol{B}$ is the magnetic flux density and is measured by the force it exerts on a wire carrying an electric current. $\boldsymbol{H}$ is the magnetic field intensity and is calcualted from the electric current which is generating the magnetic field. In the magnetic material, the relationship between the two is given by

$$\boldsymbol{B} = \mu_0 \boldsymbol{H} + \boldsymbol{M} \tag{1.1}$$

where μ_0 is the permeability constant and $\boldsymbol{M}$ is the magnetisation in the material. It will be noticed that I have chosen to measure $\boldsymbol{M}$ in units of flux density rather than in units of field intensity for reasons given elsewhere.[26] In terms of permeability, eqn. 1.1 becomes

$$\boldsymbol{B} = \mu \boldsymbol{H} = \mu_0 \mu_r \boldsymbol{H} \tag{1.2}$$

and the susceptibility is given by

$$\chi = \boldsymbol{M}/\boldsymbol{H} \tag{1.3}$$

In classical terms, the magnetic properties of materials are usually divided into the three categories: diamagnetic, paramagnetic, and ferromagnetic. Ferromagnetism can then be further divided into the categories of ferromagnetism, antiferromagnetism and ferrimagnetism. Diamagnetism is a consequence of the orbital motion of the electron and is a very weak effect. For the purposes of electromagnetic radiation fields in materials, it may be ignored. Paramagnetism arises from the magnetic effect of the orbital motion of the electron and its spin. From classical theory of the orbiting electron, we can equate the electrostatic force of attraction to the nucleus with the centripetal force, giving

$$\frac{e^2}{4\pi\varepsilon_0 r^2} = mr\omega^2 \tag{1.4}$$

The current of the circulating electron is given by $I = e\omega/2\pi$. Then the magnetic moment due to one electron in a circular orbit is given by

$$M = \mu_0 I(\text{area}) = \tfrac{1}{2}\mu_0 e r^2 \omega \tag{1.5}$$

The angular momentum of a particle in a circular orbit is $J = mr^2\omega$ so that the magnetic moment becomes

$$M = \tfrac{1}{2}\mu_0 \frac{e}{m} J \tag{1.6}$$

From quantum theory and the wave nature of an electron[27]

$$J = \frac{nh}{2\pi} \tag{1.7}$$

where n is the quantum number appropriate to the electron in its orbit. Substituting from eqn. 1.7 into eqn. 1.6 gives

$$M = \tfrac{1}{2}\mu_0 \frac{enh}{m2\pi} = \mu_0 n \frac{eh}{4\pi m} \tag{1.8}$$

It is seen from eqn. 1.8 that the magnetic moment of the electron in its orbit is a simple multiple of a quantity which is solely a property of the fundamental constants of an electron. This quantity is called the *Bohr magneton* and is defined by

$$\beta = \frac{eh}{4\pi m} = 9{\cdot}274 \times 10^{-24}\,\text{Am}^2$$

or the equivalent magnetic moment is $\beta\mu_0 = 1{\cdot}165 \times 10^{-29}$ Wb m. Then the electron in its orbit has an angular momentum of n units and a magnetic moment of $n\beta\mu_0$.

The electron also has spin. The spinning electron behaves like a small electric charge moving in a circular orbit or like a charged conducting sphere spinning about its axis. The magnetic moment of an electron due to its spin is 1 Bohr magneton. The angular momentum of the electron due to its spin is given by

$$J = \frac{1}{2}\left(\frac{h}{2\pi}\right) \tag{1.9}$$

so that the spin quantum number is $\frac{1}{2}$. Then for the electron spin

$$M = 2n\mu_0\beta \tag{1.10}$$

In general terms, eqns. 1.8 and1.10 may be combined:

$$M = gn\mu_0\beta \tag{1.11}$$

where g is called the Landé spectroscopic splitting factor. It is 1 when the magnetic moment is due to orbital motion alone and is 2 for the spin alone, but if coupling exists between the spin and orbital moments it has a value larger than 2.

In most materials, the effect of orbital magnetic moments is small because the

Table 1.1 *Spin magnetic moment of the transition elements*

Element	Number of electrons	3d shell (10 max)	4s electrons	Magnetic moment Bohr magnetons
Sc	21	↑	2	1
Ti	22	↑ ↑	2	2
V	23	↑ ↑ ↑	2	3
Cr	24	↑ ↑ ↑ ↑ ↑	1	5
Mn	25	↑ ↑ ↑ ↑ ↑	2	5
Fe	26	↑↓ ↑ ↑ ↑ ↑	2	4
Co	27	↑↓ ↑↓ ↑ ↑ ↑	2	3
Ni	28	↑↓ ↑↓ ↑↓ ↑ ↑	2	2
Cu	29	↑↓ ↑↓ ↑↓ ↑↓ ↑↓	1	0

orbits are constrained by the material and generally are not free to align themselves with any external magnetic fields. Also the orbital magnetic moments in many materials cancel each other out owing to the random orientation of all the different electron orbits. Not so electron spin. That is free to align itself with any external magnetic field. However, in most orbitals, electrons occur in pairs with equal and opposite spins and their spin magnetic moment also cancels out. Then it is only the valence electrons and electrons in incompletely filled shells which have unbalanced spins and resultant spin magnetic moments. In compounds, the valence electrons also occur in pairs with opposite spins, but in metals there is some unbalanced spin which gives rise to a small magnetic effect. It is the materials of the transition series or the lanthanide series from the periodic table which can have the larger magnetic effect. Consider the transition elements with the incompletely filled 3d electron shell. All other effects being equal, Hund's rule states that the electrons will be oriented so that there are a maximum number of unpaired spins. Then the electrons of the 3d shell of the first transition series of elements will be oriented as shown in Table 1.1. Each manganese and chromium atom will have a magnetic moment of 5 Bohr magnetons. The spin magnetic moments of each of the complete shells or sub-shells cancel, so that, for the elements given in Table 1.1, only the electrons in the 3d sub-shell contribute to the spin magnetic moment.

The unbalanced spin in each atom is free to align itself with any external magnetic field. The effect is to increase the total magnetic flux in the material because the internal atomic magnets are reinforcing the external magnetic field, giving rise to *paramagnetism*. Each atom acts individually and the effect is small. Any atom with unbalanced electron spin is paramagnetic as is also any atom with unconstrained unbalanced orbital magnetisation.

1.4 Ferromagnetism

Paramagnetism, like diamagnetism, is a small magnetic effect that can only be detected by sensitive scientific apparatus. In a few materials, notably iron and steel, there is a much stronger magnetic effect called *ferromagnetism*. This is caused by an interaction between the unbalanced spin directions of adjacent atoms, called exchange forces or exchange interaction. The exchange forces align the directions of the unbalanced electron spins between adjacent atoms so that, in a solid, there is a large spontaneous magnetisation in the material. However, freshly annealed iron specimens do not exhibit this spontaneous magnetisation externally because minimum energy encourages the generation of a large number of domains spontaneously magnetised in random directions. Then it is only under the influence of an external magnetic field that the magnetisation of all the domains is aligned and the material is said to be magnetised. Ferromagnetism requires both unbalanced electron spins and exchange forces to align the spins within a domain. Unbalanced electron spins give rise to paramagnetism; each atom acts individually and the magnetic effect is small. Unbalanced electron spins and exchange forces give rise to ferromagnetism; the atoms all act as one large magnet and the magnet effect is large.

Exchange forces are near-neighbour forces, thus ferromagnetism is very sensitive to interatomic distance in the solid. It is found that the positive alignment between electron spins in adjacent atoms only occurs for a certain range of interatomic spacing. If the spacing is too large, exchange forces become ineffective, and, if the spacing drops below a certain value, the sign of the exchange forces changes and an antiparallel configuration of spins is energetically preferable. In such a material, the total spin magnetic effect cancels out within each domain and the material appears to be non-magnetic. This is called *antiferromagnetism*. For example, body centered cubic iron and many body centred cubic iron alloys are ferromagnetic, whereas face centered cubic iron and face centered cubic iron alloys are not magnetic. Pure manganese is antiferromagnetic, but when manganese is alloyed with suitable proportions of allminium or bismuth, the interatomic distance is modified so that the alloy becomes ferromagnetic. In many metallic compounds, such as oxides and chlorides, the metal ions are separated by non-metallic ions which act as intermediaries. This gives rise to a *superexchange interaction* where the M–O–M antiparallel arrangement is stronger than the M–M parallel arrangement. An example is MnO, where each manganese ion is surrounded by oxygen ions and each oxygen ion in turn is surrounded by manganese ions. There are two equal interlaced lattices of ions whose magnetisations are in opposite directions so that manganese oxide is antiferromagnetic.

There is another kind of ferromagnetism which is similar to antiferromagnetism in having two interlaced lattices of ions whose magnetisations are in opposite directions. However, the total spin magnetisations of the two lattices are unequal and there is some resultant magnetisation in the domain. It usually describes the magnetic state of ferrites and is called *ferrimagnetism*. It explains why the saturation magnetisation of ferrites is considerably less than the sum of the spin magnetisations of its constituent ions. The original analysis was developed by Néel.[28]

In any ferromagnetic substance the spontaneous magnetisation in the domain is reduced by the thermal vibrations of the atoms in the crystal as the temperature is increased. Above a certain temperature, the exchange forces can no longer resist the thermal vibrations of the lattice and the material becomes paramagnetic. For ferromagnetic and ferrimagnetic materials the critical temperature is called the *Curie temperature* and for antiferromagnetic materials it is called the *Néel temeprature*.

1.5 Spinel ferrites

A recent survey of the properties of oxide spinels has been given by Krupicka and Novak.[29] Ferrites having the spinel structure have the general formula MFe_2O_4 where M is any divalent metal. They all have the crystal structure of the mineral spinel, $MgAl_2O_3$. The only naturally occurring ferrite is the magnetic iron ore, magnetite or ferrous ferrite, $FeFe_2O_4$. All other ferrites are man-made substances. They are ceramic solids and do not exist as individual

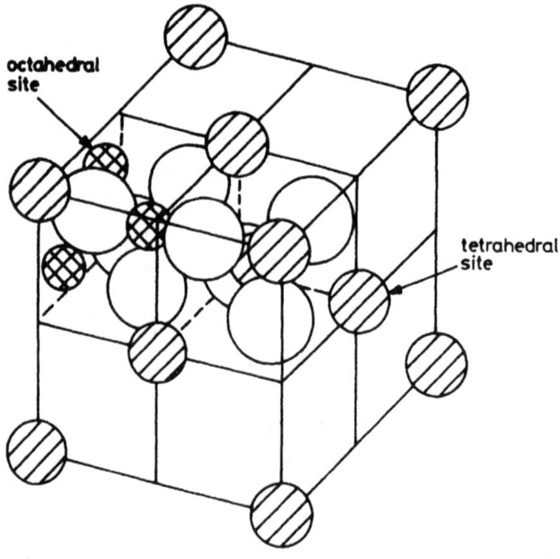

Fig. 1.3 *Unit cell of the spinel structure*
The large spheres are oxygen, the small hatched spheres are ions on tetrahedral sites and the small cross-hatched spheres are ions on octahedral sites. Ions are drawn in only 2 of the 8 octants. There are 32 oxygen atoms, 16 atoms on octahedral sites and 8 atoms on tetrahedral sites

molecules. They are usually made by sintering an intimate mixture of two oxides, MO and Fe_2O_3 which then interact in the solid to form the compound. Consequently, mixed ferrites are easy to make. Some magnetic ferrites are:

$MgFe_2O_4$; $MnFe_2O_4$; $NiFe_2O_4$; $CuFe_2O_4$ and $(LiFe)_{1/2}Fe_2O_4$. $ZnFe_2O_4$ and $CdFe_2O_4$ are non-magnetic.

The spinel structure is a cubic structure with a unit cell of 32 oxygen ions. It is shown in Fig. 1.3. The metal ions occupy the spaces between the oxygen ions in two different kinds of lattice site. There is the tetrahedral site, where the metal ion is surrounded by four oxygen ions and is at the centre of a tetrahedron as shown by Fig. 1.4*a*, and the octahedral site, where the metal ion is surrounded by six oxygen ions and is at the centre of the octahedron that has oxygen ions at its apices, as shown by Fig. 1.4*b*. In one unit cell of the spinel structure there are 16 octahedral sites, eight tetrahedral sites and 32 oxygen ions. In the normal spinel structure, the divalent metal ions occupy the tetrahedral sites while the trivalent iron ions occupy the octahedral sites. Zinc and cadmium ferrites have the normal spinel structure and are antiferromagnetic. The structure of all the other ferrites is inverted spinel, so called because the divalent metal ions occupy some octahedral sites, while the trivalent ions occupy the remaining octahedral sites and the tetrahedral sites.

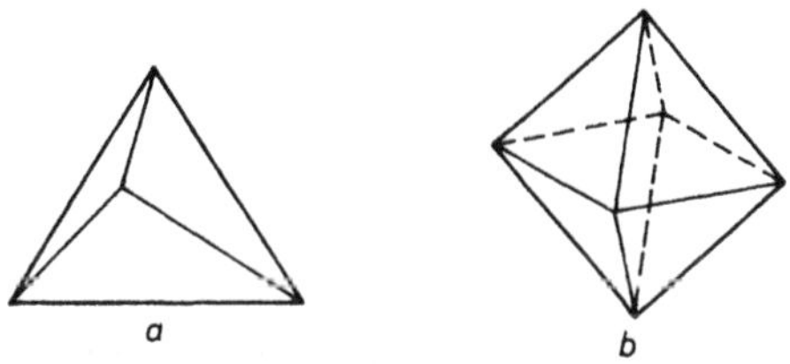

Fig. 1.4 *Geometry of sites in the spinel structure*
a Tetrahedron *b* Octahedron

If all the magnetic moments of the ions were aligned parallel, it would be possible to calculate the saturation magnetisation of a ferrite in Bohr magnetons per molecule simply by adding together the magnetic moments of the ions. However, if this is done, the values obtained are larger than those obtained experimentally. Néel[28] gave the reasons in his explanation of ferrimagnetism. He suggested that a magnetic dipole on a tetrahedral site is aligned antiparallel to a magnetic dipole on an octahedral site. Then the total magnetisation is given by the difference in magnetisations of the ions on the two sites, which agrees well with experiments. In the absence of any other magnetic ions, the ferric ions form into two opposed sublattices even if they are all on octahedral sites. The effect is shown diagrammatically in Fig. 1.5. As shown in Table 1.1, the spin magnetic moment of the divalent nickel ion is 2 Bohr magnetons, that of the divalent ferrous ion is 4 Bohr magnetons and that of the trivalent ferric ion is 5 Bohr magnetons, because it has the same electron structure as manganese. If nickel ferrite were ferromagnetic, then it could be expected to have a saturation magnetisation of 12 Bohr magnetons per molecule, whereas its actual saturation magnetisation is about 2 Bohr magnetons per molecule, as given by the theory of ferrimagnetism and shown in Fig. 1.5. Similarly, ferrous ferrite has a saturation magnetisation of about 4 Bohr magnetons per molecule and zinc ferrite is antiferromagnetic.

The saturation magnetisation of many ferrites may be increased by mixing with the non-magnetic zinc ferrite. The example of nickel ferrite is shown in

	Octahedral sites			Tetrahedral sites	
$NiFe_2O_4$	Ni^{2+}		Fe^{3+}	Fe^{3+}	
resultant	↑		↑	↓	
2 β/molecule	2		5	-5	β/ion
$FeFe_2O_4$ (Fe_3O_4)	Fe^{2+}		Fe^{3+}	Fe^{3+}	
resultant	↑		↑	↓	
4 β/molecule	4		5	-5	β/ion
$ZnFe_2O_4$	Fe^{3+}		Fe^{3+}	Zn^{2+}	
	↑		↓		
antiferromagnetic	5		-5	0	β/ion
$(LiFe)_{1/2}Fe_2O_4$	$\frac{1}{2}Li^{+}$	$\frac{1}{2}Fe^{3+}$	Fe^{3+}	Fe^{3+}	
resultant		↑	↑	↓	
2·5 β/molecule	0	$2\frac{1}{2}$	5	-5	β/ion

Fig. 1.5 *Saturation magnetisation of four different spinel ferrites*

Figs. 1.6 and 1.7. Zinc ferrite has the normal spinel structure and is mixed with nickel ferrite of the inverted spinel structure. Zinc replaces some ferric ions on tetrahedral sites and thus increases the number of Bohr magnetons contributed by ions on octahedral sites as shown in Fig. 1.6. The saturation magnetisation increases initially with zinc content at a slope which leads towards 10 Bohr magnetons per molecule for 100% zinc ferrite. However, at about 40% zinc ferrite, the saturation magnetisation then begins to fall towards zero, as shown in Fig. 1.7. An analysis of Néel's theory applied to ferrites is given by Gorter[30] and an extended discussion by Lax and Button.[14] As most of the cations used in ferrites have about the same ionic radii, completely miscible solid solutions of the different ferrites are easily possible. Mixed ferrites may be made to specification and are of much greater technical importance than the single ferrites.

Historically the spinel ferrites were the first to be used for microwave applications. Initially those developed for the high radio frequency applications were investigated for microwave use. However, they were found to have unacceptably high absorption loss at microwave frequencies and special low microwave loss ferrites were developed. The spinel ferrites normally used for microwave applications are magnesium-manganese ferrite, nickel ferrite and lithium ferrite. Magnesium-manganese ferrite has the composition $Mn_xMg_yFe_zO_4$ with $x + y + z = 3$· Its microwave properties are a function of the heat treatment of the material because the way the metallic ions are distributed between the two crystallographic sites depends upon the heat treatment. The chemical formula actually used is determined by experiment. Useful compositions lie in the ranges $0{\cdot}10 < x < 0{\cdot}15$, $0{\cdot}4 < y < 0{\cdot}5$, $0{\cdot}7 < z < 0{\cdot}9$. Saturation magnetisations range from $M_s = 0{\cdot}20$ to $0{\cdot}24$ T at room temperature. Lower saturation magnetisations are achieved by substituting some of the iron with aluminium. These lower saturation magnetisation ferrites are suitable for use at the lower microwave frequencies, thus eliminating low-field loss as described in Section 11.4. The hysteresis loop is rectangular for these ferrites so they are suitable for use in latching control applications. The magnesium-manganese ferrite is a good

Increasing proportion of zinc	Octahedral sites			Tetrahedral sites		
		$6\,Ni^{2+}$	$10\,Fe^{3+}$	$6\,Fe^{3+}$	$2\,Zn^{2+}$	
$4\,\beta$/mol		↑	↑	↓		
(average)		2	5	−5	0	β/ion
	$4\,Ni^{2+}$	$10\,Fe^{3+}$	$2\,Fe^{3+}$	$4\,Fe^{3+}$	$4\,Zn^{2+}$	
$3\tfrac{1}{2}\,\beta$/mol	↑	↑	↓	↓		
(average)	2	5	−5	−5	0	β/ion
	$2\,Ni^{2+}$	$9\,Fe^{3+}$	$5\,Fe^{5+}$	$2\,Fe^{3+}$	$6\,Zn^{2+}$	
$1\tfrac{3}{4}\,\beta$/mol	↑	↑	↓	↓		
(average)	2	5	−5	−5	0	β/ion

Fig. 1.6 *Effect of increasing proportions of zinc on the saturation magnetisation of nickel zinc ferrite*

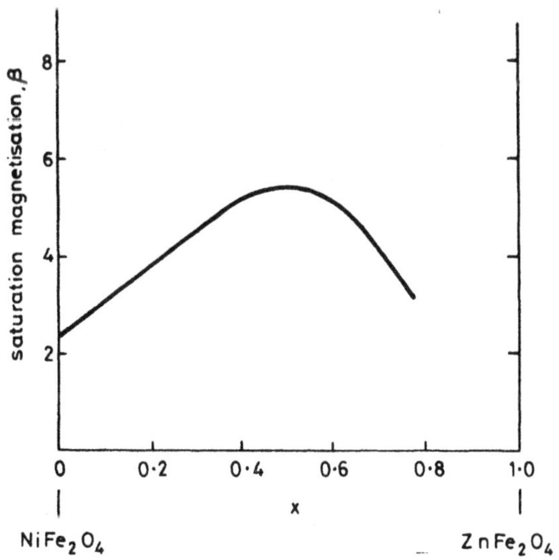

Fig. 1.7 *Saturation magnetisation of $Ni_{(1-x)}Zn_xFe_2O_4$ as x varies*

low-loss low-power microwave ferrite for use in the frequency range 7 to 15 GHz. The Curie temperature is about 300°C, which is quite low, so that its properties are temperature sensitive and it is recommended for use in situations not having wide temperature variations.

Nickel ferrites do not have such a low loss as the magnesium-manganese ferrites, but they have a higher Curie temperature of about 570°C and hence less variation of properties with temperature. They are more suitable for high-power operation. The saturation magnetisation of $NiFe_2O_4$ is 0·32 T and this can be raised by zinc substitution to 0·50 T to make a ferrite for millimetre wave applications (see Chapter 10). The saturation magnetisation can be lowered by substitution of aluminium for iron. Nickel ferrites do not have a rectangular hysteresis loop.

Lithium ferrite has the formula $(LiFe)_{1/2}Fe_2O_4$. It appears to be more complicated than the other ferrites, but it has a high Curie temperature, 645°C, a very low microwave loss and a saturation magnetisation of 0·36 T. By zinc substitution the saturation magnetisation can be raised to 0·5 T. Lithium is a monovalent ion but it is combined with the trivalent ferric ion to give an average divalent effect. Because all the iron ions are trivalent, there is no loss mechanism owing to electron transfer between divalent and trivalent iron ions, as discussed in Section 1.8, and the intrinsic microwave loss is very low. The magnetic properties of lithium ferrite microwave materials are discussed by White and Patton[31] and in less detail by Nicolas.[23] It can be seen from Fig. 1.5 that the magnetisation of lithium ferrite is $2\frac{1}{2}$ Bohr magnetons per molecule. For operation at lower microwave frequencies, the saturation magnetisation may be reduced by titanium substitution for iron. The hysteresis loop is rectangular.

1.6 Magnetic garnets

Although the spinel ferrites were the first magnetic insulators to be used for nonreciprocal or control purposes at microwave frequencies, many magnetic garnets have lower losses and are preferred in many applications. The natural silicate garnet is a source of abrasive grit and is a semiprecious stone. The best known magnetic garnet, yttrium-iron garnet, YIG, was discovered in 1956[32] together with other lanthanide series garnets. The garnet is a very complicated crystal structure and much more detailed information is given by Gilleo[24] on the properties of magnetic garnets and by Nicolas[23] on those garnets useful in microwave applications. The unit cell contains 8 formula units $A_3B_2C_3O_{12}$ where A, B and C are metal cations. In the crystal structure there are cation sites with the metal ions surrounded by oxygen ions in tetrahedron, octahedron and dodecahedron co-ordination. The A ions occupy the dodecahedron sites, B ions the octahedron sites and the C ions the tetrahedron sites. In YIG, $Y_3Fe_5O_{12}$, the yttrium ions occupy the dodecahedron sites while the iron ions occupy the tetrahedron and octahedron sites. In garnet all the cations are trivalent. Because there are no positions for divalent iron ions in the lattice, iron garnets are essentially low loss at high frequencies.

Magnetic garnets are ferrimagnetic. Magnetic ions on the different lattice sites form into two oppositely directed magnetisations. Then in YIG the resultant is one unbalanced trivalent iron ion and the magnetisation of YIG is about 5 Bohr magnetons per formula unit. There are two possible substitutions which can be used to reduce the magnetisation of YIG for low microwave frequency use:

1. Substitute trivalent aluminium ions for iron ions on tetrahedral sites, thus reducing the resultant difference in magnetisation between magnetic ions on the two sites.
2. Substitute, for yttrium ions, magnetic gadolinium ions whose magnetisation opposes the resultant magnetisation of the iron ions.

The Curie temperature of YIG is 286°C and its saturation magnetisation at room temperature is about 0·18 T. YIG has less variation of properties with temperature than magnesium-manganese ferrite and it also has a lower loss, thus for low-power nonreciprocal and control applications it is preferred to the ferrite. Ferrites have the advantage of much higher saturation magnetisation when this is required for millimetre wave devices. A wide range of metallic cations may be substituted for the cations in YIG and a wide range of magnetic garnets can be tailored for particular purposes.

Further information can be obtained from the paper by Nicolas[23] and manufacturers' catalogues. The hysteresis loop of magnetic garnets is rectangular and they are suitable for use in latching control applications.

Small spheres of single-crystal YIG have a particularly low microwave loss[33] and can be used as high-Q resonant cavities in filters,[20,34] as described in Section 9.9. YIG is suitable for use at frequencies above 3·3 GHz. Aluminium, or gadolinium substituted YIG, is used at lower frequencies.

1.7 Permanent magnetic ferrites

There is a large family of permanent magnet ferrite materials with the general formula, $x(BaO)y(MeO)z(Fe_2O_3)$, where Me is a divalent ion from the first series of transition elements or a combination of ions whose valency is two. Because of the complication of the formulas, various combinations have been given code letters. M is $BaFe_{12}O_{19}$ and is called BaM. Similarly, $SrFe_{12}O_{19}$ is SrM and $PbFe_{12}O_{19}$ is PbM. W is $Me_2BaFe_{16}O_{27}$; Y is $Me_2Ba_2Fe_{12}O_{22}$; Z is $Me_2Ba_3Fe_{24}O_{41}$; X is $Me_2Ba_2Fe_{28}O_{46}$ and U is $Me_2Ba_4Fe_{36}O_{60}$. They are described by Smit and Wijn[4] and in more detail by Kojima[35] and Sugimoto.[36]

The M-type and W-type permanent magnet ferrites have been used in microwave applications. They are used, particularly at millimetre wave frequencies (see Chapter 10), where a very large external bias magnetic field is required, or occasionally in order to eliminate the need for an external bias magnetic field.

Various partial substitutions of other cations into BaM and SrM are used to reduce the magnetisation and the anisotropy field. Other substitutions may be used to increase the magnetisation and the anisotropy field. The other ferrite used in microwave applications is NiW with either a cobalt substitution for nickel lowering the anisotropy field or an aluminium substitution for iron increasing both the magnetisation and the anisotropy field. For microwave purposes, the ferrite has to be produced with a low microwave loss which is achieved by having all the iron ions in the trivalent state. Many permanent magnet ferrites for microwave use are produced with a slight deficiency of iron compared with the stoichiometric quantity, in order to ensure that there are no iron ions in the divalent state.

1.8 Dielectric loss mechanisms

One loss mechanism in dielectric materials is the resonance of electric dipoles in the atom or molecule of the material. At a particular frequency, the dipole motion becomes resonant and will absorb power from the exciting electromagnetic field. At high frequencies, that particular dipole is unable to follow the exciting electric field and the permittivity of the material is reduced. This is a dielectric loss mechanism and occurs in the microwave frequency range for materials having a permanent electric dipole in a covalently bonded material. This means that some materials having a very high resistivity will not necessarily have a low microwave loss. However, high resistivity is also essential to produce a low microwave loss.

Conduction in ferrites occurs through the migration of electrons through the crystal. In the spinel ferrites, there are divalent and trivalent cations each occupying a fixed position in the crystal lattice. Ferrous ferrite has a low resistivity because both the trivalent and divalent cations are iron ions. At any dislocation in the crystal structure, it is easy for an iron ion to change from the divalent to the trivalent state, or vice versa, which provides a comparatively easy source of conduction. In other ferrites, ideally the iron ions are all trivalent, but unless the number of iron ions is less that the stoichiometric ratio some divalent iron ions can occur. Most ferrites are either iron rich or iron lean depending on the properties requried. The magnetic garnets have a particularly low loss because there are no divalent ions in the crystal structure, leading to less likelihood of divalent iron ions occurring. For low-frequency use, the saturation magnetisation and the initial permeability may be increased by having an iron rich ferrite at the expense of a much lower resistivity. The initial permeability and saturation magnetisation is reduced as the excess iron is reduced and the resistivity increased.

Impurities can have a very detrimental effect on the resistivity. There is a semiconductor effect with a monovalent ion occupying the position of a divalent atom in the crystal structure. Similarly, divalent ions may replace trivalent atoms leaving floating electrons or supervalent ions. Therefore it is generally necessary to have a pure homogeneous material with exactly the stoichiometric proportion of ions according to the chemical formula. As mentioned above, it is sometimes beneficial to have a slight deficiency of trivalent iron ions. Thus the production of a good low-loss ferrite is partly a technological problem of ensuring adequate purity and correct temperature conditions.

Magnetic loss mechanisms are discussed in terms of their applications in Sections 2.4 and 11.1. The magnetic loss is usually measured in terms of the resonance linewidth as discussed in Section 2.4. In order to reduce this loss, it is necessary to reduce the damping of the movement of the atomic magnetic moments as much as possible. The spin-orbit coupling often plays a significant role in the damping of the spin magnetic moments. The trivalent iron ions have the lowest damping because their electronic configuration, as shown by Table 1.1, consists of a half-full 3d shell having 5 electron spins and no orbital moment which eliminates spin-orbit coupling. This means that ferrites and garnets having trivalent iron ions as their only magnetic ions can be expected to have the lowest magnetic losses. Lithium ferrite, $(LiFe)_{1/2}Fe_2O_4$, and YIG, $Y_3Fe_5O_{12}$, fulfil this condition, and are currently used for situations requiring the lowest microwave losses.

1.9 Preparation

Ferrites and garnets are manufactured by conventional ceramic techniques. An intimate mixture of oxides is heated together to make an incompletely reacted powder which is then pressed into the required shape and sintered. The intimate mixture of oxides may be produced by co-precipitation or by mixing the oxide powders. Co-precipitation provides the most intimate mixing; nitrates or carbonates are dissolved in solution in the correct proportions and the mixture evaporated to dryness. The precipitate is heated to decompose the salts into oxides and with continued heating the oxides react together to form the ferrite. The alternative is to start with oxides which are mixed together by grinding in a ball mill and the mixture is heated to form the ferrite. Sometimes, at the first heating, the mixture only reacts incompletely and the resulting brittle powder is further ball milled or tumbled to provide fresh unreacted surfaces of the oxides and further heated. At each stage the reaction occurs by a process of diffusion in the solid. This pre-sintering often occurs at a temperature about 300 K less than the final sintering temperature but the reaction between the constituents is usually substantially complete. The ideal is to produce a homogeneous well-reacted material which is then further ball milled to a particle size of about one micron or less ready for final sintering.

The powder is pressed to its final shape before sintering. It may be extruded, pressed in a die or hydrostatically pressed. A little binder is added, which may be some organic compound or just water, but the minimum should be used because an important contribution to low microwave loss is a high density and no voids. Permanent magnet ferrites may be pressed in a magnetic field. In the ideal situation, each ferrite particle will be a single crystal and a single domain, which can then be aligned by the magnetic field. The highest grades are made by wet compression in the magnetic field. An aqueous suspension is poured into a mould in the presence of the magnetic field, giving a preferred direction of crystals in the suspension. Compression then takes place, removing most of the water. However made, the pressed shape is heat treated at about 500–600°C to eliminate the binder, and sintered, usually in an oxygen atmosphere, for 5 to 10 hours. The sintering temperature is in the range 1000 to 1500°C depending on the material being produced. There is a 20% shrinkage on sintering so that close dimensional tolerances on as-fired specimens are unlikely. Ferrite is a hard brittle material which can only be machined by grinding. Microwave materials need to be high density and thus tend to be harder than ferrites which are manufactured for use at lower frequencies.

As discussed in the last section, low microwave loss is obtained with good stoichiometry and having a slightly iron lean composition. Attention has to be taken of iron introduced to the powder from the apparatus of the mixing and milling operations. Similarly, during sintering and pre-firing, the atmosphere has to be chosen with care so as to eliminate oxygen loss from the ferrite. Therefore it is essential to control the reproducibility of the manufacturing processes. In some mixed ferrites, the distribution of some of the ions between the different crystallographic sites depends on the heat treatment of the material after sintering, therefore it is often necessary to control the subsequent cooling sequence just as much as the sintering process. Good microwave ferrites and garnets are the products of attention to detail in the manufacturing processes.

1.10 References

1 SNOEK, J. L.: 'Magnetic and electrical properties of ferrites', *Physica*, 1936, **3**, pp. 463–483
2 SNOEK, J.L.: 'Non-metallic magnetic material for high frequencies', *Philips Tech. Rev.*, 1946, **8**, pp. 353–360
3 SNOEK, J.L.: 'New developments in ferromagnetic materials' (Elsevier, 2nd. edn., 1949)
4 SMIT, J. and WIJN, H. P. J.: 'Ferrites' (Wiley, 1959)
5 GRIFFITHS, J. H. E.: 'Anomalous high-frequency resistance of ferromagnetic metals', *Nature*, 1946, **158**, pp. 670–671
6 POLDER, D.: 'On the theory of ferromagnetic resonance', *Phil. Mag.*, 1949, **40**, pp. 99–115
7 HOGAN, C. L.: 'The ferromagnetic Faraday effect at microwave frequencies and its applications', *Bell Syst. Tech. J.*, 1952, **31**, pp. 1–31
8 SUHL, H. and WALKER, L. R.: 'Topics in guided wave propagation through gyromagnetic media', *Bell Syst. Tech. J.*, 1954, **33**, pp. 579–659, 939–986 and 1133–1194
9 FOX, A. G., MILLER, S. E. and WEISS, M. T.: 'Behaviour and applications of ferrites in the microwave region', *Bell Syst. Tech. J.*, 1955, **34**, pp. 5–103
10 'Ferrites Issue', Proc. IRE, 1956, **44**, pp. 1233–1480
11 SOOHOO, R. F.: 'Theory and application of ferrites' (Prentice-Hall, 1960)
12 CLARRICOATS, P. J. B.: 'Microwave ferrites' (Chapman & Hall, 1961)

13 WALDRON, R. A.: 'Ferrites' (Van Nostrand, 1961)
14 LAX, B. and BUTTON, K. J.: 'Microwave ferrites and ferrimagnetics' (McGraw-Hill, 1962)
15 GUREVICH, A. G.: 'Ferrites at microwave frequencies' (in Russian), (State Press, 1960), (English translation Heywood, 1963)
16 THOUREL, L.: 'Emploi des ferrites en hyperfréquence' (Sofradel, 1962) (English translation 'The use of ferrites at microwave frequencies' Pergamon Press, 1964)
17 VON AULOCK, W. H. and FAY, C. E.: 'Linear ferrite devices for microwave applications. Advances in electronics and electron physics. Supplement No. 6' (Academic Press, 1968)
18 HELSZAJN, J.: 'Principles of microwave ferrite engineering' (Wiley, 1969)
19 HELSZAJN, J.: 'Nonreciprocal microwave juncions and circulators' (Wiley, 1975)
20 HELSZAJN, J.: 'YIG resonators and filters' (Wiley, 1985)
21 SOOHOO, R. F.: 'Microwave magnetics' (Harper & Row, 1985)
22 VON AULOCK, W. H.: 'Handbook of microwave ferrite materials' (Academic Press, 1965)
23 NICOLAS, J.: 'Microwave ferrites' *in* WOHLFARTH, E. P. 'Ferromagnètic materials – Vol. 2' (North-Holland, 1980), pp. 243–296
24 GILLEO, M. A.: 'Ferromagnetic insulators: Garnets', *ibid*, pp. 1–53
25 BUTTON, K. J.: 'Microwave ferrite devices: the first ten years', *IEEE Trans.*, 1984, **MTT-32**, pp. 1088–1096
26 BADEN FULLER, A. J.: 'Units of magnetisation', *Proc. IEE.*, 1982, **129A**, p.203
27 See, for example, BEISER, A.: 'Concepts of modern physics', 2nd edn. (McGraw-Hill, 1973), pp. 185 and 207
28 NÉEL, L.: 'Propriétés magnétiques des ferrites, ferrimagnétisme et antiferromagnétisme', *Ann. Phys.*, 1948, **3**, pp. 137–198
29 KRUPICKA, S. and NOVAK, P.: 'Oxide spinels' *in* WOHLFARTH, E. P. 'Ferromagnetic materials – Vol. 3' (North-Holland, 1982), pp. 189–304
30 GORTER, E. W.: 'Saturation magnetisation and crystal chemistry of ferrimagnetic oxides', *Philips Res. Rep.*, 1954, **9**, pp. 295–365
31 WHITE, G. O. and PATTON, C. E.: 'Magnetic properties of lithium ferrite microwave materials', *Jour. Magn. & Magn. Mat.*, 1978, **9**, pp. 299–317
32 BERTAUT, F. and FORRAT, F.: 'Structure des ferrites ferrimagnetique des terres rares', *Compt. Rend. Acad. Sci. de Paris.*, 1956, **242**, pp. 382–384
33 LECRAW, R. C., SPENCER, E. G. and PORTER, C. S.: 'Ferromagnetic resonance line width in yttrium iron garnet single crystals', *Phys. Rev.*, 1958, **110**, pp. 1311–1313
34 CARTER, P. S.: 'Magnetically-tunable microwave filters using single-crystal yttrium-iron-garnet resonators', *IRE Trans.*, 1961, **MTT-9**, pp. 252–260
35 KOJIMA, H.: 'Fundamental properties of hexagonal ferrites with magnetoplumbite structure' *in* WOHLFARTH, E. P. 'Ferromagnetic materials, Vol. 3' (North-Holland, 1982), pp. 305–391
36 SUGIMOTO, M.: 'Properties of ferroxplana-type hexagonal ferrites', *ibid*, pp. 393–440

Chapter 2

Plane waves in an infinite ferrite medium

2.1 Descriptive introduction

In all ferromagnetic and ferrimagnetic materials, there is interaction between microwave fields and the spinning electron, but it is only in ferrites and garnets, which are electrically insulating, that useful interaction between the magnetic properties of the material and electromagnetic waves can be obtained. A classical description of the spinning electron will be used to give a pictorial explanation of the way magnetic materials act on microwave fields. The electron behaves as if it were a negatively charged sphere which is spinning about its own axis with a fixed angular momentum. The rotation of charge gives the electron a magnetic moment which is a function of its charge, angular velocity and size. Because of its angular momentum, the electron behaves as if it were a spinning magnetic top whose magnetic moment lies along its axis of rotation. It behaves like a gyroscope, but, instead of moving under the influence of gravitation, it moves due to the influence of magnetic forces which are those of any internal or applied magnetic field.

Initially the electron will align itself with any magnetic field for minimum potential energy. If the electron is disturbed from this equilibrium position, it will not return to the position of minimum energy but will precess about the axis of the magnetic field. In the absence of losses, the equilibrium state is now a uniform precession about an axis parallel to the direction of the magnetic field. In a real material there are forces other than the magnetic field acting on the movement of the electron spin axis and these forces tend to oppose the precessional motion of the electrons. Then any gyroscope, when set into motion and left to precess, will slowly spiral to its equilibrium position. The loss of precessional energy is due to frictional and other damping mechanisms in the ferromagnetic material.

This classical description of magnetism can be used to describe the motion of the electrons in a ferrite. Assume that the ferrite is magnetically saturated by a magnetic field H so that all the electrons contributing to the ferromagnetic effect are aligned. If an alternating magnetic field acting in a plane perpendicular to

H is superimposed on to the field H, the resultant field will alternate between the two directions A and B shown in Fig. 2.1. Assume initially that the electron spin axis is vertical under the influence of the magnetic force H. If the direction of the force H is suddenly altered to the position A, the electron, behaving like a gyroscope, will precess about the axis A along the circular path a–b. If, when the electron axis has reached the position b, the direction of the force H changes to the position B, the electron axis will precess along the new circular path b–c.

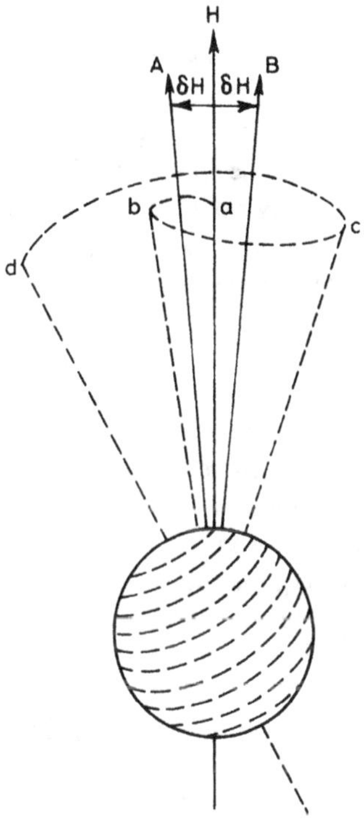

Fig. 2.1 *Precessional motion of a spinning electron in a magnetic field which oscillates between the directions A and B*

If the force then moves back to the position A, the electron axis will continue in the new circle c–d. If the alternating motion of the force H continues in step with the gyroscopic motion of the electron axis, the precession angle will continue increasing indefinitely. However, in a physical substance, damping will limit the precessional motion of the axis of the spinning electron; the precession angle will reach some equilibrium position where the losses in the material exactly offset the driving effect of the alternating force.

If a ferrite material is saturated by a steady magnetic field, the electrons will come to rest with their spin axes and magnetic moments parallel to the field H. If an additional alternating magnetic field at the correct frequency is applied perpendicular to the static field H, the electrons will begin to precess in larger and larger orbits until they finally reach an equilibrium precession orbit under the influence of the magnetic fields and the internal damping. There is a transfer of power from the alternating magnetic field to the precessing electrons in the ferrite. They dissipate their energy in internal friction which appears as heat in the material. Because the energy transfer occurs only if the frequency of the alternating magnetic field coincides with the precession frequency of the electrons in the ferrite, the phenomenon is called *resonance absorption.* If the alternating magnetic field is part of an electromagnetic wave, the ferrite absorbs power from the electromagnetic wave.

It has been assumed that the disturbing magnetic field jumped between the two positions A and B and that δH took the form of a square wave. In a linearly polarised wave, the disturbing magnetic field will be in the form of a sine wave and the effect of the disturbance will be similar to that already described. If the disturbing magnetic field is circularly polarised, however, there will be an even greater interaction between the field and the precessing electrons. Provided that the direction of rotation of the circularly polarised field is the same as the direction of rotation of the precession orbit, the circularly polarised field will be acting to increase the precession angle all the time instead of just twice in each cycle. A normal resonance curve describes the relationship between the power absorbed in the ferrite and the frequency of the disturbing magnetic field. Because the precession frequency of the electrons in the ferrite depends on the strength of the static magnetic field, a resonance curve also describes the relationship between the power absorbed in the ferrite and the strength of the static magnetic field for a disturbing magnetic field of constant frequency. If the direction of rotation of the circularly polarised field is the opposite to the direction of rotation of the precession orbit, there is only a small interaction between the electromagnetic wave and the magnetised ferrite, and no resonance absorption. For maximum interaction, the electromagnetic wave must have a circularly polarised magnetic field component in the plane perpendicular to the direction of the static magnetic field in the ferrite, whose hand of circular polarisation is the same as the direction of the precession in the ferrite. If the frequency is wrong, or if the plane in which the circularly polarised field acts is wrong, or if the hand of circular polarisation is wrong, there will be little or no resonance absorption.

Because the direction of precession in the ferrite is governed by the direction of the static magnetic field and the direction of rotation of the circularly polarised wave is governed by the direction of propagation of the wave, the interaction between a statically magnetised ferrite material and a circularly polarised electromagnetic wave is *nonreciprocal.* If a forward wave is absorbed, a reverse wave having the same hand of circular polarisation will not be absorbed because the field in the reverse wave is rotating in the opposite direction to the precession in the ferrite. A linearly polarised wave can be represented as the sum of two equal circularly polarised waves of opposite hand. The magnetised ferrite separates these two circularly polarised waves by treating them differently from one another. Only one of the circularly polarised components of the linearly polarised wave will be absorbed by resonance absorption. The other will be unaffected.

A long way away from resonance, there is a weak interaction between the magnetised ferrite material and an electromagnetic wave. If the electromagnetic wave is circularly polarised with the rotating magnetic field in the plane perpendicular to the direction of the static magnetic field in the ferrite, one hand of circular polarisation will couple to the precession in the ferrite and will be accelerated. However, a wave having the opposite hand of circular polarisation will be retarded by the precessional motion of the electrons in the ferrite. Linear polarisation is the sum of two equal circularly polarised waves of opposite hand. One hand of circular polarisation will be accelerated and the other will be retarded so that the direction of the resultant linear polarisation will be rotated as the wave propagates through the ferrite. This rotation is nonreciprocal as the direction of rotation is determined by the direction of the static magnetic field in the ferrite and not by the direction of propagation of the electromagnetic wave. For example, if, in traversing a finite length of ferrite material, a linearly polarised wave is rotated through 45° and then reflected, it will be rotated a further 45° in the same direction as before and will arrive at the beginning with the plane of polarisation at 90° to where it started.

2.2 Magnetisation equation

Under the influence of high frequency magnetic fields, the spinning electrons and their magnetic moments do not necessarily move in phase with the applied magnetic fields. Under these circumstances, any ferromagnetic material will have an internal intensity of magnetisation $\boldsymbol{M}$ which is not necessarily parallel to, or in phase with, the applied magnetic field $\boldsymbol{H}$. The total magnetic field is then given by

$$\boldsymbol{B} = \mu_0 \boldsymbol{H} + \boldsymbol{M}$$

Each minute element in the material may be considered to be a magnetic top, spinning so that the magnetic moment and the angular momentum are parallel vectors. Their ratio is a constant for any material and is called the *gyromagnetic ratio* γ. It is given by

$$\gamma = \frac{\boldsymbol{m}}{\boldsymbol{J}} \tag{2.1}$$

where $\boldsymbol{m}$ and $\boldsymbol{J}$ are the magnetic moment and the angular momentum vectors, respectively. From eqns. 1.6 to 1.11, it can be seen that

$$\gamma = \tfrac{1}{2} g \mu_0 \frac{\mathrm{e}}{\mathrm{m}}$$

where g is the Landé spectroscopic splitting factor, and, in this equation, e and m are the charge and mass of the electron. For most microwave ferrite materials, the magnetisation is due to electron spin alone and $g = 2$. Therefore we can calculate that $\gamma = 2{\cdot}21 \times 10^5$ m/C.

Normally in a ferromagnetic material, each atomic magnetic top will be orientated so that there is no torque on it. If it is disturbed, however, the top will precess as a gyroscope and the torque will be given by the equation of motion of any gyroscope

$$\text{torque} = \frac{d\boldsymbol{J}}{dt} \tag{2.2}$$

Under the influence of a disturbing magnetic field, the disturbing torque on the top is the cross-product of the magnetic field and the magnetic moment. Therefore

$$\text{torque} = \boldsymbol{m} \times \left(\boldsymbol{H} + \frac{\boldsymbol{M}}{\mu_0}\right) \tag{2.3}$$

It is normally assumed that the individual magnetic tops are all aligned within a domain and that together they make up the magnetisation in the material. Therefore we assume that the magnetic moment $\boldsymbol{m}$ is proportional to the magnetisation $\boldsymbol{M}$. Then eqns. 2.1 and 2.2 can be substituted into eqn. 2.3 to give the magnetisation equation

$$\frac{d\boldsymbol{M}}{dt} = \gamma \boldsymbol{M} \times \boldsymbol{H} \tag{2.4}$$

This is the classical equation of motion of magnetisation. It takes no consideration of damping of the motion of the magnetic tops, and is thus a simplification, but it is useful because it gives simple mathematical results which are adequate for analysing many low-loss microwave ferrite applications.

Damping can be introduced into the magnetisation equation by the addition of a damping term to the right-hand side of eqn. 2.4. A form originally proposed by Landau and Lifshitz[1] is

$$\frac{d\boldsymbol{M}}{dt} = \gamma \boldsymbol{M} \times \boldsymbol{H} - \frac{\lambda}{|\boldsymbol{M}^2|}\boldsymbol{M} \times (\boldsymbol{M} \times \boldsymbol{H}) \tag{2.5}$$

The damping term in eqn. 2.5 represents a torque vector which acts so as to reduce the precession angle.

2.3 Tensor permeability

When the ferrite material is magnetised to saturation with an externally applied d.c. magnetic field, the magnetic permeability of the ferrite to an a.c. magnetic field becomes anisotropic with rotational symmetry about the direction of the d.c. magnetising field. Substitution of a.c. field components into eqn. 2.4 will give the a.c. permeability of the ferrite.

Consider an orthogonal system of axes in rectangular co-ordinates x, y and z. Let the ferrite material be magnetised to saturation by a static magnetic field, H_0, in the z-direction, which generates a static magnetisation, M_0, in the material. If the material is magnetised to saturation, it is assumed that all the individual magnetic tops are aligned within the material and eqn. 2.4 applies throughout the material. Let there also be a time-varying magnetic field, of time dependence exp $j\omega t$, that is small compared with the saturating magnetic field. $\boldsymbol{z}$ is the unit vector in the direction of the saturating magnetic field. Then the total fields in the material will be

magnetic field intensity $H_0\boldsymbol{z} + \boldsymbol{H}$

which gives rise to an

internal magnetisation $M_0\boldsymbol{z} + \boldsymbol{M}$

If substitution of these field values is made into eqn. 2.4 and the second-order small term $\boldsymbol{M} \times \boldsymbol{H}$ is neglected, we obtain

$$j\omega\boldsymbol{M} = \gamma M_0\boldsymbol{z} \times \boldsymbol{H} - \gamma H_0\boldsymbol{z} \times \boldsymbol{M} \tag{2.6}$$

Taking the separate components of the field in the rectangular co-ordinate system, eqn. 2.6 becomes

$$\left.\begin{aligned} j\omega M_x &= -\gamma M_0 H_y + \gamma H_0 M_y \\ j\omega M_y &= \gamma M_0 H_x - \gamma H_0 M_x \\ j\omega M_z &= 0 \end{aligned}\right\} \tag{2.7}$$

The first two equations of 2.7 are simultaneous equations in M_x and M_y, therefore eqn. 2.7 can be written

$$\left.\begin{aligned} M_x &= \chi H_x + j\kappa H_y \\ M_y &= -j\kappa H_x + \chi H_y \\ M_z &= 0 \end{aligned}\right\} \tag{2.8}$$

where

$$\left.\begin{aligned} \chi &= \frac{\gamma^2 H_0 M_0}{\gamma^2 H_0^2 - \omega^2} \\ \kappa &= \frac{-\omega\gamma M_0}{\gamma^2 H_0^2 - \omega^2} \end{aligned}\right\} \tag{2.9}$$

The total alternating field is $\boldsymbol{B} = \mu_0\boldsymbol{H} + \boldsymbol{M}$ and its components are

$$\left.\begin{aligned} B_x &= \mu H_x + j\kappa H_y \\ B_y &= -j\kappa H_x + \mu H_y \\ B_z &= \mu_0 H_z \end{aligned}\right\} \tag{2.10}$$

where

$$\mu = \mu_0 + \chi \tag{2.11}$$

The relationship between $\boldsymbol{B}$ and $\boldsymbol{H}$ in eqn. 2.10 can be written in vector form

$$\boldsymbol{B} = \boldsymbol{\mu H} \tag{2.12}$$

where the permeability is the tensor

$$\boldsymbol{\mu} = \begin{vmatrix} \mu & j\kappa & 0 \\ -j\kappa & \mu & 0 \\ 0 & 0 & \mu_0 \end{vmatrix} \tag{2.13}$$

This tensor form of the permeability was first derived by Polder[2] and is often called the Polder tensor.

The tensor permeability has been derived from a simple classical model of a ferrite material magnetised to saturation. It has been assumed that the whole body of the ferrite material behaves like one large magnetic top, although this is probably not true even within a domain. However, the relationship given in eqn. 2.10 is not restricted to an ideal classical material; it has no restrictions except for a symmetry about the direction of static magnetisation. Provided that no reference is made to a specific physical model of the material, μ and κ may be arbitrary quantities which will be constant when the operating frequency and the applied static magnetic field are constant. The relations are generally applicable to any isotropic substance since the only condition needed to be satisfied by the permeability tensor is the rotational symmetry about the axis of the static magnetic field.

Therefore, for conditions when eqn. 2.9 does not apply, μ and κ may be measured and their values inserted into eqn. 2.10. However, it has been found experimentally that eqns. 2.9 and 2.11 do apply when the ferrite is magnetised to saturation, and values for μ and κ can be calculated using these equations. For most ferrite materials, γ/μ_0 is $1{\cdot}76 \times 10^{11}$ rad/s T (or 28 GHz/T). Values are quoted for γ/μ_0 because, in most practical systems, H_0 is an applied magnetic flux density rather than a magnetic field strength and will be measured in units of Tesla rather than A/m. Since eqn. 2.9 only involves ratios, it is possible to perform the complete calculation in gigahertz.

If the applied magnetic field is represented by the angular frequency ω_0 and the internal magnetisation is represented by ω_m, where

$$\omega_0 = \gamma H_0 \tag{2.14}$$

and

$$\omega_m = \frac{\gamma M_0}{\mu_0}$$

then μ and κ become

$$\left.\begin{aligned} \mu &= \mu_0\left(1 + \frac{\omega_0\omega_m}{\omega_0^2 - \omega^2}\right) \\ \kappa &= -\mu_0\frac{\omega\omega_m}{\omega_0^2 - \omega^2} \end{aligned}\right\} \qquad (2.15)$$

ω_0 is the resonant frequency in the ferrite, which is the frequency at which the components of the tensor permeability become infinite and the frequency at which resonance absorption occurs. In a real ferrite material, damping limits the amplitude attained by the components of the tensor permeability at resonance.

2.4 Damping

Derivation of the permeability tensor in the previous section ignored the effect of any damping in the ferrite or of any absorption of a.c. power by the ferrite. A possible expression for the magnetisation equation including a damping term is given in eqn. 2.5. The constant in the damping term in this equation can be replaced by[3]

$$\frac{\lambda}{|M^2|} = \frac{\gamma\alpha}{|M|} \qquad (2.16)$$

so that α is a dimensionless damping constant. If the damping is small, eqn. 2.4 can be substituted into eqn. 2.5 as modified by eqn. 2.16 to give

$$\frac{d\boldsymbol{M}}{dt} = \gamma\boldsymbol{M} \times \boldsymbol{H} - \frac{\alpha}{|\boldsymbol{M}|}\boldsymbol{M} \times \frac{d\boldsymbol{M}}{dt} \qquad (2.17)$$

Considering the magnetically biased ferrite together with the a.c. magnetic field introduced in the previous section, then in the last term

$$|\boldsymbol{M}| \approx M_0$$

$$\boldsymbol{M} \approx \boldsymbol{z}M_0$$

$$\frac{d\boldsymbol{M}}{dt} = j\omega\boldsymbol{M}$$

and eqn. 2.17 becomes

$$j\omega\boldsymbol{M} = \gamma M_0\boldsymbol{z} \times \boldsymbol{H} - \gamma H_0\boldsymbol{z} \times \boldsymbol{M} - j\omega\alpha\boldsymbol{z} \times \boldsymbol{M}$$

or

$$j\omega \boldsymbol{M} = \gamma M_0 \boldsymbol{z} \times \boldsymbol{H} - (\gamma H_0 + j\omega\alpha)\boldsymbol{z} \times \boldsymbol{M} \tag{2.18}$$

which is similar to eqn. 2.6. It is seen that γH_0 in eqn. 2.6 is replaced by $(\gamma H_0 + j\omega\alpha)$ in eqn. 2.18, and the same substitution may be made in eqn. 2.9 to give the components of the tensor permeability, including losses

$$\chi = \frac{\gamma M_0(\gamma H_0 + j\omega\alpha)}{(\gamma H_0 + j\omega\alpha)^2 - \omega^2} \tag{2.19}$$

and

$$\kappa = \frac{-\omega\gamma M_0}{(\gamma H_0 + j\omega\alpha)^2 - \omega^2} \tag{2.20}$$

with

$$\mu = \mu_0 + \chi \tag{2.11}$$

If χ and κ are written as complex quantities to give a field representation of the damping

$$\chi = \chi' - j\chi'' \tag{2.21}$$

$$\kappa = \kappa' - j\kappa'' \tag{2.22}$$

then

$$\chi' = \frac{\gamma^2 M_0 H_0\{\gamma^2 H_0^2 - \omega^2(1 - \alpha^2)\}}{\{\gamma^2 H_0^2 - \omega^2(1 + \alpha^2)\}^2 + 4\gamma^2 H_0^2\omega^2\alpha^2} \tag{2.23}$$

$$\chi'' = \frac{-\gamma M_0\omega\alpha\{\gamma^2 H_0^2 + \omega^2(1 + \alpha^2)\}}{\{\gamma^2 H_0^2 - \omega^2(1 + \alpha^2)\}^2 + 4\gamma^2 H_0^2\omega^2\alpha^2} \tag{2.24}$$

and

$$\kappa' = \frac{-\omega\gamma M_0\{\gamma^2 H_0^2 - \omega^2(1 + \alpha^2)\}}{\{\gamma^2 H_0^2 - \omega^2(1 + \alpha^2)\}^2 + 4\gamma^2 H_0^2\omega^2\alpha^2} \tag{2.25}$$

$$\kappa'' = \frac{-2\gamma^2 M_0 H_0\omega^2\alpha}{\{\gamma^2 H_0^2 - \omega^2(1 + \alpha^2)\}^2 + 4\gamma^2 H_0^2\omega^2\alpha^2} \tag{2.26}$$

Without any damping, eqn. 2.9 shows that χ and κ go through a resonance at $\omega = \gamma H_0$. This resonant value of the angular frequency has already been defined by

$$\omega_0 = \gamma H_0 \tag{2.14}$$

Most ferrites used for microwave applications have a low loss at microwave frequencies. Then $\alpha \ll 1$, and the resonant frequency is also given by eqn. 2.14. The imaginary components of χ and κ give a measure of the power absorbed in the ferrite due to resonance. The resonant linewidth is the width of the resonance curve where the magnitude of χ'' or κ'' is half its peak value. Resonance curves are obtained by varying either the frequency, ω, or the applied magnetic field, H_0, so that the resonance linewidth is specified as $\Delta\omega$ or ΔH. Substituting values into eqn. 2.24 or 2.26 gives

$$\Delta\omega = \gamma\Delta H = 2\omega_0\alpha \tag{2.27}$$

so that the damping constant can be related to measured quantities. In the approximation for small α the peak values of χ'' and κ'' at resonance, when $\omega = \omega_0$, are the same

$$\chi'' = \kappa'' = \frac{\omega_m}{2\omega_0\alpha} \tag{2.28}$$

Some curves of the values of the real and imaginary parts of the components of the susceptibility tensor are given in Figs. 2.2, 2.3, and 2.4. The horizontal axis

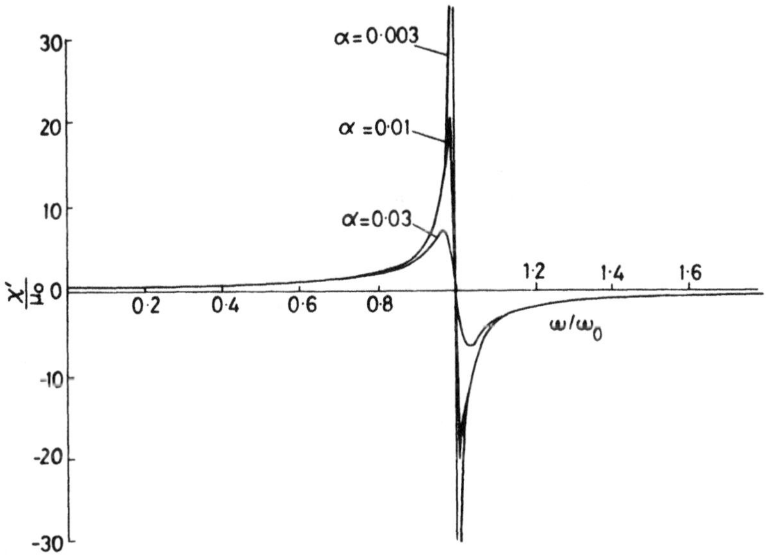

Fig. 2.2 *Real part of the diagonal component of the tensor susceptibility, χ', plotted against normalised frequency for materials having different values of damping factor, α, and for a static magnetic field such that $\omega_0 = 1{\cdot}25\,\omega_m$*

is scaled in frequency normalised to the resonant frequency, and the curves are drawn for a ferrite material where the magnetisation M_0 is given by

$$M_0 = \mu_0 \frac{0{\cdot}8\,\omega_0}{\gamma} = 0{\cdot}8\,\mu_0 H_0$$

or

$$\omega_0 = 1{\cdot}25\,\omega_m$$

Three different values of damping constant have been used and the curves are labelled with the value of the damping constant.

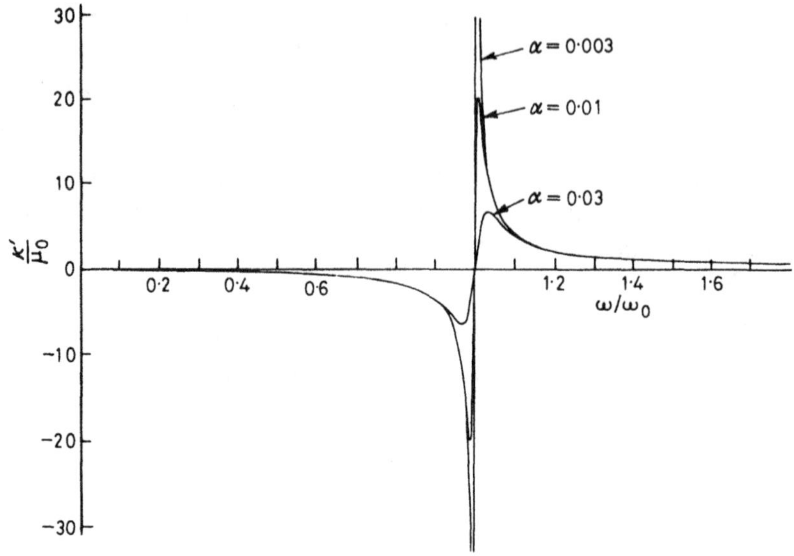

Fig. 2.3 *Real part of the cross-diagonal component of the tensor susceptibility, κ', plotted against normalised frequency for materials having different values of damping factor, α, and for a static magnetic field such that $\omega_0 = 1{\cdot}25\,\omega_m$*

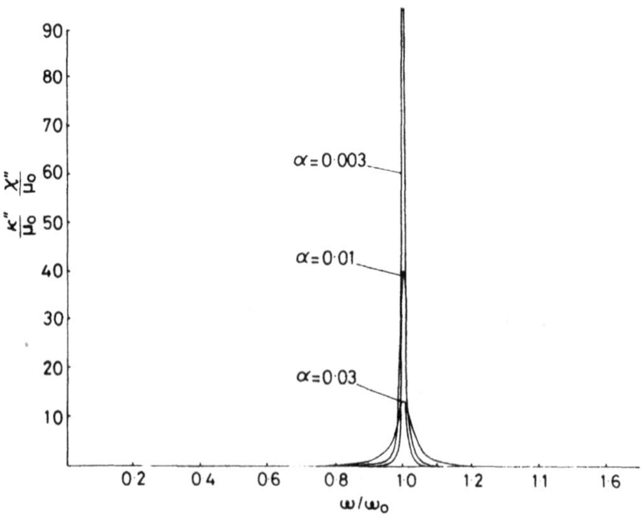

Fig. 2.4 *Imaginary part of both components of the tensor susceptibility, $\chi'' \approx \kappa''$, plotted against normalised frequency for materials having different values of damping factor, α, and for a static magnetic field such that $\omega_0 = 1{\cdot}25\,\omega_m$*

2.5 Longitudinal magnetisation

Subsequent chapters of this book analyse various ferrite loaded guided wave structures. However, it is mathematically simpler to investigate the properties of plane wave propagation in an infinite ferrite medium, which can also give a qualitative description of the properties of ferrite loaded guided wave structures. The rest of this chapter is devoted to the properties of plane wave propagation in ferrite media. A simple mathematical analysis is presented in this section, such as appears in the more elementary microwave textbooks.[4] We consider a lossless ferrite medium with the direction of propagation of the electromagnetic wave parallel to the direction of the static magnetising field. In a rectangular co-ordinate system, take z as the direction of the static magnetic field, then eqn. 2.10 relates the microwave $\boldsymbol{B}$ and $\boldsymbol{H}$ in the ferrite. The ferrite can be considered as a normal insulator with a permittivity ε and the permeability tensor given by eqn. 2.13. The electromagnetic fields in the ferrite are governed by Maxwell's equations in a non-conducting medium

$$\nabla \cdot \boldsymbol{D} = 0 \tag{2.29}$$

$$\nabla \cdot \boldsymbol{B} = 0 \tag{2.30}$$

$$\nabla \times \boldsymbol{E} = -\frac{\partial \boldsymbol{B}}{\partial t} \tag{2.31}$$

$$\nabla \times \boldsymbol{H} = \frac{\partial \boldsymbol{D}}{\partial t} \tag{2.32}$$

If the wave is propagating in the z-direction with a propagation constant β, and its angular frequency is ω, and if the conditions for a plane wave

$$\frac{\partial}{\partial x} = \frac{\partial}{\partial y} = 0$$

are substituted into eqn. 2.32, it becomes

$$\left.\begin{aligned} \beta H_y &= \omega\varepsilon E_x \\ \beta H_x &= -\omega\varepsilon E_y \\ 0 &= E_z \end{aligned}\right\} \tag{2.33}$$

If eqn. 2.10 is substituted into eqn. 2.31 together with the conditions of the plane wave, we obtain

$$\left.\begin{aligned} \beta E_y &= -\omega\mu H_x - j\omega\kappa H_y \\ \beta E_x &= -j\omega\kappa H_x + \omega\mu H_y \\ 0 &= H_z \end{aligned}\right\} \tag{2.34}$$

Elimination of the magnetic field components between eqns. 2.33 and 2.34 gives

$$\left.\begin{aligned}(\beta^2 - \omega^2\mu\varepsilon)E_y &= j\omega^2\varepsilon\kappa E_x \\ (\beta^2 - \omega^2\mu\varepsilon)E_x &= -j\omega^2\varepsilon\kappa E_y\end{aligned}\right\} \tag{2.35}$$

Eliminating the electric field components from eqn. 2.35 gives the propagation conditions of a plane wave

$$(\beta^2 - \omega^2\mu\varepsilon)^2 = \omega^4\varepsilon^2\kappa^2 \tag{2.36}$$

whence

$$\beta^2 - \omega^2\mu\varepsilon = \pm\, \omega^2\varepsilon\kappa$$

or

$$\beta^2 = \omega^2\varepsilon(\mu \pm \kappa) \tag{2.37}$$

There are two solutions for β showing that two modes of propagation are possible which still have all their components of the fields in the transverse plane. Now define the two propagation constants

$$\beta^+ = \omega\sqrt{[\varepsilon(\mu + \kappa)]} \tag{2.38}$$

$$\beta^- = \omega\sqrt{[\varepsilon(\mu - \kappa)]} \tag{2.39}$$

There are two possible modes of propagation through a magnetised ferrite material. They have been labelled 'positive' and 'negative' modes as shown by the notation used in eqns. 2.38 and 2.39. These modes do not show the independence between the two transverse components of the fields possessed by a plane wave propagating in a normal dielectric medium. Both components of the electric and magnetic fields exist in the transverse plane. Substitution for β into eqn. 2.35 gives the relationship between the components of the electric field, showing that

$$E_y = \mp\, jE_x$$

Continuing the notation of eqns. 2.38 and 2.39 for the components of the fields of the two modes

$$E_y^+ = -jE_x^+ \tag{2.40}$$

$$E_y^- = jE_x^- \tag{2.41}$$

The relationship given in eqns. 2.40 and 2.41 is that of a circularly polarised wave. Eqn. 2.40 describes a positive circularly polarised wave or a right-hand circularly polarised wave and eqn. 2.41 describes a negative circularly polarised wave or a left-hand circularly polarised wave. Therefore β^+ is the propagation constant of a positive circularly polarised wave and β^- is the propagation constant of a negative circularly polarised wave. It is seen that, for propagation through a magnetically biased ferrite material, circularly polarised modes are the fundamental modes of propagation.

From eqn. 2.33 the components of the magnetic field can be obtained, giving

$$\left.\begin{aligned} H_y &= \frac{\omega\varepsilon}{\beta} E_x \\ H_x &= -\frac{\omega\varepsilon}{\beta} E_y \end{aligned}\right\} \qquad (2.42)$$

The constant of proportionality can be shown to be equivalent to an effective free-space impedance of the ferrite material

$$\frac{\beta}{\omega\varepsilon} = \sqrt{\left(\frac{\mu \pm \kappa}{\varepsilon}\right)} = \eta^{\pm} \qquad (2.43)$$

In the expressions for the phase constant and effective impedance of a circularly polarised wave propagating through a magnetised ferrite material, the same expression occurs representing an effective permeability of the ferrite medium to the circularly polarised wave. The *effective permeability* of the ferrite material is defined by

$$\left.\begin{aligned} \mu^+ &= \mu + \kappa \\ \mu^- &= \mu - \kappa \end{aligned}\right\} \qquad (2.44)$$

Substituting from eqns. 2.9 and 2.11 into eqn. 2.44 shows that the effective permeability is dependent on the frequency of operation, the size of the static magnetic field, H_0, and the properties of the particular ferrite material

$$\left.\begin{aligned} \mu^+ &= \mu_0 + \frac{\gamma M_0}{\gamma H_0 + \omega} \\ \mu^- &= \mu_0 + \frac{\gamma M_0}{\gamma H_0 - \omega} \end{aligned}\right\} \qquad (2.45)$$

It is also different for the two hands of circular polarisation and is different for a different relationship between the direction of static magnetisation and the direction of propagation of the plane wave. The variation of effective permeability with change of static magnetic field at a fixed frequency is shown in Fig. 2.5. Below saturation there is an approximately linear relationship between the values of μ^+ and μ^- and the applied magnetic field, which can be seen from a study of eqn. 2.45, since γM_0 is proportional to applied field. Alternatively, eqns. 2.9 and 2.11 show that, to a first approximation for low fields, μ is constant and κ is proportional to the static magnetic field. Above saturation, there is a much slower variation in properties, owing to the fact that M_0 has become constant, until the resonance condition is reached. At resonance, both μ and κ become infinite but eqn. 2.45 shows that only μ^- goes through resonance. When the applied magnetic field is such that the operating frequency is smaller than the resonant frequency, eqn. 2.9 shows that κ is negative and it is only when the operating frequency is larger than the resonant frequency that κ is positive.

Since there is an approximately linear variation of permeability below saturation, numerical values can be used for μ and κ even when the ferrite is not magnetised to saturation.

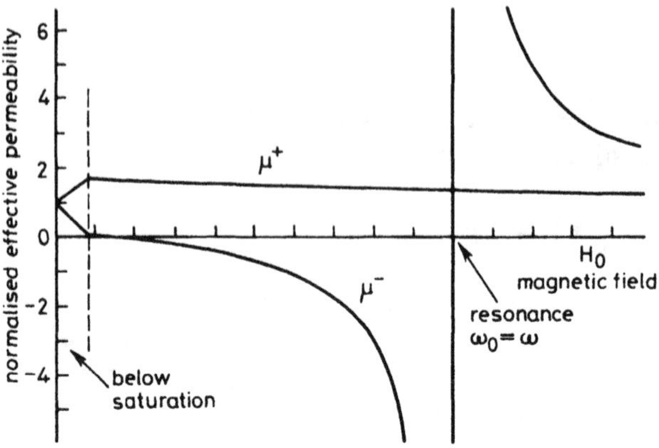

Fig. 2.5 *Variation of the microwave normalised effective permeability, μ/μ_0, with change of magnetic field of an infinite ferrite medium for two opposite hands of circularly polarised plane waves at the fixed frequency $\omega = 1{\cdot}25\,\omega_m$*

2.6 Attenuation

For a plane wave propagating in a longitudinally magnetised ferrite material, the effects of damping in the ferrite may be considered by substituting $(\gamma H_0 + j\omega\alpha)$ for γH_0 in eqn. 2.45, or by substituting expressions for the components of the susceptibility tensor from eqns. 2.23 to 2.26 into eqn. 2.44, which give the real and imaginary parts of the effective permeability

$$\mu^{+\prime} = \mu_0 + \frac{\gamma M_0(\gamma H_0 + \omega)}{(\gamma H_0 + \omega)^2 + \omega^2\alpha^2} \tag{2.46}$$

$$\mu^{+\prime\prime} = \frac{\gamma M_0 \omega\alpha}{(\gamma H_0 + \omega)^2 + \omega^2\alpha^2} \tag{2.47}$$

$$\mu^{-\prime} = \mu_0 + \frac{\gamma M_0(\gamma H_0 - \omega)}{(\gamma H_0 - \omega)^2 + \omega^2\alpha^2} \tag{2.48}$$

$$\mu^{-\prime\prime} = \frac{\gamma M_0 \omega\alpha}{(\gamma H_0 - \omega)^2 + \omega^2\alpha^2} \tag{2.49}$$

It is seen that these expressions are much simpler than eqns. 2.23 to 2.26, confirming the simplicity of the circularly polarised modes in longitudinally magnetised ferrite.

Expressions for the phase constant and attenuation constant of a plane wave propagating through a lossy material are both functions of the real and imaginary parts of the permeability and permittivity. The full expressions are given by Clarricoats.[5] For small losses, the calculation is usually performed as a perturbation on the lossless condition where the propagation constant is unchanged and the attenuation constant is a function of the loss. When the losses are appreciable, however, the phase constant is no longer the same as for the lossless condition. For most materials used in microwave ferrite devices, the dielectric loss is negligible compared to the magnetic loss, thus the imaginary part of the permittivity can be assumed to be zero. This leads to simpler expressions for the phase and attenuation constants, which are given by Lax and Button.[6] Expressed in a slightly different form, and in our notation, they are,

$$\left.\begin{aligned}\text{phase constant } \beta &= \tfrac{1}{2}\omega(\varepsilon\mu')^{1/2}\,[\{1 + (\mu''/\mu')^2\}^{1/2} + 1]^{1/2}\\ \text{attenuation constant} &= \tfrac{1}{2}\omega(\varepsilon\mu')^{1/2}\,[\{1 + (\mu''/\mu')^2\}^{1/2} - 1]^{1/2}\end{aligned}\right\} \quad (2.50)$$

where μ' and μ'' are the real and imaginary parts of the effective permeability given by eqns. 2.46 to 2.49. Suitable superscripts for the positive or negative hands of circular polarisation have been omitted from the above expressions for simplicity, and need to be assumed, i.e. to find the phase and attenuation constants of the positive circularly polarised wave, permeability values given by eqns. 2.46 and 2.47 are used.

It is seen from Fig. 2.5 that the effective permeability for the negative circularly polarised wave can have a negative value. When this is so, the wave is unable to propagate through the medium and becomes cut off. The phase constant becomes small and the attenuation constant becomes large. However, except at resonance, the power in the electromagnetic wave is not absorbed in the ferrite material but rejected from entering the material by the very low value of phase constant. Some values of the phase and attenuation constant with variation of frequency are plotted in Fig. 2.6. They have been normalised to the value of the phase constant in unmagnetised ferrite at the resonant frequency

$$\beta_0 = \omega_0\sqrt{(\varepsilon\mu_0)}$$

It will be seen that there is a range of frequencies where the phase constant is very small and the attenuation constant is appreciable. At the resonant frequency, both phase and attenuation constants are large, but at a higher cut-off frequency the phase constant becomes small and the attenuation constant increases, a true cut-off effect. For the lossless condition when the damping constant α is zero, eqn. 2.48 shows that $\mu^{-\prime}$ becomes zero at the resonant frequency, $\omega_0 = \gamma H_0$, and at a higher cut-off frequency

$$\omega_c = \gamma H_0 + \gamma M_0/\mu_0$$

Normal propagation cannot occur in the magnetised ferrite between these two frequencies. The curves on Fig. 2.6 show that the higher cut-off frequency value is slightly modified when the damping constant is no longer zero. There is a small similarity to the effect in unmagnetised plasma where propagation is not possible below the plasma frequency, which also acts as a cut-off frequency.

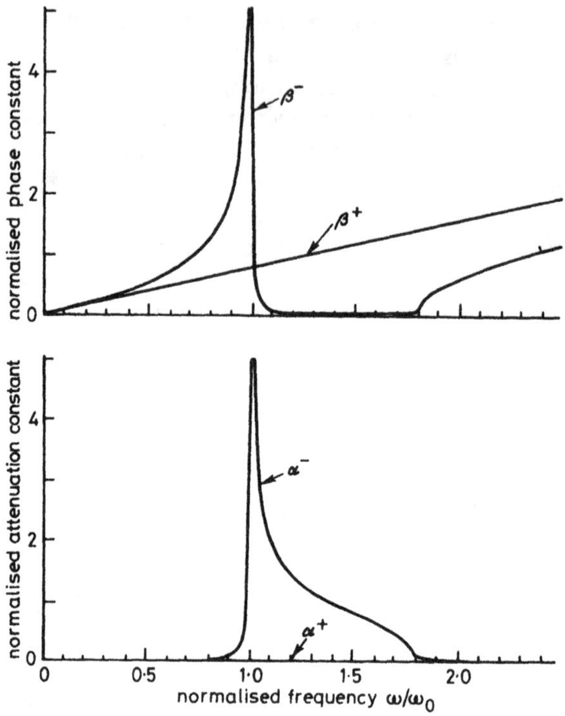

Fig. 2.6 *Normalised phase and attenuation constants, $\beta/\omega_0\sqrt{(\varepsilon\mu_0)}$, versus change in frequency for two opposite hands of circularly polarised plane waves propagating through an infinite ferrite medium with a static magnetic field such that $\omega_0 = 1{\cdot}25\,\omega_m$ and a loss factor $\alpha = 0{\cdot}01$*

2.7 Magnetisation at any angle

Having considered the relatively simple analysis of the propagating fields of a plane wave propagating through an infinite ferrite medium which is magnetised in the direction of propagation of the microwave field, it is now useful to proceed to find the propagation constant of a wave which is propagating at an arbitrary angle to the direction of static magnetisation in the ferrite. Other textbooks[7,8] and the original paper by Polder[2] take the plane wave to be propagating at an angle to the rectangular system of axes and the tensor permeability to be given by eqn. 2.13. To give a different approach to the same problem and to highlight a useful alternative form of the Polder tensor in eqn. 2.13, we consider the ferrite material to be magnetised at an arbitrary angle to the rectangular system of axes and the direction of propagation of the wave to be in the z-direction, as shown in Fig. 2.7. Assuming that the direction of magnetisation is in the x-z-plane, the tensor permeability is

$$\begin{vmatrix} \mu\cos^2\theta + \mu_0\sin^2\theta & j\kappa\cos\theta & (\mu_0 - \mu)\sin\theta\cos\theta \\ -j\kappa\cos\theta & \mu & j\kappa\sin\theta \\ (\mu_0 - \mu)\sin\theta\cos\theta & -j\kappa\sin\theta & \mu\sin^2\theta + \mu_0\cos^2\theta \end{vmatrix} \quad (2.51)$$

A similar expression is given by Hlawiczka.[9]

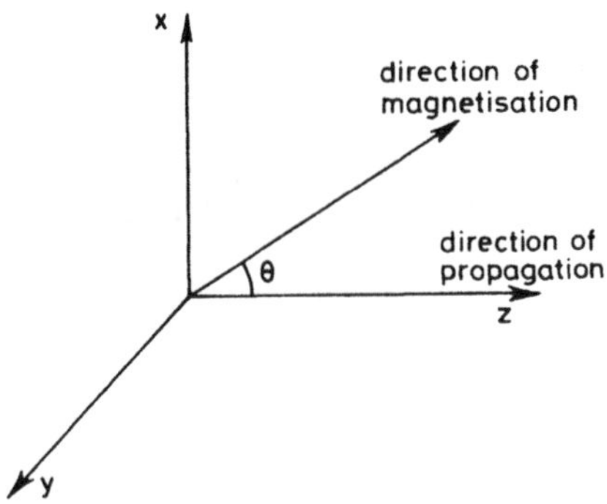

Fig. 2.7 *Co-ordinates of a wave propagating at an arbitrary angle to the direction of static magnetisation in a ferrite medium*

An even more general form of the permeability tensor is given by Tyras,[10,11] where the direction of static magnetisation of the ferrite is at an arbitrary angle to both the x- and z-axes. He concludes his paper[10] by showing that the mathematics involved in having the direction of propagation of a plane wave at an arbitrary angle to the axes of the co-ordinate system is much simpler than using the permeability tensor given in eqn. 2.51.

However, for completeness, we will use eqn. 2.51 to find the propagating conditions for a plane wave propagating at an arbitrary angle to the direction of static magnetisation in the ferrite. With the conditions of a plane wave propagating in the z-direction, Maxwell's equation, 2.31, becomes eqn. 2.33 as before, and eqn. 2.32 becomes

$$\left.\begin{aligned} j\beta E_y &= -j\omega B_x \\ -j\beta E_x &= -j\omega B_y \\ 0 &= -j\omega B_z \end{aligned}\right\} \quad (2.52)$$

It is interesting to note that, because it is a plane wave, the longitudinal component of the magnetic flux density, B_z, is zero, but there will still be a finite longitudinal component of the magnetic field intensity, H_z. Solution of eqns. 2.33, 2.51 and 2.52 to eliminate the electric field components gives

$$[\beta^2 + \omega^2\varepsilon(\mu\cos^2\theta + \mu_0\sin^2\theta)]H_x - j\omega^2\varepsilon\kappa\cos\theta\, H_y + \omega^2\varepsilon(\mu_0 - \mu)\sin\theta\cos\theta\, H_z = 0$$

$$j\omega^2\varepsilon\kappa\cos\theta\, H_x + (\beta^2 + \omega^2\varepsilon\mu)H_y - j\omega^2\varepsilon\kappa\sin\theta\, H_z = 0$$

$$(\mu_0 - \mu)\sin\theta\cos\theta\, H_x + j\kappa\sin\theta\, H_y + (\mu\sin^2\theta + \mu_0\cos^2\theta)H_z = 0$$

Eliminating the magnetic field components from the above equations gives the propagating conditions

$$(\beta^2/\omega^2\varepsilon)^2(\mu\sin^2\theta + \mu_0\cos^2\theta) + (\beta^2/\omega^2\varepsilon)[(\kappa^2 - \mu^2 + \mu\mu_0)\sin^2\theta - 2\mu\mu_0] + \mu_0(\mu^2 - \kappa^2) = 0 \qquad (2.53)$$

When solved, this gives an expression for the propagation constant of a plane wave in an infinite ferrite medium which is the same as that given in the original work by Polder[2]

$$\frac{\beta_\pm^2}{\omega^2\varepsilon} = \{(\mu^2 - \mu\mu_0 - \kappa^2)\sin^2\theta + 2\mu\mu_0 \pm [(\mu^2 - \mu\mu_0 - \kappa^2)^2\sin^4\theta + 4\mu_0^2\kappa^2\cos^2\theta]^{1/2}\}/2(\mu_0\cos^2\theta + \mu\sin^2\theta) \qquad (2.54)$$

The same result is derived by Clarricoats[7] and Lax and Button[8] but using the mathematically simpler process of having the permeability tensor described by eqn. 2.13 and the direction of propagation of the electromagnetic wave at an arbitrary angle to the direction of the rectangular system of co-ordinates.

Although there are very few, if any, applications where the direction of propagation of an electromagnetic wave in a magnetised ferrite material is at an arbitrary angle to the direction of static magnetisation, it is still useful to study some of the implications of eqn. 2.54. Taking typical values for μ and κ for a ferrite material magnetised to saturation but a long way away from resonance, $\mu = \mu_0$ and $\kappa = 0{\cdot}75\,\mu_0$, Fig. 2.8 shows the variation of the normalised effective permeability, $\mu^\pm/\mu_0$, as the angle θ is varied. The square of the normalised propagation constant is the same as the normalised effective permeability

$$\frac{\mu^\pm}{\mu_0} = \frac{\beta_\pm^2}{\omega^2\varepsilon\mu_0}$$

Fig. 2.8 shows that, working well away from resonance, the propagation constant varies in a well behaved manner between its values when the direction of propagation is parallel to or perpendicular to the direction of the static magnetic field. However, eqn. 2.54 shows that the effective permeability is a complicated function of angle and the frequency for resonance is dependent on angle. Substituting for μ from eqn. 2.9 into the denominator of the expression on the right-hand side of eqn. 2.54 shows that the denominator becomes

$$2(\gamma^2 H_0^2 - \omega^2 + \gamma^2 H_0 M_0 \sin^2\theta) \qquad (2.55)$$

which goes to zero at a frequency depending on θ. Soohoo[12] states that this expression gives the resonant frequency, but this is too simple. The variation of effective permeability is a much more complicated function of angle, which is

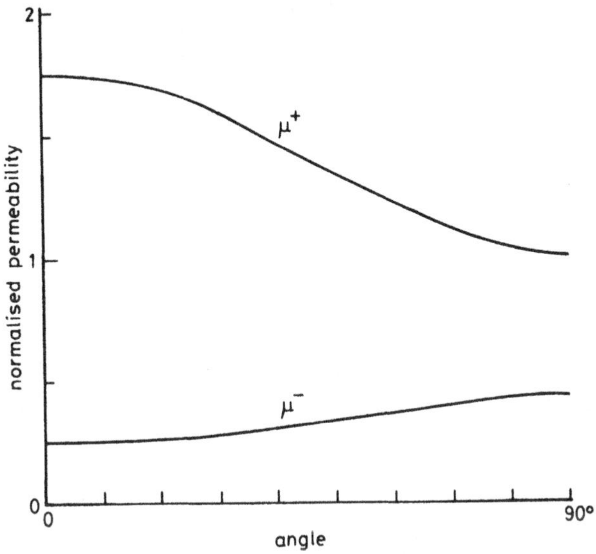

Fig. 2.8 *Normalised effective permeability, μ/μ_0, of an infinite ferrite medium as the angle varies between the direction of the static magnetisation and the direction of propagation of two oppositely rotating elliptically polarised plane waves for $\mu = \mu_0$ and $\kappa = 0{\cdot}75\,\mu_0$*

shown in Fig. 2.9 for a number of frequencies close to the resonant frequency ω_0 as previously defined by eqn. 2.14. There is an angular dependent resonance but it is not that given by a zero value for the denominator expression in eqn. 2.55. The values of μ and κ used to compute the curves shown in Fig. 2.9 are those given by the lowest loss curves in Figs. 2.2 and 2.3. As in Fig. 2.5, only the negative circularly polarised wave experiences a resonance condition. Hlawiczka[13] shows that these waves are really elliptically polarised. They are circularly polarised when $\theta = 0$ and linearly polarised when $\theta = 90°$. He gives curves of ellipticity of the modes as the angle is varied when $\mu = \mu_0$ and $\mu = 0{\cdot}9\,\mu_0$ for various values of κ.

2.8 Faraday rotation

Although, as we have shown, circularly polarised waves are the fundamental modes of propagation in a ferrite material which is magnetised in the direction of propagation, linearly polarised waves occur more frequently in practice. Then any linearly polarised wave will be separated into its circularly polarised components while in the ferrite material. Any attempt to detect a linearly polarised wave in the ferrite will measure the sum of the two circularly polarised waves. The propagation constants for the two opposite hands of circular polarisation are different, as shown by eqns. 2.38 and 2.39, so that one hand of circular polarisation will rotate further than the other in a fixed length of ferrite material.

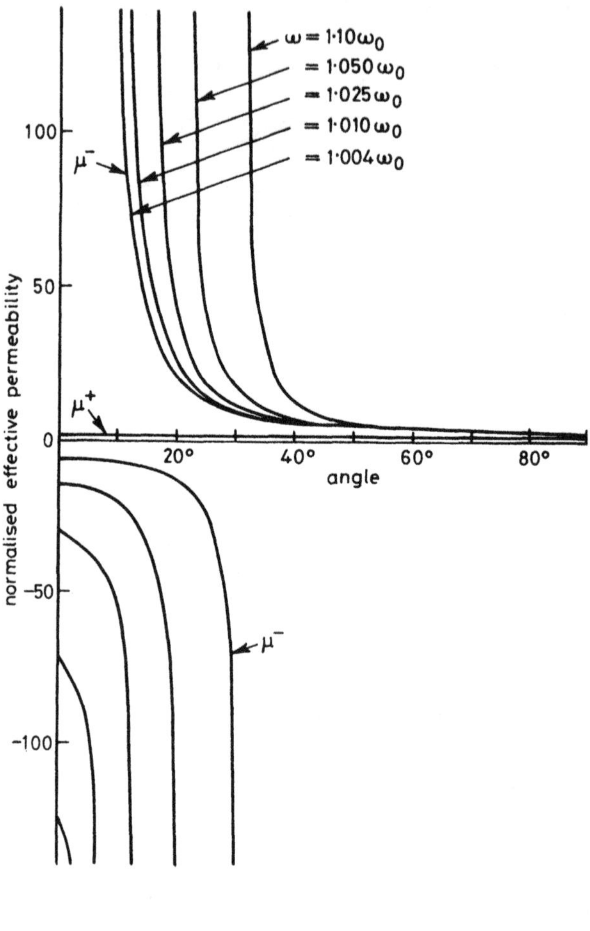

Fig. 2.9 *Normalised effective permeability, μ/μ_0, of an infinite ferrite medium as the angle varies between the direction of static magnetisation and the direction of propagation of two oppositely rotating elliptically polarised plane waves with a static magnetic field such that $\omega_0 = 1{\cdot}25\,\omega_m$, at a number of frequencies near to normal resonance. The positive effective permeability is unaffected by resonance*

Therefore the plane of polarisation of the detected linearly polarised wave will be rotated compared with the incident wave. The rotation was first observed by Faraday propagating light through paramagnetic liquids and is called *Faraday rotation*.

In propagating through a unit length, the phase change of the circularly polarised waves is given by

$$\phi^{\pm} = \beta^{\pm} \tag{2.56}$$

Fig. 2.10 shows that the angle of rotation is the difference between the angles through which each circularly polarised mode rotates so that the angle of rotation per unit length of the equivalent linearly polarised wave is given by

$$\psi = \tfrac{1}{2}(\beta^{+} - \beta^{-}) \tag{2.57}$$

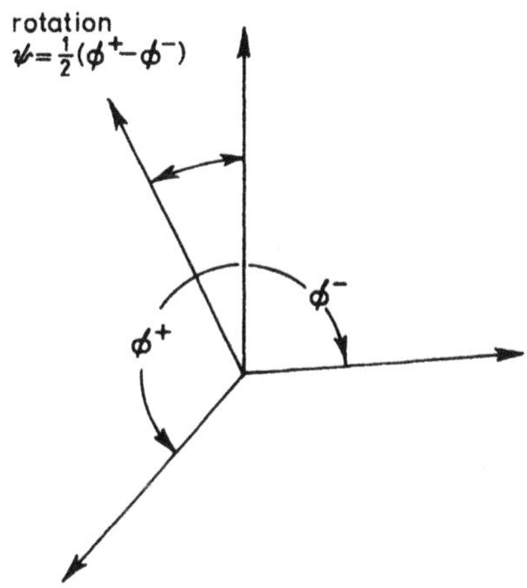

Fig. 2.10 *Angle of rotation of the positive and negative circularly polarised waves and the consequent angle of rotation of the linearly polarised wave*

The rotations first observed by Faraday of the plane of polarisation of light through a paramagnetic liquid were very small, only a few degrees in many wavelengths, but with magnetic substances such as ferrites at microwave frequencies, rotations of 90° are easily obtainable in fractions of a wavelength. It can be seen from eqns. 2.47 and 2.49 that there is a difference between the magnetic loss component of the effective permeability of the positive and negative circularly polarised waves. If the magnetic loss is very small, any difference between the losses for the two waves will be negligible. If the loss cannot be neglected, the attenuation of the two waves will be different and the resultant wave will be elliptically polarised. In many practical situations, when the losses are small, the small amount of ellipticity in the resultant linear polarisation can be neglected, but there may be situations when such ellipticity can cause poor performance of a ferrite device.

In many control situations, the ferrite material is operated below saturation. Fig. 2.5 shows that there is an approximately linear relationship between the values of μ^{+} and μ^{-} and the applied magnetic field. Then there is a simple approximation which shows that rotation is linearly proportional to the applied magnetic field. With the ferrite operating well below resonance in the unsaturated region, as shown in Fig. 2.5

$$\gamma H_0 \ll \omega, \quad \text{and} \quad \frac{\gamma M_0}{\mu_0} \ll \omega$$

If these approximations are substituted into eqn. 2.45, we obtain,

$$\mu^+ = \mu_0 + \frac{\gamma M_0}{\omega}$$

$$\mu^- = \mu_0 - \frac{\gamma M_0}{\omega}$$

Substitution into eqns. 2.38 and 2.39 give

$$\beta^{\pm} = \omega\sqrt{(\varepsilon\mu_0)}\left(1 \pm \frac{\gamma M_0}{2\omega\mu_0}\right)$$

The rotation is given by substitution in eqn. 2.57

$$\psi = \tfrac{1}{2}\gamma M_0\sqrt{(\varepsilon/\mu_0)} \tag{2.58}$$

Because for small fields M_0 is proportional to H_0, the rotation is proportional to the field. This linear relationship between the applied magnetic field and the rotation has been shown experimentally to be true for ferrite in circular waveguide,[14] and is shown in Fig. 5.2.

2.9 Transverse magnetisation

A system that occurs frequently in ferrite devices in rectangular waveguide is for the ferrite to be magnetised in a direction perpendicular to the direction of propagation. The ferrite is said to be magnetised transversely to the direction of propagation. It is also useful to analyse plane wave propagation through an infinite ferrite medium which is magnetised perpendicular to the direction of propagation. The propagation constants may be obtained most easily by putting $\theta = 90°$ in eqn. 2.54 but we will derive them by solving Maxwell's equations, as this gives an insight into the properties of the modes of propagation. In a rectangular system of co-ordinates, let the z-direction be the direction of propagation of the plane wave and the y-direction be the direction of the static magnetisation in the ferrite. Then the tensor permeability is, similar to eqn. 2.13

$$\boldsymbol{\mu} = \begin{vmatrix} \mu & 0 & j\kappa \\ 0 & \mu_0 & 0 \\ -j\kappa & 0 & \mu \end{vmatrix} \tag{2.59}$$

Applying the conditions of a plane wave to the two Maxwell curl equations, eqn. 2.32 becomes eqn. 2.33 and eqn. 2.31 together with eqn. 2.59 becomes

$$\left.\begin{aligned}\beta E_y &= -\omega\mu H_x - j\omega\kappa H_z \\ \beta E_x &= \omega\mu_0 H_y \\ 0 &= j\omega\kappa H_x - \omega\mu H_z\end{aligned}\right\} \tag{2.60}$$

Eliminating the electric field components from eqns. 2.33 and 2.60 gives

$$\beta^2 H_x = \omega^2\varepsilon\mu H_x + j\omega^2\varepsilon\kappa H_z \tag{2.61}$$

$$\beta^2 H_y = \omega^2\varepsilon\mu_0 H_y \tag{2.62}$$

$$\mu H_z = j\kappa H_x \tag{2.63}$$

These results show that the fields separate into two different modes of propagation, one involving the magnetic field component H_y and the other involving the magnetic field components H_x and H_z although B_z is zero because it is a plane wave. The plane wave described by eqn. 2.62 is a normal linearly polarised plane wave having the propagation constant

$$\beta^- = \omega\sqrt{(\varepsilon\mu_0)} \tag{2.64}$$

It has the same properties as a plane wave propagating in a simple dielectric medium. It is the mode in which the magnetic field component is parallel to the direction of the applied static magnetic field, H_0. The other mode is also a plane wave described by eqns. 2.61 and 2.63. Eliminating the magnetic field components between eqns. 2.61 and 2.63 gives the propagation constant

$$\beta^+ = \omega\sqrt{\left(\varepsilon\frac{\mu^2 - \kappa^2}{\mu}\right)} \tag{2.65}$$

The designation + or − for these two modes is obtained from eqn. 2.54. The positive wave is also linearly polarised with magnetic field components perpendicular to the direction of the static magnetic field in the transverse plane and also in the direction of propagation, although the magnetic flux density, B_z, in the direction of propagation is zero. For the case of transverse magnetisation, there are also effective permeabilities for the linearly polarised plane waves

$$\mu^- = \mu_0$$

$$\mu^+ = (\mu^2 - \kappa^2)/\mu$$

although they are often designated

$$\mu^- = \mu^{\parallel}$$

$$\mu^+ = \mu^{\perp}$$

because the + and − no longer apply. Parallel and perpendicular as superscripts apply better to directions of linear polarisation. The propagation constant of the parallel wave as shown by eqn. 2.64 is the same as that of a plane wave propagating in a dielectric medium and its field components are also the same, thus this is called the *ordinary ray*. The perpendicular wave is affected by the gyromagnetic effect of the ferrite material and although it is also a plane wave it has a component of the magnetic field intensity in the direction of propagation so it is called the *extraordinary ray*. The equivalent phenomenon in optics presents a different refractive index to two rays with perpendicular linear directions of polarisation and is called *birefringence*. It is analogous to the Cotton–Mouton effect in optics. However, magneto-optic birefringence in paramagnetic liquids is small and is easily masked by any Faraday rotation of the ray which can arise owing to any slight misalignment of the magnetic field. Similar to Faraday rotation, birefringent phase differences of 90° or larger are easily obtainable using ferrite materials at microwave frequencies.

To further investigate the properties of the propagation constant of the perpendicular wave, we will look at the effective permeability

$$\mu^{\perp} = (\mu^2 - \kappa^2)/\mu$$

Substituting for μ and κ from eqn. 2.15 shows that

$$\mu^{\perp} = 1 + \frac{\omega_m(\omega_0 + \omega_m)}{\omega_0^2 - \omega^2 + \omega_0\omega_m} \tag{2.66}$$

This has a resonance at

$$\omega^2 = \omega_0(\omega_0 + \omega_m)$$

which is the geometric mean of the resonant frequency ω_0 and the higher cut-off frequency ω_c for the longitudinally magnetised ferrite. Unlike the longitudinally magnetised situation, the resonant frequency is a function of the saturation magnetisation of the ferrite material and not just a function of the applied magnetic field. Values for the real and imaginary parts of the components of the tensor permeability can be obtained from eqns. 2.23 to 2.26 to give the real and imaginary parts of the effective permeability

$$\mu^{\perp\prime} = \frac{\mu'(\mu'^2 - \mu''^2 - \kappa'^2 + \kappa''^2) + 2\mu''(\mu'\mu'' - \kappa'\kappa'')}{\mu'^2 + \mu''^2} \tag{2.67}$$

$$\mu^{\perp\prime\prime} = \frac{\mu'(\mu'^2 - \mu''^2 - \kappa'^2 + \kappa''^2) - 2\mu''(\mu'\mu'' - \kappa'\kappa'')}{\mu'^2 + \mu''^2} \tag{2.68}$$

The real and imaginary parts of this effective permeability are plotted in Figs. 2.11 and 2.12 against change of frequency for the condition $\alpha = 0{\cdot}01$ and for a number of ferrite materials having a range of saturation magnetisations. It is seen that, similar to propagation of a plane wave in longitudinally magnetised ferrite, there is a range of frequencies for which the wave is cut off. Analysis of eqn. 2.66 shows that $\mu^{\perp}$ has a resonance at $\omega^2 = \omega_0(\omega_0 + \omega_m)$ as already mentioned, and also goes to zero at $\omega_c = \omega_0 + \omega_m$. The normalised permeability is negative between these two frequencies, and the wave is cut off and cannot propagate in the ferrite material.

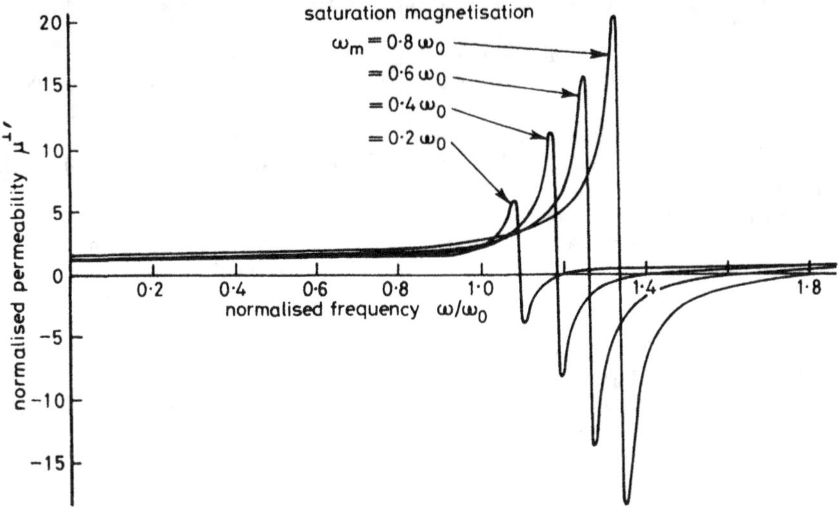

Fig. 2.11 *Real part of the normalised effective permeability of a transversely magnetised ferrite medium for the extraordinary wave versus change of frequency for a number of different ferrite materials having different saturation magnetisations but all having the loss factor α = 0·01*

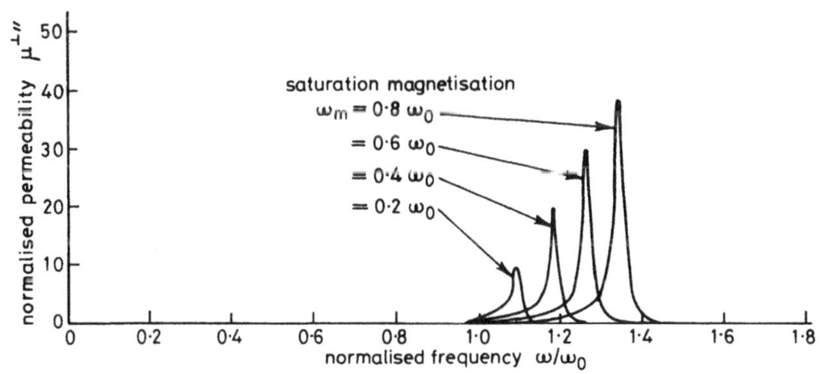

Fig. 2.12 *Imaginary part of the normalised effective permeability of a transversely magnetised ferrite medium for the extraordinary wave versus change of frequency for a number of different ferrite materials having different saturation magnetisations but all having the loss factor α = 0·01*

2.10 References

1 LANDAU, L. and LIFSHITZ, E.: 'On the theory of dispersion of magnetic permeability in ferromagnetic bodies', *Physik Z. Sowjetunion*, 1935, **8**, pp. 153–169 (in English)
2 POLDER, D.: 'On the theory of ferromagnetic resonance', *Phil. Mag.*, 1949, **40**, pp. 99–115
3 YAGER, W. A., GALT, J. K., MERRITT, F. R. and WOOD, E. A.: 'Ferromagnetic resonance in nickel ferrite', *Physical Rev.*, 1950, **80**, pp. 744–748

4 See, for example, BADEN FULLER, A. J.: 'Microwaves', 2nd edn. (Pergamon Press, 1979) Chapter 7
5 CLARRICOATS, P. J. B.: 'Microwave ferrites' (Chapman & Hall, 1961), p. 100
6 LAX, B. and BUTTON, K. J.: 'Microwave ferrites and ferrimagnetics' (McGraw-Hill, 1962), p. 300
7 CLARRICOATS, P. J. B.: 'Microwave ferrites' (Chapman & Hall, 1961), p. 98
8 LAX, B. and BUTTON, K. J.: 'Microwave ferrites and ferrimagnetics' (McGraw-Hill, 1962), p. 298
9 HLAWICZKA, P.: 'Gyrotropic waveguides' (Academic Press, 1981), p. 13
10 TYRAS, G.: 'The permeability matrix for a ferrite medium magnetized at an arbitrary direction and its eigenvalues', *IRE Trans.*, 1959, **MTT-7**, pp. 176–177
11 TYRAS, G.: 'Radiation and propagation of electromagnetic waves' (Academic Press, 1969), p. 60
12 SOOHOO, R. F.: 'Theory and application of ferrites' (Prentice-Hall, 1960), p. 118
13 HLAWICZKA, P.: 'Gyrotropic waveguides' (Academic Press, 1981), pp. 15–18
14 BADEN FULLER, A. J.: 'Microwave propagation through round waveguide partially filled with ferrite', *Proc. IEE*, 1961, **108C**, pp. 339–348

Chapter 3

Longitundinally magnetised ferrite in circular waveguide

3.1 Introduction

Although a useful appreciation of the properties of ferrite materials at microwave frequencies can be obtained from a study of electromagnetic wave propagation through an infinite extent of ferrite material, it is not a system that can be realised in practice. To be practical, the maximum size of any single piece of ferrite is about 0·1 m side or 0·0002 m^3 volume, thus it is only at optical frequencies that an infinite ferrite is a valid approximation. At microwave frequencies, any accurate analysis has to take account of the boundary of the ferrite material.

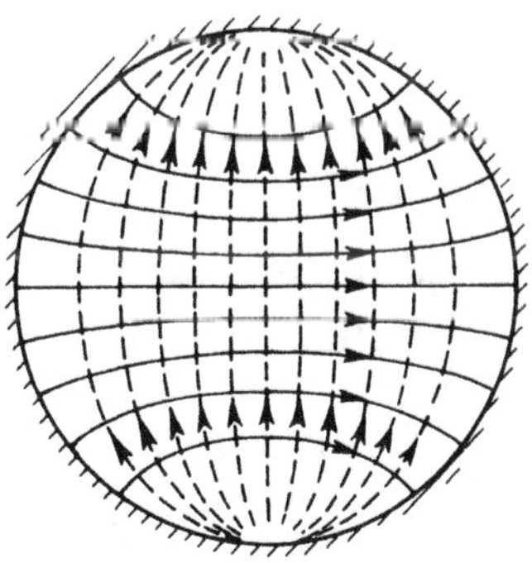

Fig. 3.1 *Fields in a cross-section of circular waveguide*
—— electric field
---- magnetic field

In a microwave system, the electromagnetic field is guided along a transmission line, such as microstrip, or enclosed in hollow metal waveguide pipe. In this chapter we consider microwave transmission inside circular metal waveguide. Fig. 3.1 shows the transverse electric and magnetic fields for the dominant TE_{11}-mode in circular waveguide. From the diagram it is seen that the field at the centre of the waveguide is almost the same as that of a plane wave, thus a rod of ferrite at the centre of the waveguide might be expected to behave similarly to that of an infinite ferrite material with a plane wave. In particular, the effective permeability of the ferrite material to the microwave field when the ferrite is magnetised in the direction of the axis of the waveguide is given in eqn. 2.45, which is shown in Fig. 2.5. A practical device will use a ferrite rod at the centre of the waveguide, as shown in Fig. 3.2. The ferrite will be magnetised so that it is operating either at resonance, or in the cut-off condition (i.e. where μ^- is negative) or below saturation where the effective permeability is strongly dependent on the strength of the applied magnetic field. Such a device can then be used as a resonance isolator or a field displacement isolator or a phase changer on circularly polarised waves, or as a rotator of linearly polarised waves. However, there are problems in supporting and cooling a ferrite rod at

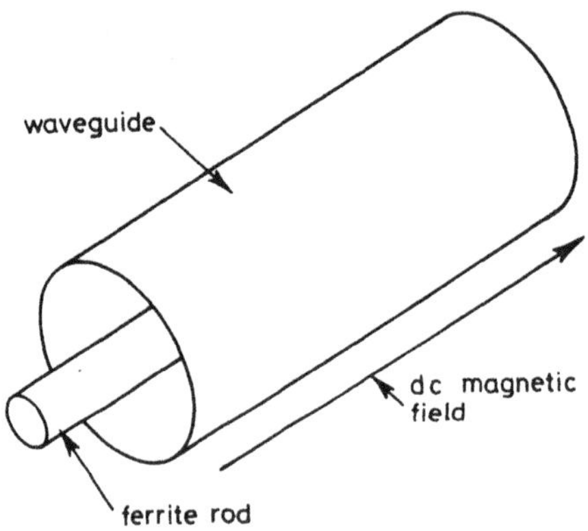

Fig. 3.2 *Ferrite rod at the centre of circular waveguide*

the centre of circular waveguide and ferrite tubes have also been considered for ferrite devices. Although the substitution of the plane wave permeability of the ferrite for the properties of the rod in the circular waveguide will give an approximate result for the effect of the magnetised ferrite rod in the waveguide, more accurate results are usually required, and a plane wave approximation is not valid for a ferrite tube near to the outside wall of the waveguide. For this reason, a full theoretical analysis is given for longitudinally magnetised ferrite tubes or rods inside circular waveguide.

In the 1950s, the Faraday rotation device, consisting of a longitudinally magnetised ferrite rod or tube in circular waveguide, was an important amplitude or phase control device to microwave engineers. At about this time, many theoretical investigations of microwave propagation through ferrite loaded circular waveguide were published. The first theoretical treatment of circular waveguide containing gyromagnetic media was given by Gamo,[1] who restricted himself to a completely filled waveguide but considered a material that was gyroelectric as well as gyromagnetic. He gives a study of some cut-off frequencies and shows the variation of propagation coefficient with waveguide radius. Kales[2] gives a similar theoretical treatment of ferrite-filled waveguide and also indicates the solution for a ferrite rod at the centre of the waveguide, but he does not give any results computed from theory. Suhl and Walker[3] have undertaken a more detailed study of the completely filled waveguide. Like Gamo, they begin by considering a medium which is both gyroelectric and gyromagnetic; because of the complexity of the solutions, they simplify the problem to the gyromagnetic material only, before finding expressions for the propagation coefficient. They give a large number of results of calculations of cut-off frequencies and phase coefficients for the completely filled guide. They also discuss perturbation methods of calculating the propagation coefficients for the partly filled guide, which are only valid for very thin ferrite rods or for very dilute ferrite materials. Clarricoats[4] gives a perturbation theory of microwave propagation in circular waveguide with a ferrite rod in the centre. He makes the assumption that the ferrite material has the same effective permeability to circularly polarised waves inside the waveguide as an infinite ferrite medium has to plane circularly polarised waves, and he shows that this method gives ample accuracy for the design of ferrite devices. This analysis is similar to his work on propagation in a circular waveguide containing a dielectric rod.[5] A good summary of the theoretical analysis of dielectric loaded circular waveguide is also given in the review paper by Chatterjee and Chatterjee.[6]

Acknowledging the limitation of all previously published results, Waldron[7] gives an exact theoretical analysis of an infinite length of circular waveguide with a ferrite rod at the centre. He has computed a large number of curves of cut-off conditions and phase coefficients for various sizes of rod and waveguide and various ferrite parameters. He has also computed[8] values of the microwave field components and density of energy flow for the HE_{11}-mode as functions of transverse position in the waveguide. Tompkins[9] has given similar results for one physical configuration. Waldron[10] and Mirimanov and Anisimova[11] have also analysed microwave propagation in circular waveguide with a tube of ferrite adjacent to the waveguide wall. Waldron gives an expression linking cut-off conditions for ferrite tubes and rods and has computed a few cut-off conditions, but no phase coefficients have been computed. Rizzi[12] gives theoretical results for a ferrite rod or tube, and also a few results of rotation which were calculated for the design of a particular ferrite device. I have given[13] exact expressions for the propagation constants for four configurations of ferrite loaded circular waveguide:

1 completely filled with ferrite
2 with a ferrite rod at the centre
3 with a ferrite tube adjacent to the waveguide wall
4 with a ferrite tube in the centre of the waveguide.

Some specimen results are given of the variation of propagation constant as the volume of ferrite varies for fixed ferrite properties and a fixed waveguide diameter. Waldron and Bowie[14] also show how to calculate the loss due to the ferrite in the waveguide.

These results are picked up in subsequent textbooks. Clarricoats[15] gives the exact analytical expressions but concentrates on perturbation methods. Waldron[16] gives an exact analytical analysis for both the ferrite rod and the ferrite tube adjacent to the waveguide wall but the bulk of his computed results are given in his earlier papers.[7,8,10] Helszajn[17] follows Waldron in his exact analysis of the ferrite rod at the centre of circular waveguide. The textbooks give a good introduction to the subject but the best design information is probably given in the original papers by Waldron.[7,8,10] I have given[13] a good statement of the analytical results in a form that is easily interpreted so as to compute propagation constants, and the analysis in the next sections of this chapter follow this earlier paper.[13]

3.2 Wave equations in the ferrite

The geometry that is being considered is a length of circular waveguide with either a rod or tube of ferrite inside it. In order to simplify the mathematical analysis, any boundaries that occur in the transverse plane are ignored and an infinite length of constant cross-section ferrite loaded waveguide is considered.

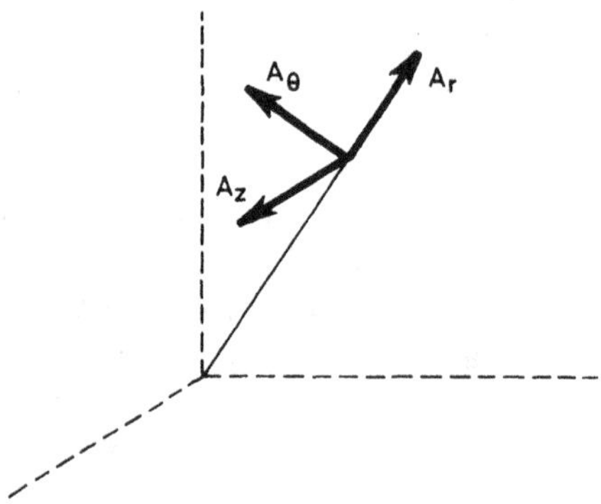

Fig. 3.3 *r, θ, z components of a vector*

The cylindrical polar co-ordinates, r, θ and z are used, having the axis of the co-ordinate system coincident with the axis of the waveguide; any field quantity will be expressed in terms of its r, θ and z components as shown by Fig. 3.3. In particular, it is convenient to separate the field quantities into two components, one parallel to the z-axis along the direction of propagation down the waveguide, and one perpendicular to it in the transverse plane of the waveguide. Then the general vector $\boldsymbol{A}$ will have the components $\boldsymbol{z}A_z$ and $\boldsymbol{A}_t$ where the subscript t denotes the transverse component of the vector quantity. In order to simplify the algebraic manipulation of Maxwell's equations following Suhl and Walker[3] and as used by myself,[18] a star denotes another vector of the same amplitude as $\boldsymbol{A}_t$ but rotated through 90° in the transverse plane. Therefore $\boldsymbol{A}_t^* = -z \times \boldsymbol{A}_t$, $(\boldsymbol{A}_t^*)^* = -\boldsymbol{A}_t$ and $\boldsymbol{A}_t \cdot \boldsymbol{A}_t = \boldsymbol{A}_t^* \cdot \boldsymbol{A}_t^*$. Similarly, the differential operator ∇ can be divided into the components $z\partial/\partial z$ and ∇_t. ∇_t and ∇_t^* have properties similar to those given for $\boldsymbol{A}_t$ and $\boldsymbol{A}_t^*$.

The wave equation will now be derived. Following the analysis of Section 2.5, despite the fact that later the cylindrical polar co-ordinate system is necessary, we start using the rectanglar co-ordinate system, x, y and z. It is assumed that the wave is propagating in the z-direction with a propagation constant β, and its angular frequency is ω. The gyromagnetic properties of the ferrite, given by eqn. 2.10, are substituted into Maxwell's equations, eqns. 2.31 and 2.32, and are written in their component form to give

$$\frac{\partial E_z}{\partial y} + j\beta E_y = -j\omega\mu H_x + \omega\kappa H_y \tag{3.1}$$

$$-j\beta E_x - \frac{\partial E_z}{\partial x} = -\omega\kappa H_x - j\omega\mu H_y \tag{3.2}$$

$$\frac{\partial E_y}{\partial x} - \frac{\partial E_x}{\partial y} = -j\omega\mu_0 H_z \tag{3.3}$$

$$\frac{\partial H_z}{\partial y} + j\beta H_y = j\omega\varepsilon E_x \tag{3.4}$$

$$-j\beta H_x - \frac{\partial H_z}{\partial y} = j\omega\varepsilon E_y \tag{3.5}$$

$$\frac{\partial H_y}{\partial x} - \frac{\partial H_x}{\partial y} = j\omega\varepsilon E_z \tag{3.6}$$

Writing these equations in the transverse vector form, eqns. 3.1 and 3.2 become

$$-j\omega\mu \boldsymbol{H}_t - \omega\kappa \boldsymbol{H}_t^* = \nabla_t^* E_z + j\beta \boldsymbol{E}_t^* \tag{3.7}$$

and eqn. 3.3 becomes

$$-j\omega\mu_0 H_z = \nabla_t \cdot \boldsymbol{E}_t^* \tag{3.8}$$

Eqns. 3.4 and 3.5 become

$$j\omega\varepsilon \boldsymbol{E}_t = \nabla_t^* H_z + j\beta \boldsymbol{H}_t^* \tag{3.9}$$

and eqn. 3.6 becomes

$$j\omega\varepsilon E_z = \nabla_t \cdot \boldsymbol{H}_t^* \tag{3.10}$$

Following Suhl and Walker,[3] the starred quantities are now eliminated from these equations. Operating on eqns. 3.7 and 3.9 with $\nabla_t \cdot$ and combining with eqns. 3.8 and 3.10 gives

$$\nabla_t \cdot \boldsymbol{E}_t = j\beta E_z \tag{3.11}$$

and

$$\mu \nabla_t \cdot \boldsymbol{H}_t + \omega\kappa\varepsilon E_z = j\beta\mu_0 H_z \tag{3.12}$$

Operating similarly with $\nabla_t^* \cdot$ and using the identity $\nabla_t^* \cdot \boldsymbol{H}_t^* = \nabla_t \cdot \boldsymbol{H}_t$ gives

$$-\omega^2 \varepsilon\mu_0 H_z = \nabla_t^2 H_z + j\beta\nabla_t \cdot \boldsymbol{H}_t \tag{3.13}$$

and

$$\omega^2 \varepsilon\mu E_z + \omega\kappa\nabla_t \cdot \boldsymbol{H}_t = -\nabla_t^2 E_z - j\beta\nabla_t \cdot \boldsymbol{E}_t \tag{3.14}$$

Eliminating $\nabla_t \cdot \boldsymbol{E}_t$ and $\nabla_t \cdot \boldsymbol{H}_t$ from eqns. 3.11 to 3.14 gives

$$\mu\nabla_t^2 H_z + \mu_0(\omega^2\varepsilon\mu - \beta^2)H_z = j\beta\omega\varepsilon\kappa E_z \tag{3.15}$$

$$\mu\nabla_t^2 E_z + [\omega^2\varepsilon(\mu^2 - \kappa^2) - \mu\beta^2]\, E_z = -j\beta\omega\mu_0\kappa H_z \tag{3.16}$$

These equations show that the gyromagnetic properties of the ferrite couple the longitudinal components of the electric and magnetic fields in the waveguide, so that the modes of propagation in ferrite-filled waveguide cannot be pure TE or TM. For a waveguide mode to exist in the magnetised ferrite, the longitudinal components of both the electric and magnetic fields must be present. For dielectric loaded waveguide, such as a dielectric rod at the centre of circular waveguide, the boundary causes the coupling so that any propagating mode has both longitudinal components of the electric and magnetic fields present. However, a waveguide filled with a uniform dielectric material behaves in exactly the same way as hollow metal waveguide and supports TE and TM modes, whereas a waveguide completely filled with a uniform ferrite material magnetised longitudinally still needs the longitudinal components of both the electric and magnetic fields.

Eliminating H_z and E_z in turn from eqns. 3.15 and 3.16 gives

$$\begin{aligned}\{(\mu\nabla_t^2)^2 &+ [(\omega^2\varepsilon\mu - \beta^2)(\mu + \mu_0) - \omega^2\varepsilon\kappa^2]\,(\mu\nabla_t^2) \\ &+ \mu\mu_0\,[(\omega^2\varepsilon\mu - \beta^2)^2 - (\omega^2\varepsilon\kappa)^2]\}E_z = 0\end{aligned} \tag{3.17}$$

and a similar expression for H_z.

A special notation is now used. Eqn. 3.17 is a fourth-order differential equation which is not as complicated as it looks. However, with the application of the relationships for the boundaries of the ferrite, numerical methods are needed to obtain values of the propagation constant. In order to aid the computing of solutions, some special units will be used – following Waldron[7] – to simplify the expressions used in the computation. In the design of microwave components using hollow metal waveguide, dimensions may be scaled proportional to the free-space wavelength. In this chapter, all dimensions will be measured in units of the free-space wavelength, λ_0, of the microwave energy, so that if the actual dimension is R, the normalised dimension used here will be

$$r = R/\lambda_0 = R\omega(\mu_0\varepsilon_0)^{1/2}/2\pi$$

Similarly, the propagation constant, β, is normalised so that $\beta_n = \lambda_0/\lambda_g$ where λ_g is the waveguide wavelength inside the ferrite loaded waveguide. The permittivity and permeability of the ferrite material are represented by their *relative* values where

$$\varepsilon_r = \varepsilon/\varepsilon_0, \qquad \mu_r = \mu/\mu_0, \qquad \text{and} \qquad \kappa_r = \kappa/\mu_0$$

Moreover, the permeability and permittivity of the ferrite are taken relative to the medium filling the rest of the waveguide, whether this is air (vacuum) or any other dielectric material. If the space apart from the ferrite in the waveguide is filled with a dielectric other than air, then the free-space wavelength is taken to be the wavelength in an unbounded medium of that dielectric. Deducing the mathematics in terms of these normalised units means that the frequency of the microwave signal and the permittivity and permeability of free space cancel out in the characteristic equations, thus simplifying the computation.

To return to eqn. 3.17, it is a quadratic in $(\mu\nabla_t^2)$ and it may be put into the form

$$(\nabla_t^2 + \gamma_1^2)(\nabla_t^2 + \gamma_2^2)E_z = 0 \tag{3.18}$$

and

$$\gamma^2 = \frac{2\pi^2\varepsilon_r}{\mu_r}(\mu_r(\mu_r + 1)(1 - \beta') - \kappa_r^2$$
$$\pm \sqrt{\{[(1 - \beta')(\mu_r - 1)\mu_r - \kappa_r^2]^2 + 4\kappa_r^2\mu_r\beta'\}}) \tag{3.19}$$

where γ_1 takes the positive sign and γ_2 the negative and

$$\beta' = \frac{\beta_n^2}{\varepsilon_r\mu_r}$$

E_z has two complementary sets of solutions given by the solutions to the equation

$$\nabla_t^2 E_z + \gamma_{1\cdot 2}^2 E_z = 0 \tag{3.20}$$

To move from the generalised transverse form of the differential operator to cylindrical polar co-ordinates, eqn. 3.20 becomes

$$\frac{\partial^2 E_z}{\partial r^2} + \frac{1}{r}\frac{\partial E_z}{\partial r} + \frac{1}{r^2}\frac{\partial^2 E_z}{\partial \theta^2} + \gamma_{1\cdot 2}^2 E_z = 0 \tag{3.21}$$

Using the method of separation of variables in the solution to eqn. 3.21 shows that there is a sinusoidal variation of the fields with θ. Any practical system will consist of a circularly symmetric system of ferrite material inside circular waveguide; a sinusoidal distribution of fields in the circumferential direction will repeat itself on turning through an angle of $\theta = 2\pi$. This can be depicted mathematically by an exponential dependence, so that the θ-dependence of the fields inside the waveguide is assumed to be exp $-jn\theta$, where n is a positive or negative integer. Then eqn. 3.21 becomes

$$\frac{\partial^2 E_z}{\partial r^2} + \frac{1}{r}\frac{\partial E_z}{\partial r} - \left(\frac{n^2}{r^2} - \gamma_{1\cdot2}^2\right)E_z = 0 \tag{3.22}$$

Eqn. 3.22 is Bessel's equation and its solution is

$$E_z = K_{1\cdot2}J_n(\gamma_{1\cdot2}r) + L_{1\cdot2}Y_n(\gamma_{1\cdot2}r) \tag{3.23}$$

where J_n and Y_n are Bessel functions of the first and second kind, respectively, of order n and $K_{1\cdot2}$ and $L_{1\cdot2}$ are constants determined by the boundary conditions. Because n^2 occurs in eqn. 3.22, the order of the Bessel function is $|n|$, always positive, although signed values of n are used in all other analytical expressions. γ may be real or imaginary, and so may the arbitrary constants K and L, so that, in any actual calculation, modified Bessel functions are as likely to occur as not. The full solution to eqn. 3.18 is then

$$E_z = [K_1J_n(\gamma_1r) + L_1Y_n(\gamma_1r) + K_2J_n(\gamma_2r) + L_2Y_n(\gamma_2r)]\exp jn\theta \tag{3.24}$$

3.3 Field Components

Eqn. 3.24 is a solution to eqn. 3.17 and gives a general expression for the longitudinal component of the electric field in the ferrite. It would have been equally possible to start with the differential equation in H_z similar to eqn. 3.17 having a solution for H_z which is the same as eqn. 3.24. However, having specified E_z, H_z is now defined in termsof E_z by eqn. 3.16, and the other components of the electric and magnetic field in the ferrite are related to them through Maxwell's equations. $\nabla_t^2 E_z$ is derived from eqn. 3.20 and is given by

$$\nabla_t^2 E_z = [-K_1\gamma_1^2J_n(\gamma_1r) - L_1\gamma_1^2Y_n(\gamma_1r) - K_2\gamma_2^2J_n(\gamma_2r) - L_2\gamma_2^2Y_n(\gamma_2r)]\exp jn\theta \tag{3.25}$$

Substituting the values for E_z from eqns. 3.24 and 3.25 into eqn. 3.16 gives an expression for H_z which is quoted as eqn. 3.35. To find expressions for the other components of the field, it is necessary to return to Maxwell's curl equations, eqns. 2.31 and 2.32. The transverse vector form of eqns. 2.31 and 2.32 are given in eqns. 3.7 and 3.9. Putting these equations into their vector components in the r and θ direction gives

$$j\eta\mu_rH_r + \eta\kappa_rH_\theta = -\frac{1}{2\pi r}\frac{\partial E_z}{\partial\theta} - j\beta_rE_\theta \tag{3.26}$$

$$j\eta\mu_r H_\theta - \eta\kappa_r H_r = -\frac{1}{2\pi}\frac{\partial E_z}{\partial r} + j\beta_n E_r \tag{3.27}$$

$$j\frac{\varepsilon_r}{\eta}E_r = \frac{1}{2\pi r}\frac{\partial H_z}{\partial \theta} + j\beta_n H_\theta \tag{3.28}$$

$$j\frac{\varepsilon_r}{\eta}E_\theta = -\frac{1}{2\pi}\frac{\partial H_z}{\partial r} - j\beta_n H_r \tag{3.29}$$

$\eta = \sqrt{(\mu_0/\varepsilon_0)}$ enters into these equations through the normalising process. Substitution for E_θ and E_r from eqns. 3.28 and 3.29 into eqns. 3.26 and 3.27 gives a pair of simultaneous equations in H_r and H_θ in terms of E_z and H_z. Solving these expressions for the transverse components of the field gives

$$2\pi\eta[\kappa_r^2 - \mu_r^2(1-\beta')^2]\,H_r = j\mu_r(1-\beta')\left(-\varepsilon_r\frac{1}{r}\frac{\partial E_z}{\partial \theta} + \beta_n\eta\frac{\partial H_z}{\partial_r}\right)$$
$$-\kappa_r\left(\varepsilon_r\frac{\partial E_z}{\partial r} + \beta_n\eta\frac{1}{r}\frac{\partial H_z}{\partial \theta}\right) \tag{3.30}$$

$$2\pi\eta[\kappa_r^2 - \mu_r^2(1-\beta')^2]H_\theta = \kappa_r\left(-\varepsilon_r\frac{1}{r}\frac{\partial E_z}{\partial \theta} + \beta_n\eta\frac{\partial H_z}{\partial r}\right)$$
$$+ j\mu_r(1-\beta')\left(\varepsilon_r\frac{\partial E_z}{\partial r} + \beta_n\eta\frac{1}{r}\frac{\partial H_z}{\partial \theta}\right) \tag{3.31}$$

$$E_r = \frac{\eta\beta}{\varepsilon_r}H_\theta - \frac{j\eta}{2\pi\varepsilon_r}\frac{1}{r}\frac{\partial H_z}{\partial \theta} \tag{3.32}$$

$$E_\theta = -\frac{\eta\beta}{\varepsilon_r}H_r + \frac{j\eta}{2\pi\varepsilon_r}\frac{1}{r}\frac{\partial H_z}{\partial r} \tag{3.33}$$

Appropriate substitution into eqns. 3.30 to 3.33 gives the following expressions for the components of the field in the unbounded ferrite medium:

$$E_z = [K_1J_n(\gamma_1 r) + L_1Y_n(\gamma_1 r) + K_2J_n(\gamma_2 r)$$
$$+ L_2Y_n(\gamma_2 r)]\exp j(n\theta + \omega t - \beta_n z) \tag{3.34}$$

$$H_z = \frac{j}{\eta}\{(A_1 - A_2\gamma_1^2)[K_1J_n(\gamma_1 r) + L_1Y_n(\gamma_1 r)]$$
$$+ (A_1 - A_2\gamma_2^2)[K_2J_n(\gamma_2 r)$$
$$+ L_2Y_n(\gamma_2 r)]\}\exp j(n\theta + \omega t - \beta_n z) \tag{3.35}$$

$$H_r = \frac{1}{\eta}\left\{(A_3 - A_4\gamma_1^2)\frac{n}{r}[K_1 J_n(\gamma_1 r) + L_1 Y_n(\gamma_1 r)]\right.$$
$$+ (A_3 - A_4\gamma_2^2)\frac{n}{r}[K_2 J_n(\gamma_2 r) + L_2 Y_n(\gamma_2 r)]$$
$$- (A_5 - A_6\gamma_1^2)\gamma_1 [K_1 J_n'(\gamma_1 r) + L_1 Y_n'(\gamma_1 r)]$$
$$\left. - (A_5 - A_6\gamma_2^2)\gamma_2 [K_2 J_n'(\gamma_2 r) + L_2 Y_n'(\gamma_2 r)]\right\}$$
$$\times \exp j(n\theta + \omega t - \beta_n z) \tag{3.36}$$

$$H_\theta = \frac{j}{\eta}\left\{-(A_5 - A_6\gamma_1^2)\frac{n}{r}[K_1 J_n(\gamma_1 r) + L_1 Y_n(\gamma_1 r)]\right.$$
$$- (A_5 - A_6\gamma_2^2)\frac{n}{r}[K_2 J_n(\gamma_2 r) + L_2 Y_n(\gamma_2 r)]$$
$$+ (A_3 - A_4\gamma_1^2)\gamma_1[K_1 J_n'(\gamma_1 r) + L_1 Y_n'(\gamma_1 r)]$$
$$\left. + (A_3 - A_4\gamma_2^2)\gamma_2[K_2 J_n'(\gamma_2 r) + L_2 Y_n'(\gamma_2 r)]\right\}$$
$$\times \exp j(n\theta + \omega t - \beta_n z) \tag{3.37}$$

$$E_r = j\left\{-(A_9 - A_{10}\gamma_1^2)\frac{n}{r}[K_1 J_n(\gamma_1 r) + L_1 Y_n(\gamma_1 r)]\right.$$
$$- (A_9 - A_{10}\gamma_2^2)\frac{n}{r}[K_2 J_n(\gamma_2 r) + L_2 Y_n(\gamma_2 r)]$$
$$+ (A_7 - A_8\gamma_1^2)\gamma_1[K_1 J_n'(\gamma_1 r) + L_1 Y_n'(\gamma_1 r)]$$
$$\left. + (A_7 - A_8\gamma_2^2)\gamma_2[K_2 J_n'(\gamma_2 r) + L_2 Y_n'(\gamma_2 r)]\right\}$$
$$\times \exp j(n\theta + \omega t - \beta_n z) \tag{3.38}$$

$$E_\theta = \left\{-(A_7 - A_8\gamma_1^2)\frac{n}{r}[K_1 J_n(\gamma_1 r) + L_1 Y_n(\gamma_1 r)]\right.$$
$$- (A_7 - A_8\gamma_2^2)\frac{n}{r}[K_2 J_n(\gamma_2 r) + L_2 Y_n(\gamma_2 r)]$$
$$+ (A_9 - A_{10}\gamma_1^2)\gamma_1[K_1 J_n'(\gamma_1 r) + L_1 Y_n'(\gamma_1 r)]$$
$$\left. + (A_9 - A_{10}\gamma_2^2)\gamma_2[K_2 J_n'(\gamma_2 r) + L_2 Y_n'(\gamma_2 r)]\right\}$$
$$\times \exp j(n\theta + \omega t - \beta_n z) \tag{3.39}$$

where

$$
\left.
\begin{aligned}
A_1 &= \frac{\varepsilon_r}{\beta_n \kappa_r}[\mu_r^2(1-\beta') - \kappa_r^2] \\
A_2 &= \frac{\mu_r}{4\pi^2 \beta_n \kappa_r} \\
A_3 &= \frac{\mu_r(\mu_r+1)(1-\beta') - \kappa_r^2}{2\pi[\kappa_r^2 - \mu_r^2(1-\beta')^2]} \\
A_4 &= \frac{\mu_r}{8\pi^3 \varepsilon_r[\kappa_r^2 - \mu_r^2(1-\beta')^2]} \\
A_5 &= \frac{\mu_r[\mu_r^2(1-\beta')^2 - \kappa_r^2(1-\beta') + \kappa_r^2/\mu_r]}{2\pi\kappa_r[\kappa_r^2 - \mu_r^2(1-\beta')^2]} \\
A_6 &= \frac{\mu_r^2(1-\beta')}{8\pi^3 \varepsilon_r \kappa_r[\kappa_r^2 - \mu_r^2(1-\beta')^2]} \\
A_7 &= \frac{\beta_n}{\varepsilon_r} A_3 \\
A_8 &= \frac{\beta_n}{\varepsilon_r} A_4 \\
A_9 &= \frac{\beta_n}{\varepsilon_r} A_5 - \frac{1}{2\pi\varepsilon_r} A_1 \\
A_{10} &= \frac{\beta_n}{\varepsilon_r} A_6 - \frac{1}{2\pi\varepsilon_r} A_2
\end{aligned}
\right\} \qquad (3.40)
$$

and the primed Bessel functions are the differential derivatives with regard to the argument.

The above equations give expressions for the field components in a ferrite medium in terms of four arbitrary constants. It is necessary to apply the boundary conditions in order to evaluate these arbitrary constants. Eliminating the arbitrary constants from the equations arising as a result of applying the boundary conditions gives the characteristic equation which enables the propagation constant to be evaluated.

In order to apply the boundary conditions to a circular waveguide only partially filled with ferrite, it is necessary to derive expressions similar to eqns. 3.34 to 3.39 for the fields in an unbounded dielectric material. They are

$$E_z = [K_3 J_n(\gamma_a r) + L_3 Y_n(\gamma_a r)] \exp \mathrm{j}(n\theta + \omega t - \beta_n z) \qquad (3.41)$$

$$H_z = \frac{\mathrm{j}}{\eta}[K_4 J_n(\gamma_a r) + L_4 Y_n(\gamma_a r)] \exp \mathrm{j}(n\theta + \omega t - \beta_n z) \qquad (3.42)$$

$$H_r = \frac{1}{\eta}\left\{A_{11}\frac{n}{r}[K_3 J_n(\gamma_a r) + L_3 Y_n(\gamma_a r)] + A_{12}\gamma_a[K_4 J_n'(\gamma_a r) + L_4 Y_n'(\gamma_a r)]\right\}\exp j(n\theta + \omega t - \beta_n z) \quad (3.43)$$

$$H_\theta = \frac{j}{\eta}\left\{A_{12}\frac{n}{r}[K_4 J_n(\gamma_a r) + L_4 Y_n(\gamma_a r)] + A_{11}\gamma_a[K_3 J_n'(\gamma_a r) + L_3 Y_n'(\gamma_a r)]\right\}\exp j(n\theta + \omega t - \beta_n z) \quad (3.44)$$

$$E_r = j\left\{-A_{11}\frac{n}{r}[K_4 J_n(\gamma_a r) + L_4 Y_n(\gamma_a r)] - A_{12}\gamma_a[K_3 J_n'(\gamma_a r) + L_3 Y_n'(\gamma_a r)]\right\}\exp j(n\theta + \omega t - \beta_n z) \quad (3.45)$$

$$E_\theta = \left\{A_{12}\frac{n}{r}[K_3 J_n(\gamma_a r) + L_3 Y_n(\gamma_a r)] + A_{11}\gamma_a[K_4 J_n'(\gamma_a r) + L_4 Y_n'(\gamma_a r)]\right\}\exp j(n\theta + \omega t - \beta_n z) \quad (3.46)$$

where

$$\gamma_a^2 = 4\pi^2(1 - \beta_n^2) \quad (3.47)$$

and

$$\left.\begin{aligned} A_{11} &= -\frac{2\pi}{\gamma_a^2} \\ A_{12} &= -\beta_n A_{11} \end{aligned}\right\} \quad (3.48)$$

In order to apply the boundary conditions to model a practical system of ferrite loaded circular waveguide, it is necessary to equate components of the fields at each side of the boundary. The expressions for the components of the fields have become very complicated, and some new functions are defined which are used to simplify the equations for the components of the fields inside the waveguide.

$$\begin{aligned} F_1(r) &= J_n(\gamma_1 r) \\ F_2(r) &= Y_n(\gamma_1 r) \\ F_3(r) &= J_n(\gamma_2 r) \\ F_4(r) &= Y_n(\gamma_2 r) \\ F_5(r) &= (A_1 - A_2\gamma_1^2)J_n(\gamma_1 r) \\ F_6(r) &= (A_1 - A_2\gamma_1^2)Y_n(\gamma_1 r) \end{aligned}$$

$$F_7(r) = (A_1 - A_2\gamma_2^2)J_n(\gamma_2 r)$$

$$F_8(r) = (A_1 - A_2\gamma_2^2)Y_n(\gamma_2 r)$$

$$F_9(r) = -(A_5 - A_6\gamma_1^2)\frac{n}{r}J_n(\gamma_1 r) + (A_3 - A_4\gamma_1^2)\gamma_1 J_n'(\gamma_1 r)$$

$$F_{10}(r) = -(A_5 - A_6\gamma_1^2)\frac{n}{r}Y_n(\gamma_1 r) + (A_3 - A_4\gamma_1^2)\gamma_1 Y_n'(\gamma_1 r)$$

$$F_{11}(r) = -(A_5 - A_6\gamma_2^2)\frac{n}{r}J_n(\gamma_2 r) + (A_3 - A_4\gamma_2^2)\gamma_2 J_n'(\gamma_2 r)$$

$$F_{12}(r) = -(A_5 - A_6\gamma_2^2)\frac{n}{r}Y_n(\gamma_2 r) + (A_3 - A_4\gamma_2^2)\gamma_2 Y_n'(\gamma_2 r)$$

$$F_{13}(r) = -(A_7 - A_8\gamma_1^2)\frac{n}{r}J_n(\gamma_1 r) + (A_9 - A_{10}\gamma_1^2)\gamma_1 J_n'(\gamma_1 r)$$

$$F_{14}(r) = -(A_7 - A_8\gamma_1^2)\frac{n}{r}Y_n(\gamma_1 r) + (A_9 - A_{10}\gamma_1^2)\gamma_1 Y_n'(\gamma_1 r)$$

$$F_{15}(r) = -(A_7 - A_8\gamma_2^2)\frac{n}{r}J_n(\gamma_2 r) + (A_9 - A_{10}\gamma_2^2)\gamma_2 J_n'(\gamma_2 r)$$

$$F_{16}(r) = -(A_7 - A_8\gamma_2^2)\frac{n}{r}Y_n(\gamma_2 r) + (A_9 - A_{10}\gamma_2^2)\gamma_2 Y_n'(\gamma_2 r)$$

$$F_{17}(r) = J_n(\gamma_a r)$$

$$F_{18}(r) = Y_n(\gamma_a r)$$

$$F_{19}(r) = J_n'(\gamma_a r)$$

$$F_{20}(r) = Y_n'(\gamma_a r)$$

$$F_{21}(r) = A_{11}\gamma_a J_n'(\gamma_a r)$$

$$F_{22}(r) = A_{11}\gamma_a Y_n'(\gamma_a r)$$

$$F_{23}(r) = A_{12}\frac{n}{r}J_n(\gamma_a r)$$

$$F_{24}(r) = A_{12}\frac{n}{r}Y_n(\gamma_a r)$$

3.4 Ferrite-filled circular waveguide

The characteristic equation for wave propagation through circular waveguide completely filled with microwave ferrite material magnetised in the direction of the axis of the waveguide is mathematically the simplest expression and has been given by the largest number of authors.[1,2,3,7,13,15] Consider a circular waveguide of radius r_1 completely filled with ferrite. The boundary condition is that the tangential electric field is zero at the waveguide wall. Since the ferrite medium is continuous through the origin of the co-ordinate system, the constants L associated with the Bessel funtions of the second kind are zero. Then equating the longitudinal and circumferential electric fields to zero at $r = r_1$ gives

$$\left.\begin{aligned} K_1F_1(r_1) + K_2F_3(r_1) &= 0 \\ K_1F_{13}(r_1) + K_2F_{15}(r_1) &= 0 \end{aligned}\right\} \tag{3.49}$$

Elimination of K_1 and K_2 gives the characteristic equation

$$\begin{vmatrix} F_1(r_1) & F_3(r_1) \\ F_{13}(r_1) & F_{15}(r_1) \end{vmatrix} = 0 \tag{3.50}$$

where the functions are defined in Section 3.3. Even a comparatively simple expression as eqn. 3.50 does not have a simple analytical solution for values of β_n because the unknown occurs in the expressions for the argument of the Bessel functions. However, numerical methods used with a computer enable solutions to be obtained for eqn. 3.50. If only limited computing power is available, eqn. 3.50 may be further simplified by only considering the cut-off conditions for the modes propagating in the waveguide, i.e. by putting $\beta_n = 0$ in eqn. 3.50.

Waldron[7] has made an extensive investigation of the cut-off conditions for ferrite-filled waveguide and only a summary of the results will be given here. When $\beta_n = 0$, eqn. 3.50 reduces to two independent equations, but for the completely filled guide the same result may be obtained from eqns. 3.15 and 3.16 by putting $\beta = 0$. This shows that at cut off, separation of the mode types into TE- and TM-modes occurs and that the cut-off condition of the TE-modes is independent of the gyromagnetic effect of the ferrite. The cut-off relations for the two mode types is obtained by solving eqns. 3.15 and 3.16 with $\beta = 0$ and then letting $E_z = 0$ or $\partial H_z/\partial r = 0$ at $r = r_1$. As we have already seen, the normally propagating modes in ferrite loaded waveguide are not pure TE or TM, but need the existence of the longitudinal component of both the electric and magnetic field. These modes are called HE- or EH-modes, depending whether the longitudinal component of the magnetic or electric field is dominant. Identification is further assisted by the fact that the cut-off condition for the HE-modes is independent of the value of κ whereas that for the EH-modes is a function of κ.

In discussing the properties of plane waves propagating through longitudinally magnetised ferrite, it was shown that the circularly polarised waves were the fundamental modes propagating in the ferrite mateiral. In the waves propagating in ferrite loaded circular waveguide, the hand of circular polarisation is given by the sign of the integer n, which is always positive as the order of the Bessel functions, but is used with the correct sign in the relationships for the field quantities. However, in a practical system, the influence of the ferrite on the wave is governed by the relationship between the direction of rotation of the circularly polarised field and the direction of precession of the magnetic dipoles in the ferrite. This means that reversing the direction of the static magnetic field acting on the ferrite has the same effect as reversing the hand of rotation of the circularly polarised wave. Therefore, in the calculation it is easier to do the equivalent to reversing the magnetic field, that is to change the sign of κ_r, than to change the hand of rotation of the circularly polarised wave by changing the sign of n.

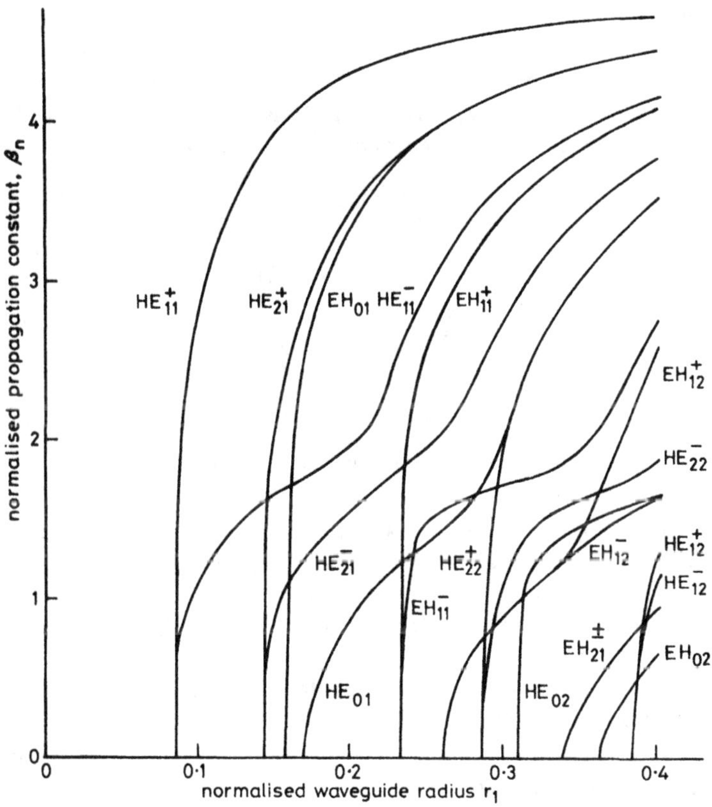

Fig. 3.4 *Propagation conditions for a circular waveguide completely filled with longitudinally magnetised ferrite. The normalised propagation constant is plotted against the normalised waveguide radius as the waveguide radius varies. The ferrite properties are* $\mu_r = 1$, $\kappa_r = 0{\cdot}75$ *and* $\varepsilon_r = 13$

Some typical results are shown in Fig. 3.4 of the variation of the normalised propagation constant β_n as the normalised radius of the completely filled waveguide, r_1, varies. The ferrite properties are $\mu_r = 1$, $\kappa_r = 0{\cdot}75$ and $\varepsilon_r = 13$. Results have been calculated for all the circularly polarised modes up to circular order $n = 2$. The positive and negative hands of circular polarisation are shown by $+$ or $-$ signs added as superscripts to the mode notation. These results show that there is a useful region where only the dominant HE_{11}-mode can propagate and that there is a large difference between the propagation constants for the two hands of circular polarisation. Therefore, useful quantities of rotation or differential phase change can be obtained using the dominant mode.

3.5 Ferrite rod at centre of circular waveguide

In the early days of ferrite devices, one of the most popular configurations used for control applications consisted of a longitudinally magnetised rod of ferrite mounted axially at the centre of circular waveguide, as shown in Fig. 3.2. The waveguide configuration is shown in cross section in Fig. 3.5. Area A is the ferrite medium where $0 \leqslant r < r_1$. Area B is the homogeneous dielectric medium which, so often in practice, is air where $r_1 \leqslant r < r_2$. Region C is the

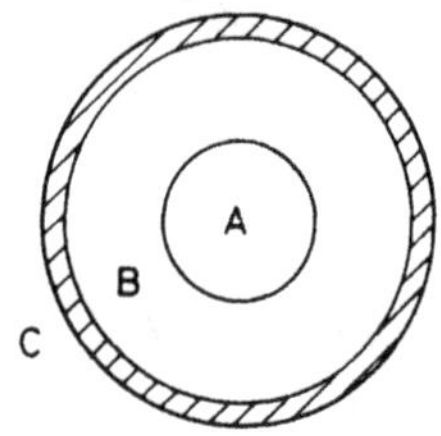

Fig. 3.5 *Cross-section of circular waveguide partially filled with ferrite*

perfectly conducting waveguide wall where $r_2 \leqslant r$. Because area A is continuous through the origin, the constants L associated with the Bessel functions of the second kind will be zero. Then the arbitrary constants occurring in the mathematical expressions for the components of the fields will be K_1 and K_2 in the ferrite region, and K_3, K_4, L_3 and L_4 in the outer dielectric region. The characteristic equation for propagation in the ferrite loaded waveguide is obtained by equating various components of the fields across the ferrite dielectric boundary. The functions used to simplify the expressions for the field components are given in Section 3.3, on pages 56 and 57.

For the boundary conditions at the waveguide walls, equating the expression for the longitudinal component of electric field to zero gives

$$K_3 F_{17}(r_2) + L_3 F_{18}(r_2) = 0 \tag{3.51}$$

and similarly the circumferential component gives

$$K_4 F_{19}(r_2) + L_4 F_{20}(r_2) = 0 \tag{3.52}$$

At the ferrite dielectric boundary, where $r = r_1$, the tangential components of the electric and magnetic fields are continuous. Equating the expressions for these field components across the boundary gives

$$\left.\begin{aligned} K_3F_{17}(r_1) + L_3F_{18}(r_1) &= K_1F_1(r_1) + K_2F_3(r_1) \\ K_4F_{17}(r_1) + L_4F_{18}(r_1) &= K_1F_5(r_1) + K_2F_7(r_1) \\ K_3F_{21}(r_1) + L_3F_{22}(r_1) + K_4F_{23}(r_1) + L_4F_{24}(r_1) &= K_1F_9(r_1) + K_2F_{11}(r_1) \\ K_3F_{23}(r_1) + L_3F_{24}(r_1) + K_4F_{21}(r_1) + L_4F_{22}(r_1) &= K_1F_{13}(r_1) + K_2F_{15}(r_1) \end{aligned}\right\} \qquad (3.53)$$

Elimination of the arbitrary constants K and L from eqns. 3.51 to 3.53 gives the characteristic equation

$$\begin{vmatrix} F_{17}(r_2) & F_{18}(r_2) & 0 & 0 & 0 & 0 \\ 0 & 0 & F_{19}(r_2) & F_{20}(r_2) & 0 & 0 \\ F_{17}(r_1) & F_{18}(r_1) & 0 & 0 & F_1(r_1) & F_3(r_1) \\ 0 & 0 & F_{17}(r_1) & F_{18}(r_1) & F_5(r_1) & F_7(r_1) \\ F_{21}(r_1) & F_{22}(r_1) & F_{23}(r_1) & F_{24}(r_1) & F_9(r_1) & F_{11}(r_1) \\ F_{23}(r_1) & F_{24}(r_1) & F_{21}(r_1) & F_{22}(r_1) & F_{13}(r_1) & F_{15}(r_1) \end{vmatrix} = 0 \qquad (3.54)$$

where the functions are defined in Section 3.3. Computer optimisation techniques may be used to find the values for β_n for which eqn. 3.54 is satisfied and these give the propagating conditions for wave propagation along the ferrite loaded waveguide. As with the ferrite-filled waveguide, Waldron[7] has shown how the cut-off condition splits into separate mathematical expressions for the TE(HE)- and TM(EH)-modes and that the cut-off conditions of the TE-modes are independent of the gyromagnetic effect of the ferrite.

The results of a specimen investigation of the solution of eqn. 3.54 taken from reference 13 is given in Fig. 3.6. It is a plot of the variation of the propagation constant β_n as the radius of the ferrite rod, r_1, varies for the same ferrite properties used in the calculation of the results in Fig. 3.4. The waveguide has a normalised radius $r_2 = 0{\cdot}36$. However, Fig. 3.6 is very cluttered so the results have been plotted again with the modes of the same circular order on the same diagram. That is, all the results for one value of n on one diagram so that it is easier to see the properties of the modes. Fig. 3.7 shows the propagating conditions for the circularly symmetric modes where $n = 0$. Fig. 3.8 shows the propagating conditions for the modes of order 1 where $n = 1$. This shows that the dominant mode in empty circular waveguide, the TE_{11}-mode, can propagate and that this mode becomes the HE_{11}-mode in the ferrite loaded waveguide. The positive and negative hands of circular polarisation are shown by + or − signs added as superscripts to the mode notation. It can be seen that the cut-off condition is the same for both hands of circular polarisation of the HE-modes whereas it is different for the EH-modes. It is also seen that the gyromagnetic effect separates the phase constants of modes with opposite hands of circular polarisation.

A study of the behaviour of the HE_{11}-mode is particularly instructive since this is the mode derived from the TE_{11}-mode in empty waveguide. It is also the dominant mode in circular waveguide, thus it is the mode most likely to be excited in the ferrite loaded waveguide if it is connected directly to a length of air or dielectric filled circular waveguide. It is seen from Fig. 3.8 that when there is only a thin rod of ferrite at the centre of the waveguide, the ferrite has very little effect on the propagation constant of the wave propagating in the

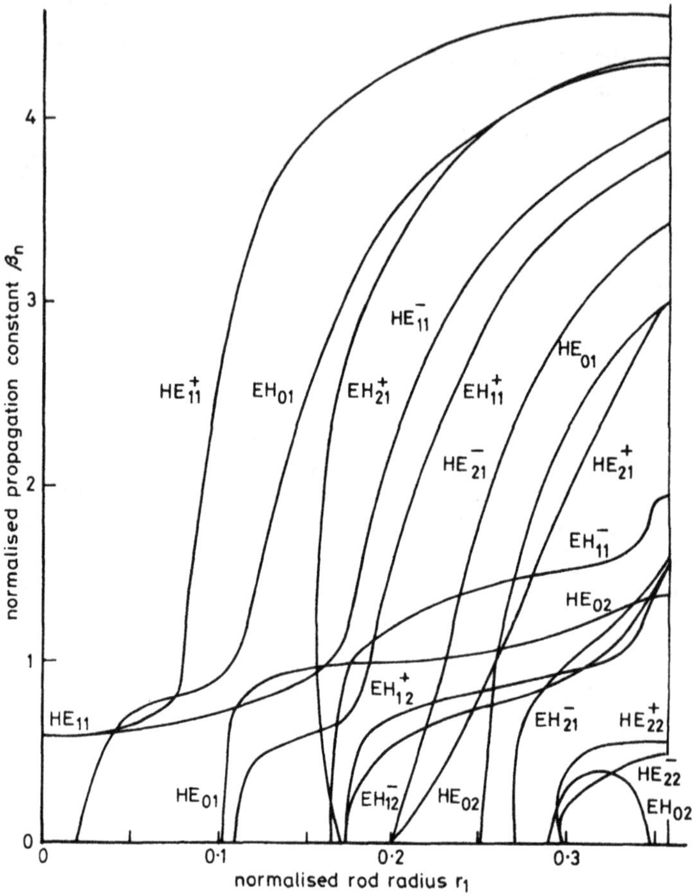

Fig. 3.6 *Propagation conditions for a longitudinally magnetised ferrite rod in circular waveguide. The normalised propagation constant is plotted against the normalised rod radius as the rod radius varies from zero (empty waveguide) to $r_1 = r_2$ (completely filled waveguide). The normalised waveguide radius $r_2 = 0{\cdot}36$ and the ferrite properties are the same as in Fig. 3.4*

waveguide. As the radius of the rod increases, the propagation constant of the positively rotating wave increases first, to be followed by that of the negatively rotating wave. There is a range of radii where there is a maximum difference between the two oppositely rotating waves and there will be a large rotation of a linearly polarised wave. Fig. 3.6 shows that there are also three other modes that can propagate with this geometry. In many designs of ferrite device, the two circularly symmetric modes are unlikely to be excited but the EH_{11}^{+}-mode might be troublesome and ought to be avoided. For some categories of Faraday rotation switch, the circularly symmetric modes might also be disadvantageous and a different geometry needs to be used where only the dominant mode can propagate. Fig. 3.4 shows that there is a range of sizes of ferrite-filled waveguide where only the dominant mode can propagate and there is also a similar result for ferrite tubes of small diameter. Waldron[7] gives very extensive results of the calculation of cut-off conditions with a wide variation of r_1, r_2, μ_r, κ_r and ε_r and further design information can be obtained from his paper.[7]

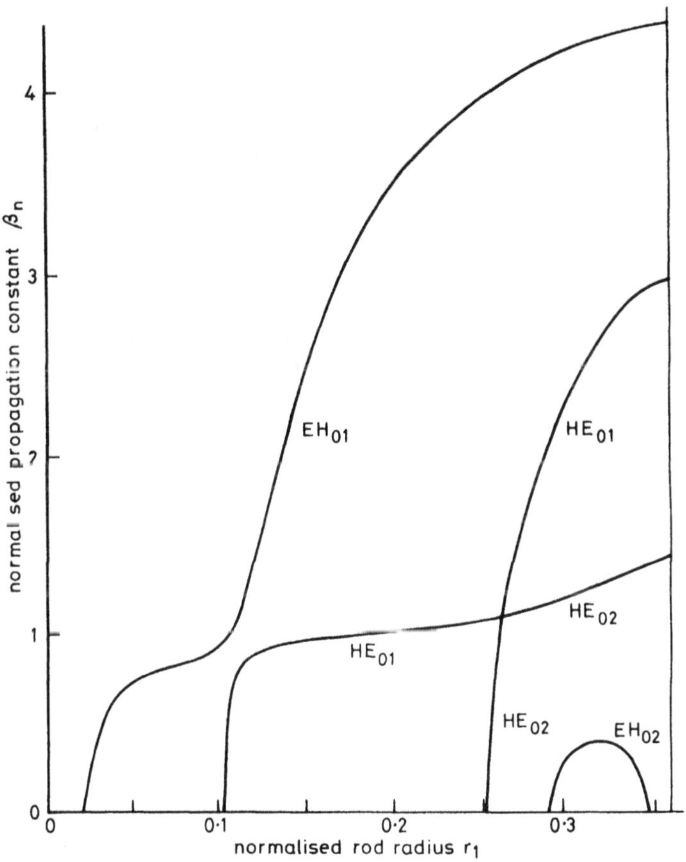

Fig. 3.7 *Results from Fig. 3.6 for n = 0*

Following a solution to eqn. 3.54 it is possible to substitute values into eqns. 3.51 to 3.53 to obtain values for all the arbitrary constants except one, and then substitution can be made into the expressions for the components of the fields, eqns. 3.34 to 3.39 and 3.41 to 3.46. In an accompanying paper to his earlier work, Waldron[8] has calculated relative values for all the components of the electric and magnetic fields across the cross-section of the waveguide for the HE_{11}-mode. In particular, he shows that the rapid change in value of the propagation constant at about $r_1 = 0{\cdot}1$ is due to a concentration of the field in the ferrite. As most ferrite materials will have a slightly larger loss than empty

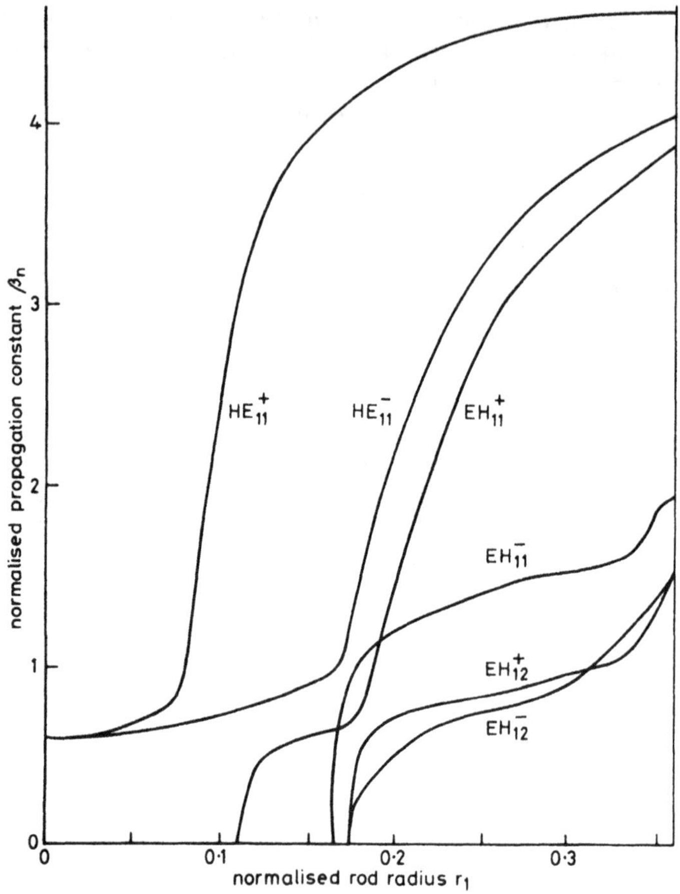

Fig. 3.8 *Results from Fig. 3.6, for n = 1*

waveguide, this means that for the condition of maximum rotation the positive circularly polarised wave will be attenuated slightly more than the negative circularly polarised wave. Waldron[8] also gives curves of the power density at different positions in the cross-section of the waveguide calculated from the Poynting vector values at these positions. He also shows that the simple expression for wave impedance that is valid for homogeneous dielectric filled waveguide

$$Z_0 = E_r/H_\theta = -E_\theta/H_r$$

is not valid in waveguide filled or partially filled with ferrite. For any waveguide partially filled with either dielectric or ferrite, E_r/H_θ is not the same as E_θ/H_r and both are functions of radius. Therefore the conventional concept of impedance, calculated from a wave impedance in the waveguide, is invalid except for simple waveguides which support only TE- or TM-modes.[19]

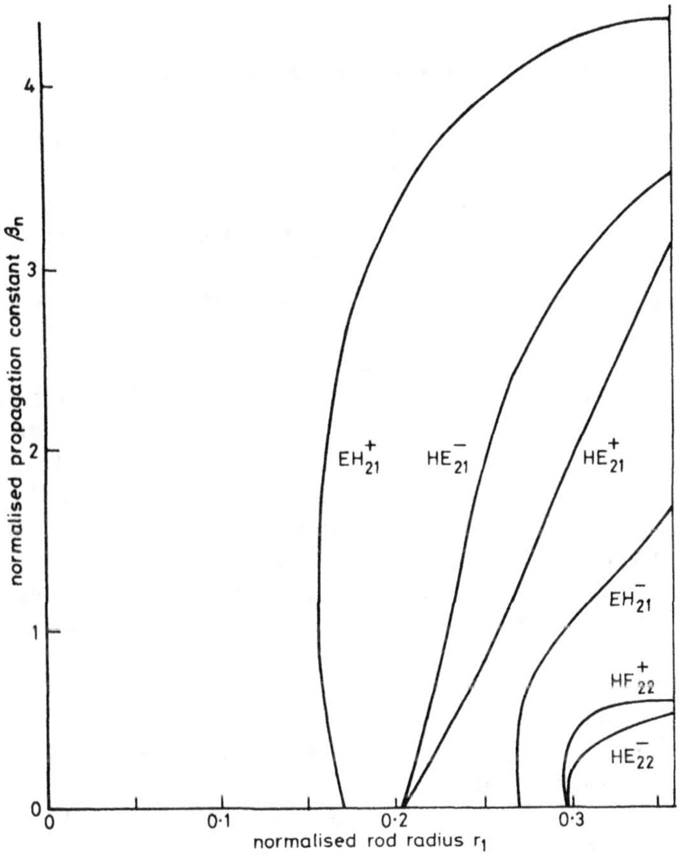

Fig. 3.9 *Results from Fig. 3.6 for n = 2*

The propagating conditions for the modes of order 2, where $n = 2$, are shown in Fig. 3.9, where it is seen that the phase constant curve for the EH_{21}^+-mode has a negative slope near to cut off. Waldron[20,21] has studied this effect extensively for a particular geometry of ferrite rod in circular waveguide where the same effect occurs for the HE_{11}-mode. He shows that the portion of the curve near to cut off with a negative slope represents a backward wave where the phase velocity and the group velocity are in opposite directions. Calculations of the power density across the cross-section of the waveguide show that, in part of the cross-section, the Poynting vector is directed in the direction of the phase velocity, and in the rest of the cross-section, the Poynting vector is directed in the opposite direction. For much of the propagating condition there is a region of apparent forward power flow and a region of apparent backward power flow. In the negative slope region, the negative power is greater than the forward power, and a backward wave results. Otherwise the forward power is greater and normal propagation occurs.

3.6 Ferrite tube adjacent to the wall of circular waveguide

There are situations where it may be disadvantageous to have a rod of ferrite at the centre of circular waveguide. It needs to be mounted in some kind of dielectric sleeve which then forms the dielectric material filling the waveguide surrounding the ferrite. All the theory considered so far assumes that there is no loss in the ferrite or the dielectric material. If the dielectric is air or vacuum, then its loss can probably be ignored, but the loss in any ferrite material is larger and may not always be ignored. Particularly for high-power applications, the ferrite may become hot and may need cooling. Forced air cooling can be used, provided a method of mounting the ferrite can be devised which allows for the air flow, but it is not easy. Another way of cooling is to move the ferrite to the wall of the waveguide where it may be cooled from outside the waveguide. A study of the field in empty circular waveguide, shown in Fig. 3.1, shows that, although the field at the centre of the waveguide approximates to that of a plane wave, the field adjacent to the walls does not. In particular, a circularly polarised TE_{11}-mode approximates to a circularly polarised field at the centre of the waveguide so that a ferrite rod at the centre might be expected to behave like an infinite ferrite to a plane wave. However, a circularly poalrised TE_{11}-wave does not provide a circularly polarised field adjacent to the walls of the waveguide, thus a ferrite tube is less likely to be able to provide a ferrite control device. Such an argument has been used to show that it is not worth considering a ferrite tube inside circular waveguide. However, useful rotation can be produced and resutls for a longitudinally magnetised ferrite tube in circular waveguide are given in this section.

The geometry is shown in Fig. 3.5 but this time area A is the dielectric medium where $0 \leqslant r < r_1$, and area B is the ferrite medium where $r_1 \leqslant r < r_2$. Region C is the perfectly conducting waveguide wall at $r_2 \leqslant r$. Because area A is continuous through the origin, the constants L associated with the Bessel functions of the second kind will be zero. Then the arbitrary constants occurring in the mathematical expressions for the components of the fields will be K_1, K_2, L_1 and L_2 in the ferrite region and K_3 and K_4 in the dielectric region. For the boundary at the waveguide walls, equating the expressions for the longitudinal and circumferential components of the electric field to zero gives

$$\left.\begin{aligned} K_1F_1(r_2) + L_1F_2(r_2) + K_2F_3(r_2) + L_2F_4(r_2) &= 0 \\ K_1F_{13}(r_2) + L_1F_{14}(r_2) + K_2F_{15}(r_2) + L_2F_{16}(r_2) &= 0 \end{aligned}\right\} \tag{3.55}$$

At the ferrite dielectric boundary, where $r = r_1$, the tangential components of the electric and magnetic fields are continuous. Equating the expressions for these field components across the boundary gives

$$\left.\begin{aligned} K_1F_1(r_1) + L_1F_2(r_1) + K_2F_3(r_1) + L_2F_4(r_1) &= K_3F_{17}(r_1) \\ K_1F_5(r_1) + L_1F_6(r_1) + K_2F_7(r_1) + L_2F_8(r_1) &= K_4F_{17}(r_1) \\ K_1F_9(r_1) + L_1F_{10}(r_1) + K_2F_{11}(r_1) + L_2F_{12}(r_1) &= K_3F_{21}(r_1) + K_4F_{23}(r_1) \\ K_1F_{13}(r_1) + L_1F_{14}(r_1) + K_2F_{15}(r_1) + L_2F_{16}(r_1) &= K_3F_{23}(r_1) + K_4F_{21}(r_1) \end{aligned}\right\} \quad (3.56)$$

Elimination of the arbitrary constants K and L from eqns. 3.55 and 3.56 gives the characteristic equation

$$\begin{vmatrix} F_1(r_2) & F_2(r_2) & F_3(r_2) & F_4(r_2) & 0 & 0 \\ F_{13}(r_2) & F_{14}(r_2) & F_{15}(r_2) & F_{16}(r_2) & 0 & 0 \\ F_1(r_1) & F_2(r_1) & F_3(r_1) & F_4(r_1) & F_{17}(r_1) & 0 \\ F_5(r_1) & F_6(r_1) & F_7(r_1) & F_8(r_1) & 0 & F_{17}(r_1) \\ F_9(r_1) & F_{10}(r_1) & F_{11}(r_1) & F_{12}(r_1) & F_{21}(r_1) & F_{23}(r_1) \\ F_{13}(r_1) & F_{14}(r_1) & F_{15}(r_1) & F_{16}(r_1) & F_{23}(r_1) & F_{21}(r_1) \end{vmatrix} = 0 \quad (3.57)$$

where the functions are defined in Section 3.3. As before computer optimisation may be used to find the values for β_n for which eqn. 3.57 is satisfied and this gives the propagating conditions for wave propagation in the ferrite loaded waveguide. Waldron[10] has given an expression equivalent to eqn. 3.57, as has Mirimanov and Anisimova[11] and Rizzi.[12] As with his related work on ferrite rods in circular waveguide, Waldron[10] has made an extensive study of the cut-off conditions. He shows that the cut-off conditions again separate into two equations, one representing HE-modes and one representing EH-modes, and that the cut-off conditions of the HE-modes are independent of the gyromagnetic effect of the ferrite. The application given in Rizzi's paper[12] shows how the tube was investigated particularly because of its possibility for high-power applications. The results of a specimen investigation of the solution of eqn. 3.57 taken from reference 13 is given in Figs. 3.10 to 3.12. As with Fig. 3.6, they are plots of the variation of the propagation constant β_n as the inside radius of the ferrite tube, r_1, is varied for the same ferrite properties and same value of r_2 as used for calculating the results given in Figs. 3.4 and 3.6.

It can be seen from Fig. 3.11 that there is not as much rotation of the dominant HE_{11}-mode using the tube of ferrite as that given by the rod of ferrite which is shown by Fig. 3.8. However, Rizzi[12] has shown that useful rotation can be obtained with a tube of ferrite inside a smaller diameter waveguide. I have conducted a similar investigation. The results are given in Fig. 3.13, which shows the propagation constant of the dominant HE_{11}-mode for both hands of circular polarisation as the inside radius of the tube, r_1, varies with the outer radius of the tube, r_2, which is also the radius of the waveguide, as a parameter. The ferrite properties used are different from those used in the investigation whose results are given in Figs. 3.4 to 3.12. These results show that useful rotation can be obtained with a tube of longitudinally magnetised ferrite in

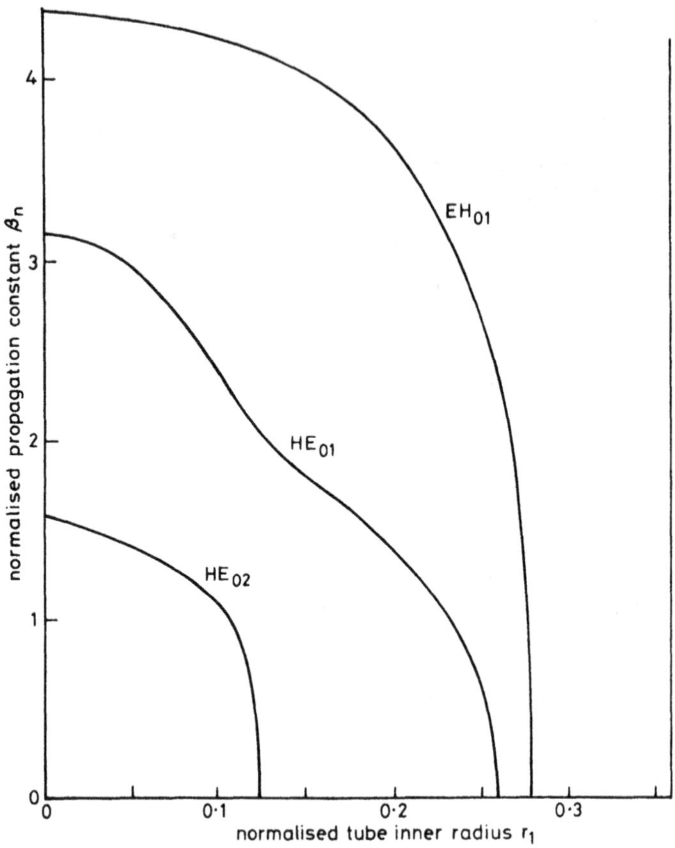

Fig. 3.10 *Propagation conditions for a longitudinally magnetised ferrite tube in circular waveguide. The normalised propagation constant is plotted against the normalised radius of the inside of the tube as the inside tube radius varies from zero (completely filled waveguide) to $r_1 = r_2$ (empty waveguide). The normalised waveguide radius $r_2 = 0{\cdot}36$, and the ferrite properties are the same as in Fig. 3.4. Circularly symmetric modes $n = 0$*

circular waveguide. However, Rizzi[12] shows that there is a much higher loss due to the ferrite with a tube adjacent to the walls than with a rod at the centre of the waveguide, but even then the dissipative properties of the tube may make it preferable.

Mode notation

It appears to be generally accepted that the modes may be identified by varying one parameter in the calculation and observing how the propagation constant varies. Then the modes are identified by relating any particular propagating mode to a mode in empty circular waveguide or to the cut-off conditions. Waldron,[7,10] in his extensive investigation of the cut-off conditions, has labelled

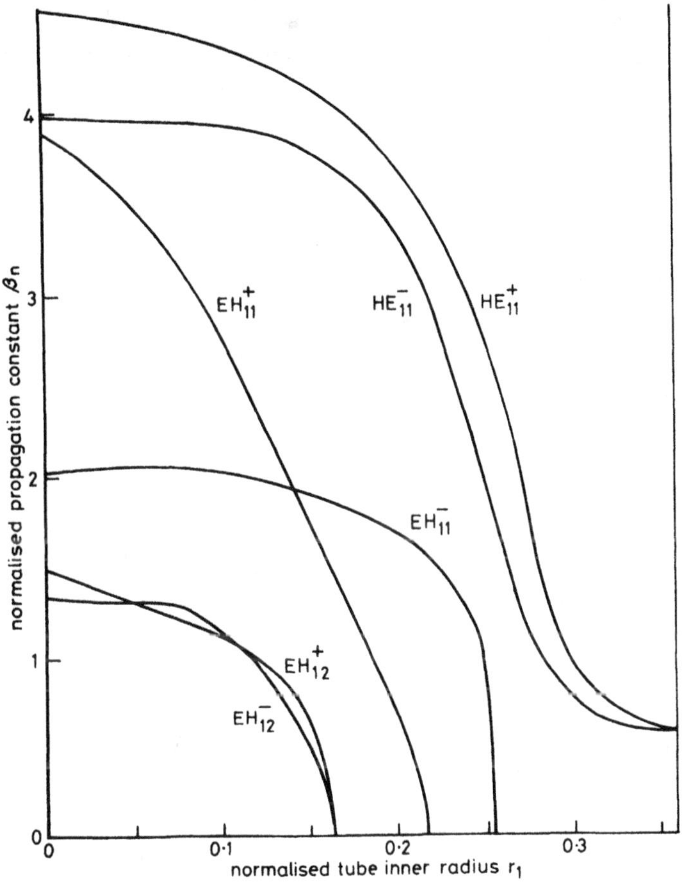

Fig. 3.11 *Propagation conditions for a ferrite tube, the same as for Fig. 3.10 for the modes* $n = 1$

most of the modes having propagating conditions. He also used the properties, already discussed above, that the cut-off equations separate into two, one identifiable with the HE-modes, which is independent of the values of μ_r and κ_r, and the other identifiable with the EH-modes. This effect may be seen in Figs. 3.8, 3.9, 3.11 and 3.12, where the cut-off of the HE-modes is the same for both hands of circular polarisation. However, this identification does not always tally when identification is made through comaprison with the completely filled waveguide.

It is possible to look at the curves for rods, where the rod completely fills the waveguide, and compare the modes with the curves for tubes, where the tube has zero inside radius. This comparison, using the results given here, provides agreement for the mode nomenclature *except* for the positive circularly polarised modes of circumferential order, $n = 2$. The relevant curves taken from Figs. 3.9 and 3.12 are plotted in juxtaposition in Fig. 3.14. This shows that the EH_{21}^{+}- and HE_{21}^{+}-modes appear to swap places in going from a rod at the centre of circular waveguide to a tube adjacent to the wall of the waveguide. It is also seen

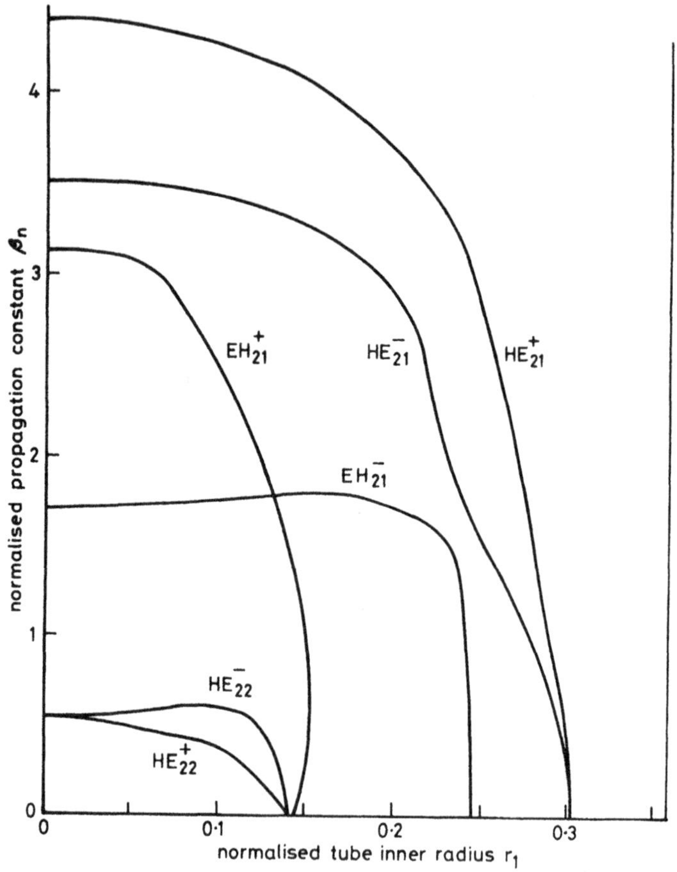

Fig. 3.12 *Propagation conditions for a ferrite tube, the same as for Fig. 3.10 for the modes* $n = 2$

from Figs. 3.9 and 3.12 that, as long as we insist that the cut-off conditions of the HE-modes is independent of the hand of circular polarisation of the mode, we have labelled the modes correctly in Fig. 3.14. As the geometry or the ferrite properties are varied it is probable that the relative sizes of the longitudinal components of the electric and magnetic field will vary. Waldron[8] shows this result in the values he gives there of the sizes of the field components of the HE_{11}-mode for a rod of ferrite in circular waveguide. Thus there may well be a

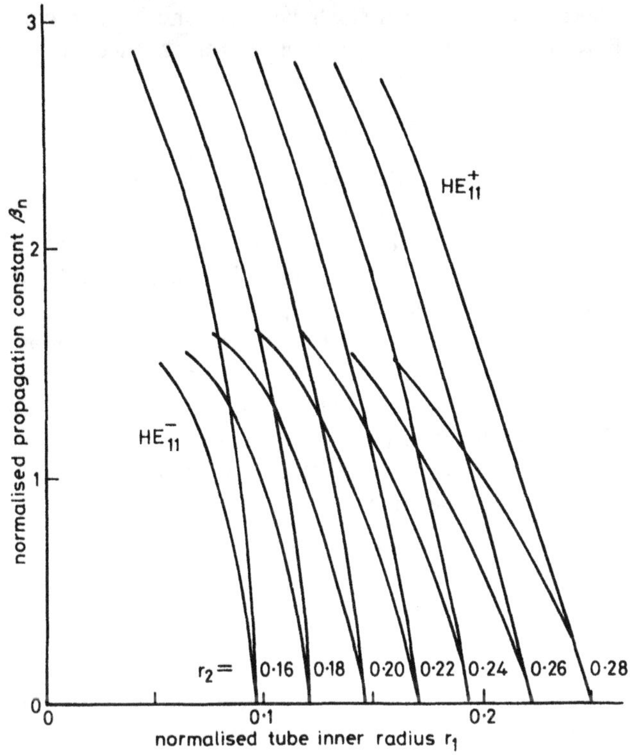

Fig. 3.13 *Propagation conditions for HE_{11}-mode for a longitudinally magnetised ferrite tube in circular waveguide. The normalised propagation constant is plotted against the variation of the normalised radius of the inside of the tube. A number of curves are given for different tube outer radii. The ferrite properties are $\mu_r = 0{\cdot}98$, $\kappa_r = 0{\cdot}63$ and $\varepsilon_r = 11{\cdot}2$*

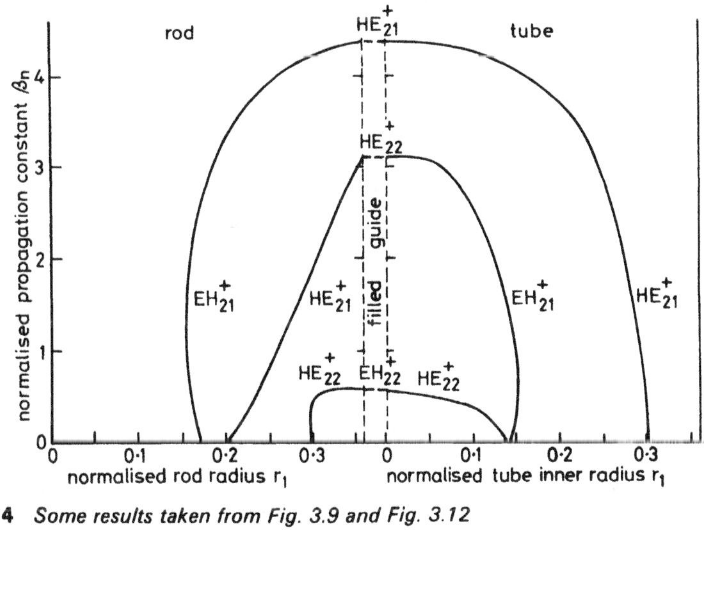

Fig. 3.14 *Some results taken from Fig. 3.9 and Fig. 3.12*

condition where a mode, which starts with HE-mode characteristics for one geometry, develops EH-mode characteristics for another geometry. This is a possible explanation for the apparent anomaly shown in Fig. 3.14.

3.7 More complicated waveguide structures

The theory given in the preceding sections may be extended to analyse any circularly symmetric system of longitudinally magnetised ferrite in circular waveguide. Expressions for the field components are given in eqns. 3.34 to 3.48 and it is only necessary to define a different set of arbitrary constants in each different region and to equate the relevant components of the electric and magnetic fields across each boundary to obtain a characteristic equation similar to eqn. 3.57. As the number of regions increases, the number of arbitrary constants will increase and the size of the determinant that has to be evaluated will increase. However, the functions comprising the elements of the deter-

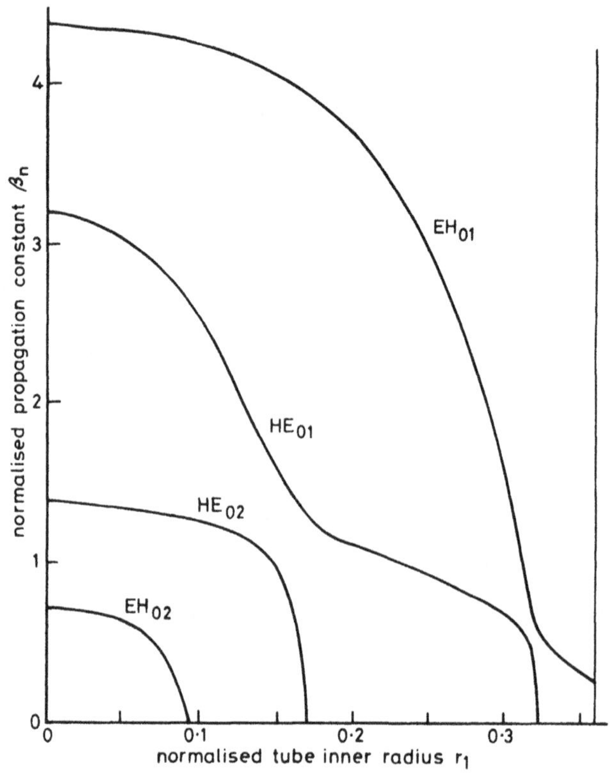

Fig. 3.15 *Propagation conditions for a longitudinally magnetised ferrite tube in circular waveguide with a dielectric (air) space between the tube and the waveguide wall. The other conditions are the same as for Fig. 3.10. The normalised radius of the outside of the ferrite tube r_2 = 0·36, and the normalised radius of the waveguide r_3 = 0·40. Circularly symmetric modes n = 0*

minant are no more complicated and are still given in Section 3.3. I have investigated[13] the geometry consisting of a tube of ferrite mounted at the centre of circular waveguide. There is air, or some other dielectric material, between the waveguide wall and the ferrite and also inside the centre of the ferrite tube.

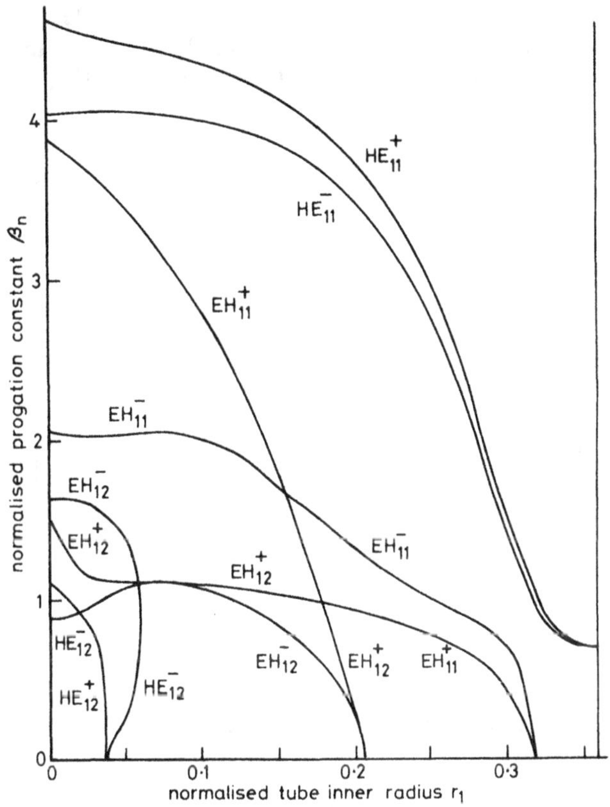

Fig. 3.16 *Propagation conditions for a ferrite tube, the same as for Fig. 3.15 for the modes $n = 1$*

The mathematical equations are given in that paper[13] and the results of a specimen investigation are given in Figs. 3.15 to 3.17, which are similar to those given in Figs. 3.7 to 3.12 and use the same values of ferrite properties. The normalised radii are: r_1, which is the radius of the inside of the tube and is the variable; $r_2 = 0{\cdot}36$ is the radius of the outside of the tube, and $r_3 = 0{\cdot}40$ is the radius of the circular waveguide. For this particular geometry and size, the TM_{01}-mode is able to propagate in the empty waveguide as well as the dominant TE_{11}-mode, and so it is unlikely to be used for practical devices. However, it is interesting to compare these results with the other results of the rod and tube of the same sizes. Compared with the results given in Fig. 3.13, these results show that the useful rotation given by the dominant HE_{11}-mode is reduced as the waveguide size is increased.

3.8 Approximate methods

Perturbation theory is discussed in Chapter 12 and the perturbation formula for the change of propagation constant owing to the perturbation of a waveguide is given in eqn. 12.73. Suhl and Walker[22] apply perturbation techniques to calculating the effect of a longitudinally magnetised ferrite rod in circular waveguide. Because, of necessity, the theory is only applicable to very weakly magnetised ferrite rods or to very thin rods, these perturbation techniques are

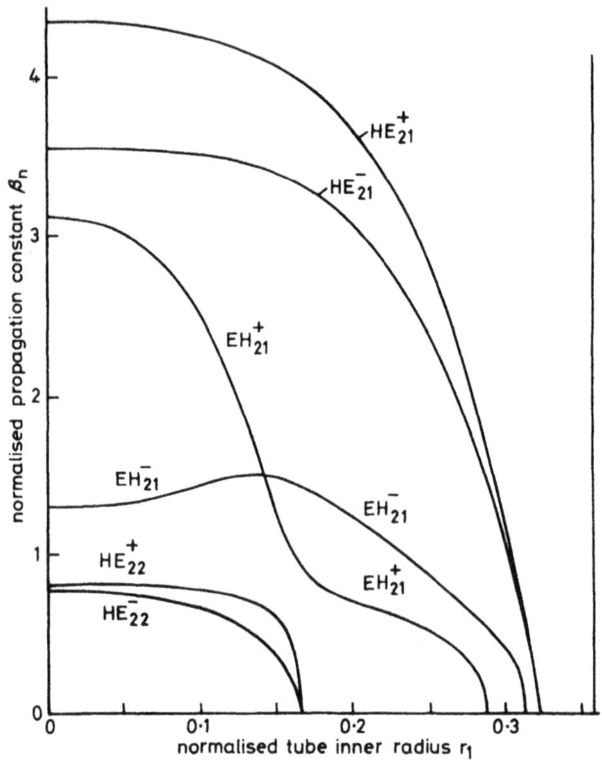

Fig. 3.17 *Propagation conditions for a ferrite tube, the same as for Fig. 3.15 for the modes $n = 2$*

of limited applicability. However, Clarricoats[4] uses the assumption that the fields inside the ferrite approximate to those of a plane wave and produces a perturbation formula that is applicable over the range of sizes of ferrite rod that are used to make Faraday rotation devices. He makes the assumption that the ferrite has the same effective permeability to circularly polarised waves inside the waveguide as an infinite ferrite medium has to plane circularly polarised waves and he shows that this method gives ample accuracy for the design of ferrite devices. In his book[15] he uses both applications of perturbation theory to give design information for a ferrite rod at the centre of circular waveguide. However, the assumption that the field in the waveguide approximates to that of a plane wave is not applicable to ferrite tubes in circular waveguide. Most of the perturbation theories were developed when the exact theories were not available and before the advent of powerful digital computers. Now that exact theories and computer calculations are available, it is doubtful whether the use of perturbation theory is advantageous. The perturbation formulas are not that simple. A very full description of the various perturbation methods is given by Clarricoats,[15] together with all the necessary formulas.

3.9 Attenuation

All the results given so far in this chapter have made the assumption that the wave is propagating through the ferrite loaded waveguide without loss. For most real situations, the losses are small and the assumption is valid for the calculation of propagation constants and relative strengths of the components of the fields. However, attenuation does occur and it is useful to be able to estimate the effect of attenuation. In a ferrite loaded waveguide there are four kinds of loss:

1 Wall loss owing to the finite conductivity of the metal of the waveguide wall to the currents flowing in it.
2 Dielectric loss in the dielectric material filling the non-ferrite part of the waveguide.
3 Dielectric loss in the ferrite
4 Magnetic loss in the ferrite.

The wall loss is usually small and may be neglected compared with the loss due to the ferrite. Even if the dielectric material is not air, most low-loss microwave dielectric materials have losses that are small compared to those of microwave ferrite materials and the loss of the dielectric material can also be neglected. Then the only significant losses are due to the ferrite material. The material losses may be represented by giving complex values to the permittivity and permeability

$$\left.\begin{aligned} \varepsilon_r &= \varepsilon'_r - j\varepsilon''_r \\ \mu_r &= \mu'_r - j\mu''_r \\ \kappa_r &= \kappa'_r - j\kappa''_r \end{aligned}\right\} \qquad (3.58)$$

Mathematically it is possible to put these complex values of the permittivity and permeability of the ferrite into the expressions used for the determination of the propagation constant and to determine a complex value for the propagation constant. Howver, the determinants used to evaluate the propagation constant are already sufficiently complicated without the additional complication of having to do all the calculation in complex numbers. Provided the losses are small, as they usually are, approximate methods may be used to evaluate the attenuation of the wave in the ferrite loaded waveguide.

Clarricoats[15] gives a method for calculating the attenuation in ferrite loaded waveguide by using the results for dielectric loaded waveguide.[5] As has already been discussed in the previous section, he shows that this approximation is particularly valid for a ferrite rod in circular waveguide where the permeability used is the plane wave permeability of an infinite ferrite medium to circularly polarised waves. Waldron[14] has proposed an approximate method which is valid for small losses whatever the geometry of the ferrite in the waveguide and for which values may be calculated using the data given earlier in this chapter. A summary of his method is given here. The object is to find an expression for the imaginary part of the complex normalised propagation constant

$$\beta_n = \beta_n' - j\beta_n''$$

in terms of the complex material constants given in eqn. 3.58. The propagation constant is a function of the dimensions and the material constants

$$\beta_n = \beta_n(r_1, r_2, r_3, \varepsilon_r, \mu_r, \kappa_r) \qquad (3.59)$$

Because the dimensions are normalised in terms of the free-space wavelength in the dielectric material, the dimensions are functions of the properties of that material. Waldron shows how allowance can be made for the effect of any loss factor in the permittivity of the dielectric material. However, in the analysis that follows, we assume that the loss in the dielectric material is negligible. Provided the attenuation is small, the attenuation is given by differentiating eqn. 3.59 assuming that the dimensions are constants. The perturbation is to introduce the loss factors in the permittivity and permeability. Therefore

$$\delta\beta_n = \frac{\partial\beta_n}{\partial\varepsilon_r}\delta\varepsilon_r + \frac{\partial\beta_n}{\partial\mu_r}\delta\mu_r + \frac{\partial\beta_n}{\partial\kappa_r}\delta\kappa_r \qquad (3.60)$$

If we let $\delta\varepsilon_r$ etc. be the small imaginary parts of the expressions given in eqn. 3.58, then eqn. 3.60 becomes

$$\beta_n'' = \frac{\partial\beta_n}{\partial\varepsilon_r}\varepsilon_r'' + \frac{\partial\beta_n}{\partial\mu_r}\mu_r'' + \frac{\partial\beta_n}{\partial\kappa_r}\kappa_r'' \qquad (3.61)$$

which is valid as long as the loss factors are small. The expressions used to evaluate the propagation constant are too complicated to give analytical results for the partial differential expressions in eqn. 3.61. However, a computer program is available to calculate values of the propagation constant in terms of the variables in eqn. 3.59 and numerical differentiation may be used to evaluate the partial differentials in eqn. 3.61. Waldron[14] gives curves of the values of these partial differentials for a rod in circular waveguide for the HE_{11}-mode, from which the attenuation of a wave in the ferrite loaded waveguide may be calculated. A similar technique may be used to calculate the attenuation due to a tube of ferrite in circular waveguide.

3.10 References

1 GAMO, H.: 'The Faraday rotation of waves in a circular waveguide', *J. Phys. Soc. Japan.*, 1953, **8**, pp. 176–182

2 KALES, M. L.: 'Modes in wave guides containing ferrites', *J. Appl. Phys.*, 1953, **24**, pp. 604–608

3 SUHL, H. and WALKER, L. R.: 'Topics in guided wave propagation through gyromagnetic media. Pt.I and Pt.III', *Bell Syst. Tech. J.*, 1954, **33**, pp. 579–659, 1133–1194

4 CLARRICOATS, P. J. B.: 'A perturbation method for circular waveguides containing ferrites', *Proc. IEE*, 1959, **106B**, pp. 335–340

5 CLARRICOATS, P. J. B.: 'Propagation along unbounded and bounded dielectric rods', *Proc. IEE*, 1961, **108C**, pp. 170–186

6 CHATTERJEE, S. K. and CHATTERJEE, R.: 'Dielectric loaded waveguides – A review of theoretical solutions', *Rad. & Elect. Eng.*, 1965, **30**, pp. 145–160, 195–205, 259–283, 353–364

7 WALDRON, R. A.: 'Electromagnetic wave propagation in cylindrical waveguides containing gyromagnetic media', *J. Brit. IRE*, 1958, **18**, pp. 597–612, 677–690, 733–746

8 WALDRON, R. A.: 'Features of cylindrical waveguides containing gyromagnetic media', *J. Brit. IRE*, 1960, **20**, pp. 695–706

9 TOMPKINS, J. E.: 'Energy distribution in partially ferrite-filled wave guides', *J. Appl. Phys.*, 1958, **29**, pp. 399–400

10 WALDRON, R. A.: 'Theory of the mode spectra of cylindrical waveguides containing gyromagnetic media', *J. Brit. IRE* , 1959, **19**, pp. 347–356

11 MIRIMANOV, R. G. and ANISIMOVA, Yu. V.: 'Circular waveguide partially filled with ferrite as a retarding system', *Rad. i Elek.*, 1957, **2**, pp. 843–855 (English translation in *Rad. Eng. & Electron*, **2**, 7, pp. 37–53)

12 RIZZI, P. A.: 'High-power ferrite circulators', *IRE Trans.*, 1957, **MTT-5**, pp. 230 237

13 BADEN FULLER, A. J.: 'Microwave propagation through waveguide partially filled with ferrite', *Proc. IEE*, 1961, **108C**, pp. 339–348

14 WALDRON, R. A. and BOWE, D. J.: 'Loss properties of cylindrical waveguides containing gyromagnetic media', *J. Brit. IRE*, 1963, **25**, pp. 321–334

15 CLARRICOATS, P. J. B.: 'Microwave ferrites', (Chapman and Hall, 1961)

16 WALDRON, R. A.. 'Ferrites' (Van Nostrand, 1961)

17 HELSZAJN, J.: 'Microwave ferrite engineering' (Wiley-Interscience, 1969)

18 BADEN FULLER, A. J.: 'Microwaves', 2nd edn. (Pergamon, 1979), p. 101

19 BADEN FULLER, A. J.: 'Microwaves', 2nd edn. (Pergamon, 1979), p. 103

20 WALDRON, R. A: 'Properties of inhomogeneous cylindrical waveguides in the neighbourhood of cut-off', *J. Brit. IRE*, 1963, **25**, pp. 547–555

21 WALDRON, R. A.: 'Properties of ferrite-loaded cylindrical waveguides in the neighbourhood of cut-off', *Proc. IEE*, 1962, **109B**, Suppl. No. 21, pp. 90–94

22 SUHL, H. and WALKER, L. E.: 'Topics in guided wave propagation through a gyromagnetic media. Pt.III', *Bell Syst. Tech. J*, 1954, **33**, pp. 1133–1194

Chapter 4

Transversely magnetised ferrite in circular waveguide

4.1 Introduction

In the last chapter we considered the effect of a longitudinally magnetised piece of ferrite material in circular waveguide. The problem is mathematically complicated but it has been solved and the results of some computations are given. Unless the static magnetic field is provided by a solenoidal coil around the outside of the waveguide, which is only convenient for control applications, there is no immediately obvious method of applying the magnetic field. However, for certain configurations of field transverse to the direction of propagation along the waveguide it is possible to place the waveguide between the poles of a magnet, or to place a number of small magnets adjacent to the wall of the waveguide outside the waveguide. Therefore, in this chapter, we consider the propagating conditions for a circularly symmetric system of ferrite in circular waveguide which is magnetised in some direction perpendicular to the direction of propagation along the waveguide. Four different systems of transverse magnetisation are shown in Fig. 4.1. The radially magnetised ferrite tube shown in Fig. 4.1*a* is almost impossible to realise in practice and will not be considered further. The circumferentially magnetised ferrite tube shown in Fig. 4.1*b* is used in the design of a latching phase changer. Using the square-loop magnetic property of microwave ferrite materials, the ferrite may be magnetised to saturation circumferentially by passing a pulse of electric current through a wire threading the tube. The ferrite will then remain magnetised to remanence until it is demagnetised. The analysis of the circumferentially magnetised ferrite tube is given in Sections 4.2 to 4.4. Both the simple transverse magnetisation, shown in Fig. 4.1*c*, and the four-pole field, shown in Fig. 4.1*d*, have been used to provide birefringence to the microwave field in the waveguide. The effect of the transverse magnetisation of Fig. 4.1*c* is described in Section 4.5, the effect of the four-pole magnetisation of Fig. 4.1*d* is described in Section 4.6 and the governing equations for a theoretical analysis are given in Section 4.7.

4.2 Circumferentially magnetised ferrite tube

In Chapter 2 we considered the interaction between a plane wave and an infinite extent of ferrite material magnetised in the direction of propagation of the wave. In particular it was seen that the chief interaction was between precessing electrons in the ferrite and a circularly polarised magnetic field pattern in the plane perpendicular to the direction of magnetisation in the ferrite. The effective permeability of the ferrite material to the wave is given by eqn. 2.44 and its variation with the strength of the static magnetic field is shown in Fig. 2.5. In

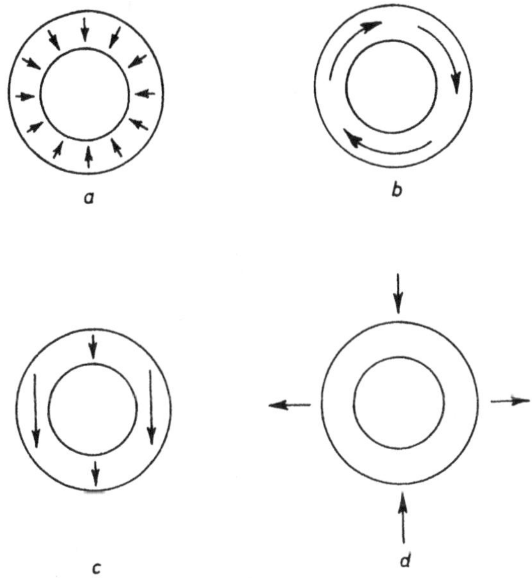

Fig. 4.1 *Systems of transverse magnetisation of a ferrite loaded circular waveguide*
a Radially magnetised tube
b Circumferentially magnetised tube
c Transverse magnetisation
d Four-pole field

Chapter 3, it was shown that, for a longitudinally magnetised ferrite rod at the centre of circular waveguide, the field due to the dominant TE_{11}-mode approximates to that of a plane wave, and the effective permeability of the ferrite material is given by eqn. 2.44. For the circumferentially magnetised ferrite tube in circular waveguide, there is also a waveguide mode which provides a circularly polarised magnetic field in a plane perpendicular to the static magnetisation in the ferrite. It is the TE_{01}-mode in the empty circular waveguide.

Following the notation used in Chapter 3, we consider the fields in an empty circular waveguide in terms of a cylindrical polar co-ordinate system, r, θ and z, having its axis coincident with the axis of the circular waveguide. The wave is propagating in the z-direction with a propagation constant β and has an angular frequency ω. Because the circularly symmetric field pattern of the TE_{01}-mode is being considered, there is no variation of the fields in the θ-direction, and the integer n will be zero. Expressions for the fields in empty circular waveguide are given in many textbooks;[1] those for the TE_{01}-mode are:

$$\left.\begin{aligned}
H_r &= -\frac{j\beta}{k} H_0 J_0'(kr) \exp j(\omega t - \beta z) \\
H_\theta &= 0 \\
H_z &= H_0 J_0(kr) \exp j(\omega t - \beta z) \\
E_r &= 0 \\
E_\theta &= \frac{j\omega\mu_0}{k} H_0 J_0'(kr) \exp j(\omega t - \beta z) \\
E_z &= 0
\end{aligned}\right\} \tag{4.1}$$

where $k^2 = \omega^2\mu_0\varepsilon - \beta^2$, J_0 is the Bessel function of the first kind and $J_0'(x) = J_1(x)$ is the differential of J_0 with regard to its argument. The field will be circularly polarised at a particular radius in the diametral plane when $jH_r = \pm H_z$. That is, when

$$J_1(kr) = \pm \frac{k}{\beta} J_0(kr) \tag{4.2}$$

There are two radii which satisfy eqn. 4.2 corresponding to the plus or the minus in the equation. This means that the field is circularly polarised with the opposite hand at each radius. It can also be seen that the radius corresponding to circular polarisation varies with frequency or with the propagation constant of the wave. However, if there are two ferrite tubes in the waveguide at approximately the correct radius and the rest of the space in the waveguide is filled with air, the large permittivity of the ferrite material tends to concentrate the fields in the ferrite and the variation of the position of circular polarisation of the magnetic field with change of frequency is much less. Two ferrite tubes magnetised as shown in Fig. 4.2 will provide differential phase change for the TE_{01}-mode in the ferrite loaded circular waveguide. Making use of the square-loop hysteresis properties of the microwave ferrite material, the ferrite tubes in Fig. 4.2 can be magnetised to remanence and will remain in that state until they are demagnetised or magnetised in the opposite direction. The technique is discussed in more detail in Section 7.2. Altering the direction of magnetisation will change the effective permeability of the ferrite from μ^+ to μ^- or vice versa. Altering the direction of propagation of the wave in the waveguide views the magnetised ferrite tubes from the opposite end so that the direction of the magnetisation is reversed, and such a device will have a different phase length for waves propagating in opposite directions. It is nonreciprocal and provides differential phase change. It is shown in Section 7.2 how to produce variable magnetisation in the ferrite below remanence, so that such a device can provide variable phase change, but normally the ferrite tubes are only magnetised to remanence and the device provides a fixed quantity of differential phase change.

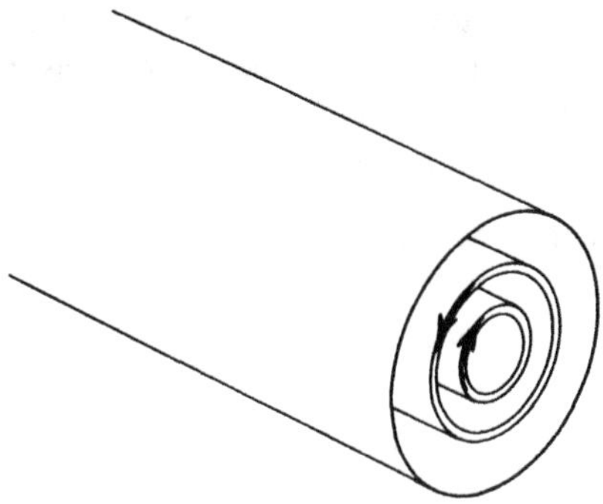

Fig. 4.2 *Circular waveguide loaded with two ferrite tubes circumferentially magnetised in opposite directions*

4.3 Circumferentially magnetised ferrite-filled waveguide

Suhl and Walker[2] were the first to attempt any solution to the problem of circumferentially magnetised ferrite completely filling circular waveguide. They analysed the fields in the ferrite in a way similar to that used in the last chapter, and showed that for the circularly symmetric field when $n = 0$ a solution could be obtained for E_z in terms of Bessel functions and that the modes were TM-modes and similar to those propagating through a circular waveguide loaded with a dielectric tube. For the circularly symmetric modes, the modes of propagation separate into TM-modes having the same properties as those that propagate in dielectric loaded waveguide and TE-modes whose propagation conditions are dependent on the gyromagnetic properties of the ferrite. This is similar to the two modes of propagation of a plane wave in an infinite transversely magnetised ferrite material given in eqns. 2.64 and 2.65. The equation for H_z given by Suhl and Walker has solutions which are confluent hypergeometric functions. They present their results in a form which gives Whittaker functions, and they chose this form because Whittaker functions are closely related to Bessel functions. For the other possible modes in the circular waveguide when $n \neq 0$, the equations are much more complicated; there are equations similar to eqns. 3.15 and 3.16 and there are no modes with just transverse electric or magnetic fields.

Following Suhl and Walker[2] we derive a wave equation for the circularly symmetric modes in a circumferentially magnetised ferrite material. Working in the r, θ and z system of polar co-ordinates, the tensor permeability of the circumferentially magnetised ferrite material is similar to eqn. 2.59

$$\boldsymbol{\mu} = \begin{vmatrix} \mu & 0 & \mathrm{j}\kappa \\ 0 & \mu_0 & 0 \\ -\mathrm{j}\kappa & 0 & \mu \end{vmatrix} \tag{4.3}$$

The gyromagnetic properties of the ferrite, given by eqn. 4.3, are substituted into Maxwell's equations, eqns. 2.31 and 2.32. As we are considering the circularly symmetric modes alone, there is no variation of the fields in the θ-direction. Then the governing field equations are

$$\mathrm{j}\beta E_\theta = -\mathrm{j}\omega(\mu H_r + \mathrm{j}\kappa H_z) \tag{4.4}$$

$$-\mathrm{j}\beta E_r - \frac{\partial E_z}{\partial r} = -\mathrm{j}\omega\mu_0 H_\theta \tag{4.5}$$

$$\frac{1}{r} E_\theta + \frac{\partial E_\theta}{\partial r} = -\mathrm{j}\omega(-\mathrm{j}\kappa H_r + \mu H_z) \tag{4.6}$$

$$\mathrm{j}\beta H_\theta = \mathrm{j}\omega\varepsilon E_r \tag{4.7}$$

$$-\mathrm{j}\beta H_r - \frac{\partial H_z}{\partial r} = \mathrm{j}\omega\varepsilon E_\theta \tag{4.8}$$

$$\frac{1}{r} H_\theta + \frac{\partial H_\theta}{\partial r} = \mathrm{j}\omega\varepsilon E_z \tag{4.9}$$

The decision to confine the investigation to circularly symmetric modes has meant that the terms in eqns. 4.4 to 4.9 which cause coupling between the longitudinal components of the electric and magnetic fields are equal to zero. The TE- and TM-modes are the fundamental forms of circularly symmetric modes propagating in the ferrite-filled waveguide. If E_r and H_θ are eliminated from eqns. 4.5, 4.7 and 4.9, the following wave equation describes the fields of the TM_0-modes:

$$\frac{\partial^2 E_z}{\partial r^2} + \frac{1}{r}\frac{\partial E_z}{\partial r} + k^2 E_z = 0 \tag{4.10}$$

where k is defined following eqn. 4.1. The solution of eqn. 4.10 is zero-order Bessel functions and gives rise to standard TM_0-modes in dielectric-filled waveguide. However, it is only mathematically that this separation into TE- and TM-modes occurs, and it only applies in practice to a waveguide completely filled with a homogeneous ferrite material. For a ferrite tube or rod, as with a dielectric tube or rod in circular waveguide, the need to match field components at the boundary between dielectric and air means that composite modes are propagated having longitudinal components of both the electric and magnetic fields. Then eqn. 4.10 on its own describes the fields of the TM_0-modes in ferrite-filled circular waveguide. For the TE_0-modes, the wave equation is given by eliminating E_θ and H_r from eqns. 4.4, 4.6 and 4.8

$$\frac{\partial^2 H_z}{\partial r^2} + \frac{1}{r}\frac{\partial H_z}{\partial r} + \left[\omega^2\varepsilon\frac{(\mu^2-\kappa^2)}{\mu} - \beta^2 - \frac{\kappa\beta}{\mu r}\right]H_z = 0 \tag{4.11}$$

The solutions to this equation are confluent hypergeometric functions. Suhl and Walker[2] show how, by change of variable, it becomes Whittaker's equation which has solutions of zero-order Whittaker functions. For the completely filled waveguide, the boundary condition is given by finding zero's for $\partial H_z/\partial r$ and these values are not tabulated for the Whittaker functions. Ivanov[3] analyses the circumferentially magnetised ferrite-filled circular waveguide in terms of confluent hypergeometric functions. He gives curves of normalised propagation constant against normalised waveguide radius for the TE_{01}-mode for various values of κ when $\mu = \mu_0$, representing the remanent condition of the ferrite. An alternative approach has been given by Bolle and Heller,[4] which is discussed in the following section.

4.4 Circumferentially magnetised ferrite tube theory

Suhl and Walker[2] appear to have investigated the circumferentially magnetised ferrite tube as part of a study of the nonreciprocal helix for use in travelling wave tubes. They do not appear to have considered the phase-change properties of such a ferrite tube. Fox, Miller and Weiss[5] have the first recorded suggestion of the use of a circumferentially magnetised ferrite tube in circular waveguide to provide nonreciprocal phase change. They give an equation similar to eqn. 4.2 but not confined to the circularly symmetric modes. However, for the circularly symmetric modes, the magnetic field in the ferrite will be circularly polarised and interacting with the gyromagnetic properties of the ferrite at all angles around the circle, whereas for any other modes the circularly polarised magnetic fields occur at only a few specific points around the circle and the ferrite is being used inefficiently.

Continuing with Suhl and Walker's derivation,[2] the other components of the fields in the waveguide may be given in terms of E_z and H_z. The expressions are derived from eqns. 4.4 to 4.9.

$$E_r = -j\frac{\beta}{k^2}\frac{\partial E_z}{\partial r} \tag{4.12}$$

$$H_\theta = -j\frac{\omega\varepsilon}{k^2}\frac{\partial E_z}{\partial r} \tag{4.13}$$

$$H_r = -j\frac{1}{k^2}\left(\omega^2\varepsilon\kappa H_z + \beta\frac{\partial H_z}{\partial r}\right) \tag{4.14}$$

$$E_\theta = j\frac{1}{k^2}\left(\omega\kappa\beta H_z + \omega\mu\frac{\partial H_z}{\partial r}\right) \tag{4.15}$$

A full solution to the ferrite loaded waveguide problem requires the application of boundary conditions between the ferrite and the dielectric filling the rest of the waveguide, and the boundary condition at the waveguide wall. Such problems have been investigated by Bernues and Bolle[6] and Potekhin and Yurgenson,[7,8] but using special functions for the solution to the wave equation. Further discussion of their results is given below.

Bolle and Heller[4] have used an alternative approach to the solution of eqns. 4.4, 4.6 and 4.8. Instead of eliminating E_θ from eqns. 4.4, 4.6 and 4.8 to give an equation in H_z, H_z and H_r are eliminated from these equations to give an equation in E_θ which comes out to be

$$\frac{\partial^2 E_\theta}{\partial r^2} + \frac{1}{r}\frac{\partial E_\theta}{\partial r} + \left[\omega^2\varepsilon\frac{(\mu^2 - \kappa^2)}{\mu} - \beta^2 + \frac{\kappa\beta}{\mu r} - \frac{1}{r^2}\right]E_\theta = 0 \tag{4.16}$$

The other components of the field are given by

$$H_r = \frac{1}{\omega(\mu^2 - \kappa^2)}\left[\left(\frac{\kappa}{r} - \beta\mu\right)E_\theta + \kappa\frac{\partial E_\theta}{\partial r}\right] \tag{4.17}$$

$$H_z = \frac{\mathrm{j}}{\omega(\mu^2 - \kappa^2)}\left[\left(\frac{\mu}{r} - \beta\kappa\right)E_\theta + \mu\frac{\partial E_\theta}{\partial r}\right] \tag{4.18}$$

Bolle and Heller[4] have investigated the solution of eqn. 4.16, which is also a confluent hypergeometric equation. They decided to construct their own solutions that relate directly to Bessel functions, because when either β or κ go to zero, eqn. 4.16 reduces to Bessel's equation. This aids in the understanding of the modes in the waveguide since the new functions become Bessel functions when investigating cut-off conditions or when the ferrite behaves like an isotropic dielectric material. Bolle and Heller introduced two new functions which are linearly independent solutions of eqn. 4.16, called the Bolle and Heller functions, $B_1(\alpha;x)$ and $H_1(\alpha;x)$.

Eqn. 4.16 may be put in the form

$$\frac{\partial^2 E_\theta}{\partial r^2} + \frac{1}{r}\frac{\partial E_\theta}{\partial r} + \left[m^2 + \frac{p}{r} - \frac{1}{r^2}\right]E_\theta = 0 \tag{4.19}$$

where

$$m^2 = \omega^2\varepsilon\frac{(\mu^2 - \kappa^2)}{\mu} - \beta^2 = k^2 - \omega^2\varepsilon\frac{\kappa^2}{\mu}$$

$$p = \frac{\beta\kappa}{\mu}$$

In the Bolle and Heller functions

$$\alpha = p/m \quad \text{and} \quad x = mr$$

and the solution to eqn. 4.16 is

$$E_\theta = AB_1(\alpha;mr) + BH_1(\alpha;mr) \tag{4.20}$$

where A and$B$ and arbitrary constants.

The functions are given by

$$B_1(\alpha;x) = \sum_{n=0}^{\infty} a_n x^{n+1} \tag{4.21}$$

where

$$a_0 = \tfrac{1}{2}, \qquad a_1 = -\alpha a_0/3$$

$$a_{n+2} = -\frac{\alpha a_{n+1} + a_n}{(n+2)(n+4)}$$

and

$$H_1(\alpha;x) = \frac{2}{\pi}\left\{[\gamma + \ln \tfrac{1}{2}x]\, B_1(\alpha;x) + \sum_{n=0}^{\infty} b_{n+2}x^{n+1} + \frac{b_0}{x} + b_1\right\} \tag{4.22}$$

where γ is Euler's constant and

$$b_0 = -1/(1+\alpha^2) \qquad b_1 = -\alpha/(1+\alpha^2) \qquad b_2 = -\tfrac{1}{4}$$

$$b_{n+2} = -\frac{2(n+1)a_n + b_n + \alpha b_{n+1}}{n(n+2)}$$

For the condition where m is imaginary, two new functions have been defined by Ince and Tsandoulas[9] which are modified Bolle and Heller functions. They are $P_1(\alpha;x)$ and $Q_1(\alpha;x)$, which have very similar expressions to $B_1(\alpha;x)$ and $H_1(\alpha;x)$. They are defined by

$$P_1(\alpha;x) = \sum_{n=0}^{\infty} a_n x^{n+1} \tag{4.23}$$

where

$$a_0 = \tfrac{1}{2}, \qquad a_1 = -\alpha a_0/3$$

$$a_{n+2} = \frac{a_n - \alpha a_{n+1}}{(n+2)(n+4)}$$

and

$$Q_1(\alpha;x) = [\gamma + \ln \tfrac{1}{2}x]P_1(\alpha;x) + \sum_{n=0}^{\infty} b_{n+2}x^{n+1} + \frac{b_0}{x} + b_1 \tag{4.24}$$

where γ is Euler's constant and

$$b_0 = 1/(1-\alpha^2) \qquad b_1 = \alpha/(1-\alpha^2) \qquad b_2 = -\tfrac{1}{4}$$

$$b_{n+2} = \frac{b_n - \alpha b_{n+1} - 2(n+1)a_n}{n(n+2)}$$

and the modified solution to eqn. 4.19 is

$$E_\theta = AP_1(\alpha;mr) + BQ_1(\alpha;mr) \qquad (4.25)$$

where A and B are arbitrary constants, and the modulus of m is used in the argument of the functions.

Both the Bolle function and the modified Bolle function go to zero for $x = 0$ and in this way are similar to Bessel functions of the first kind. The Heller functions and the modified Heller functions both go to infinity for $x = 0$ and are similar to Bessel functions of the second kind. Plots of the values of $B_1(\alpha;x)$ and $H_1(\alpha;x)$ and the derivatives of these two functions are given by Bolle and Heller.[4] Plots of the values of $P_1(\alpha;x)$ and$Q_1(\alpha;x)$ are given by Ince and Tsandoulas[9] as well as some further information on the properties of $B_1(\alpha;x)$ and $H_1(\alpha;x)$. For a completely filled waveguide of radius r_1 the propagating conditions are given by

$$E_\theta = 0$$

Therefore either

$$B_1(\alpha;x) = 0 \quad \text{or} \quad P_1(\alpha;x) = 0 \qquad (4.26)$$

However, the completely filled guide is unlikely to be used in practice since only part of the ferrite is contributing to the differential phase change and the rest is just increasing the loss. Also it is difficult to see how a solid rod of ferrite could be magnetised circumferentially. Bolle and Heller[4] did investigate the propagating conditions for the situation when the Bolle function describes the fields in the ferrite and they showed that differential phase change does occur. They then proceeded to indicate how the circumferentially magnetised ferrite tube in circular waveguide, similar to that shown in Fig. 4.2, may be analysed. They give some results from their calculations.

One of the problems of using a circumferentially magnetised ferrite tube as a differential phase changer is that it operates most efficiently on the HE_{01}-mode in the ferrite loaded waveguide, whereas the HE_{11}-mode is the dominant mode. In particular, in the presence of circumferentially magnetised ferrite loading, it is difficult to suppress conversion from the desired HE_{01}-mode to the dominant HE_{11}-mode. Tsandoulas and Ince[10] have shown that dielectric loading of the centre of the circular waveguide causes mode inversion between the HE_{01}-mode and the HE_{11}-mode. The frequency bandwidth of the mode inversion region is about 40% using a dielectric with a relative permittivity of 120. They[9] analysed the three-region dielectric/ferrite loaded waveguide shown in Fig. 4.3. In order for mode inversion to occur, region 1 needs to be filled with a dielectric material having a large relative permittivity. The circumferential components of the electric field in the three regions are given by

$$\left.\begin{aligned} E_{\theta 1} &= EJ_1(k_1r)\exp j(\omega t - \beta z) \\ E_{\theta 2} &= [AB_1(\alpha;mr) + BH_1(\alpha;mr)]\exp j(\omega t - \beta z) \\ E_{\theta 3} &= [CJ_1(k_3r) + DY_1(k_3r)]\exp j(\omega t - \beta z) \end{aligned}\right\} \tag{4.27}$$

where

$$k_1^2 = \omega^2\mu_0\varepsilon_1 - \beta^2$$
$$k_3^2 = \omega^2\mu_0\varepsilon_3 - \beta^2$$

and m and α are defined following eqn. 4.19. Applying the boundary conditions at the interfaces between the three regions and the waveguide wall serves to eliminate the arbitrary constants A, B, C, D and E and to give the characteristic equation. Its solution gives the propagating conditions for the circularly symmetric modes. Ince and Tsandoulas[9] give results of phase constant and differential phase change for the HE_{01}-mode.

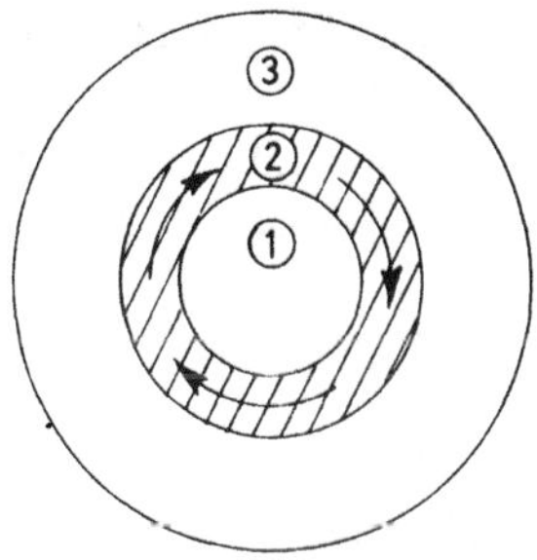

Fig. 4.3 *Three-region ferrite and dielectric loaded circular waveguide (1 and 3 are dielectric and 2 is the circumferentially magnetised ferrite tube)*

Potekhin and Yurgenson[7] tackle the same problem analysing the electromagnetic fields in a circumferentially magnetised ferrite without using the simplification of confining the solution to modes with circularly symmetric waveguide modes fields. Maxwell's curl equations then give a relationship between the longitudinal components of both the electric and magnetic fields. They eliminate the transverse components of the fields to give wave equations in both E_z and H_z. Using our notation, the governing equations come out to be

$$\frac{\partial^2 E_z}{\partial r^2} + \frac{1}{r}\frac{\partial E_z}{\partial r} + \left[\omega^2\varepsilon\mu - \beta^2 - \frac{n^2}{r^2}\right]E_z - j\frac{n\omega\kappa}{r}H_z = 0 \tag{4.28}$$

$$\frac{\partial^2 H_z}{\partial r^2} + \frac{1}{r}\frac{\partial H_z}{\partial r} + \left[\omega^2\varepsilon\frac{(\mu^2 - \kappa^2)}{\mu} - \beta^2 + \frac{\beta\kappa}{\mu r} - \frac{n^2}{r^2}\right]H_z$$
$$- j\frac{n\omega\varepsilon\kappa}{\mu r}E_z = 0 \tag{4.29}$$

and the other components of the fields in the ferrite are given by

$$\left.\begin{aligned} E_r &= -\frac{1}{k^2}\left(\mathrm{j}\beta\frac{\partial E_z}{\partial r} + \frac{n\omega\mu}{r}H_z\right) \\ E_\theta &= \frac{1}{k^2}\left(-\frac{n\beta}{r}E_z + \mathrm{j}\omega\mu\frac{\partial H_z}{\partial r} + \mathrm{j}\omega\kappa\beta H_z\right) \\ H_r &= \frac{1}{k^2}\left(\frac{n\omega\varepsilon}{r}E_z - \mathrm{j}\beta\frac{\partial H_z}{\partial r} - \mathrm{j}\omega^2\varepsilon\kappa H_z\right) \\ H_\theta &= -\frac{1}{k^2}\left(\mathrm{j}\omega\varepsilon\frac{\partial E_z}{\partial r} + \frac{n\beta}{r}H_z\right) \end{aligned}\right\} \quad (4.30)$$

Potekhin and Yurgenson[7] give expressions for the differential equations obtained by the elimination of E_z or H_z from eqns. 4.28 and 4.29. They are fourth-order differential equations and will not be reproduced here. They go on to give solutions in terms of some new functions which they define. Finally they[8] calculate the propagation constants for the HE_{11}-mode for a circumferentially magnetised ferrite tube in circular waveguide similar to that shown in Fig. 4.3. They give the results of some computations of differential phase change and phase constant as various parameters of the ferrite or waveguide are changed.

4.5 Transverse magnetisation

In this section the transversely magnetised ferrite rod or tube in circular waveguide, as shown in Fig. 4.1*c*, is discussed. The properties were first inves-

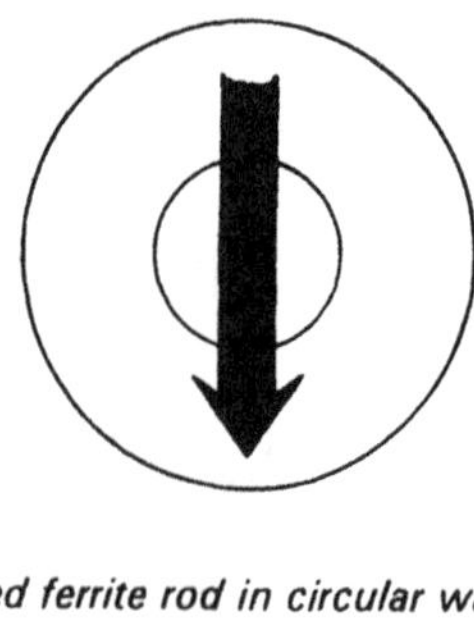

Fig. 4.4 *Transversely magnetised ferrite rod in circular waveguide*

tigated with a ferrite rod in circular waveguide, as shown in Fig. 4.4. In an initial approach, it is possible to assume that the permeability of the ferrite to the dominant TE_{11}-mode in ferrite-filled waveguide is the same as that experienced by a plane wave in an infinite ferrite medium given by eqn. 2.66. Then a wave that propagates in the waveguide with its plane of polarisation parallel to the biasing magnetic field will experience a permeability different from a wave whose plane of polarisation is perpendicular to the biasing magnetic field. This is the phenomenon of birefringence in circular waveguide which was first introduced in the context of plane wave propagation in Section 2.9. Birefringence is used in waveguide to construct half-wave and full-wave plates.[11] Using ferrite to provide the differential phase change in these plates means that they become electrically controllable. A discussion of the various waveguide devices that may be constructed using birefringence in circular waveguide is given in Sections 5.6 and 5.7.

The first recorded observation of microwave birefringence in ferrites was by Weiss and Fox,[12] who gave experimental results of differential phase change of the dominent TE_{11}-mode in circular waveguide due to transversely magnetised ferrite completely filling the circular waveguide. Magnetic birefringence was also investigated by Turner,[13] who shows how a ferrite strip, mounted longitudinally in the circular waveguide adjacent to the wall and magnetised perpendicular to the wall, causes nonreciprocal birefringence of the TE_{11}-mode in circular waveguide. These ideas are developed by Fox, Miller and Weiss[5] who investigate the theory of the optimum position of the ferrite in the waveguide. They show that a tube of ferrite ought to be more efficient in producing birefringent phase change than a rod of ferrite and that a four-pole configuration of transverse magnetisation in the ferrite, as shown in Fig. 4.1*d*, ought to be superior to the simple transverse magnetisation shown in Fig. 4.1*c*. They give experimental results which demonstrate the superiority of the four-pole field at 24 GHz and they also show how the birefringent element can be used with a continuously rotating external magnetic field as a continuous phase changer or frequency changer of circularly polarised waves. Details of the arguments leading to the superiority of the four-pole transverse field configuration are given in the next section. However, the simple transverse field is easier to produce and continued to be investigated. Cacheris[14] describes a birefringent device being used as a microwave single-sideband modulator He gives experimental results on a transversely magnetised ferrite rod or tube in circular waveguide at 9·375 GHz. Karayianis and Cacheris[15] give further experimental results on transversely magnetised ferrite rods and tubes in circular waveguide at X-band.

No full theoretical analysis of the birefringent effect due to a transversely magnetised ferrite rod or tube in circular waveguide has been given. If the birefringent ferrite element is a thin rod, its effect may be analysed by perturbation theory. Such an analysis is given by Clarricoats[16] and the results of such an analysis are given by Lax and Button.[17] However, a waveguide structure containing so little ferrite that perturbation theory is valid, is unlikely to give appreciable birefringence and the designer needs to refer to experimental results that have been published. Karayianis and Cacheris[15] give the largest amount of experimental information. At 9·5 GHz, using a transversely magnetised ferrite rod which completely filled the circular waveguide, they obtained a differential phase change of 2800°/m and 2000°/m using a ferrite diameter of 6·3 mm and 8·6 mm, respectively. There were problems matching the reduced diameter of the ferrite-filled circular waveguide to the larger diameter air-filled circular waveguide. They also give results of measurements taken on ferrite tubes with a simple transverse magnetic field as shown in Fig. 4.1*c*. Obviously the amount of differential phase change increases as the tube thickness increases, but for thicker tubes other modes will be able to propagate. The optimum thickness was found to be about 2·4 mm at 9·375 GHz. Tubes of that thickness gave differential phase changes of 3000°/m and 5000°/m when the tube outer diameter, which is also the waveguide diameter, was 24 mm and 18 mm, respectively. They[15] also give experimental results taken at various frequencies in the band 9–10 GHz. From this it is seen that the tube is more efficient in producing birefringent differential phase change than the small diameter ferrite-filled waveguide. However, experimental results given in the next section show that the four-pole form of transverse magnetisation gives much larger values of differential phase change.

4.6 Four-pole field

The description of the efficacy of the four-pole field on a ferrite tube as shown in Fig. 4.1*d* was first given by Fox, Miller and Weiss[5] and is also given by Clarricoats.[16] The field due to the dominant TE_{11}-mode in circular waveguide has been shown in Fig. 3.1. It is known that this is similar to the dominant TE_{10}-mode in rectangular waveguide. It is shown in Chapter 6 how a slab of ferrite in rectangular waveguide magnetised perpendicular to the broad face of the waveguide causes nonreciprocal phase change of the dominant mode. Two similarly magnetised ferrite slabs in circular waveguide are shown in Fig. 4.5.

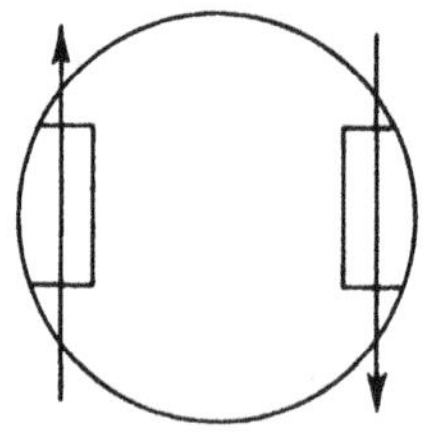

Fig. 4.5 *Two transversely magnetised ferrite strips in circular waveguide causing nonreciprocal and dielectric birefringence*

For the vertically polarised mode in the circular waveguide, the ferrite will cause nonreciprocal phase change. The horizontally polarised mode will be virtually unaffected by the ferrite slabs, so that the structure shown in Fig. 4.5 is birefringent. If a further pair of ferrite slabs are inserted into the waveguide so as to cause a nonreciprocal phase change for the horizontally polarised mode as well, as shown in Fig. 4.6, the birefringent phase change will be doubled compared with that caused by the structure shown in Fig. 4.5. There is an additional advantage for the structure shown in Fig. 4.6; even if the ferrite is unmagnetised, the structure shown in Fig. 4.5 will cause birefringence owing to the permittivity of the ferrite slabs. The symmetrical system of Fig. 4.6 removes this dielectrically caused birefringence.

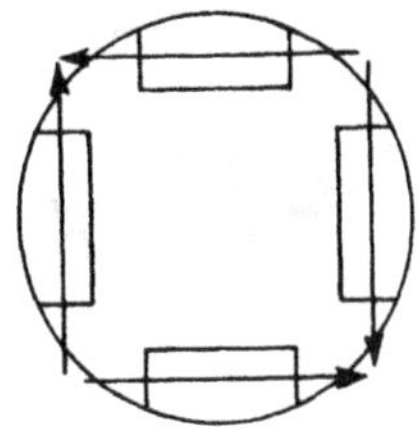

Fig. 4.6 *Symmetrical system of transversely magnetised ferrite strips in circular waveguide causing nonreciprocal birefringence*

There is also another system of ferrite slabs in circular waveguide which will cause birefringence of the dominant TE_{11}-mode. Further reference to Fig. 3.1 shows that, at a point on the circumference of the waveguide whose diameter makes an angle of 45° to the plane of polarisation of the wave, the magnetic field of the wave is elliptically polarised in the plane of the waveguide wall. Therefore, a set of four ferrite slabs mounted and magnetised as shown in Fig. 4.7 will also cause birefringence of the dominant TE_{11}-mode in circular waveguide. The two structures of Figs. 4.6 and 4.7 can be combined into the ferrite tube magnetised

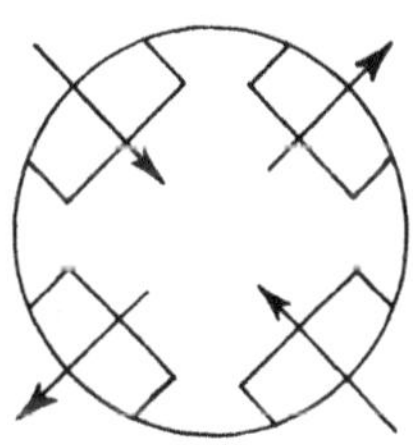

Fig. 4.7 *Another system of transversely magnetised ferrite strips in circular waveguide causing nonreciprocal birefringence*

as shown in Fig. 4.8, which makes all parts of the ferrite tube active in producing birefringence. On the basis of this qualitative analysis, a ferrite tube magnetised with a four-pole field is more efficient in producing birefringent phase change than the same tube with a simple transverse magnetic field. In the previous section, experimental results were quoted to show that a greater degree of differential phase change was obtained when the transversely magnetised ferrite tube is in reduced diameter circular waveguide. However, if the ferrite tube is too small in diameter, there may be difficulty in generating a biasing four-pole magnetic field. Owing to this difficulty, and because matching to the empty circular waveguide is simplified, the four pole field birefringent phase changer has been investivated using a ferrite tube whose outer diameter is the same as the diameter of the empty circular waveguide.

The results of some of my previously unpublished experimental work on the four-pole birefringent phase changer give some design information. Measurements were made at 8·5 GHz in 26 mm diameter circular waveguide. The birefringent phase change for various sizes of ferrite tube whose outer diameter is the same as that of the waveguide was measured as the strength of the four-pole magnetic bias field was varied. The birefringent phase change versus tube thickness is plotted in Fig. 4.9. These measurements, using a waveguide diameter of 26 mm at 8·5 GHz, scale with frequency to compare directly with those

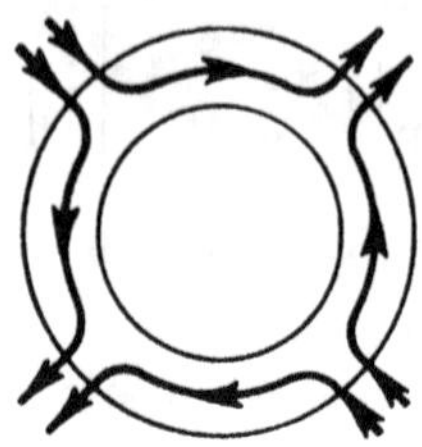

Fig. 4.8 *Systems shown in Figs. 4.6 and 4.7 combined into a ferrite tube in circular waveguide*

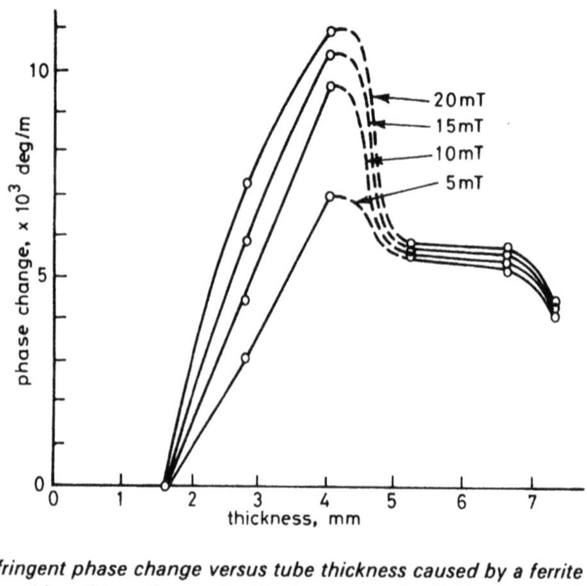

Fig. 4.9 *Birefringent phase change versus tube thickness caused by a ferrite tube in circular waveguide with a four-pole biasing magnetic field, measured at 8·5 GHz in waveguide of 26 mm diameter*

of Karyianis and Cacheris,[15] referred to in the previous section, using a transversely magnetisaed ferrite tube in waveguide with a diameter of 24 mm at 9·375 GHz. It may be seen from Fig. 4.9 that the differential phase change increases as the ferrite tube thickness increases up to 4 mm thickness and that this effect is similar to that observed by Karyianis and Cacheris. Their recommended optimum thickness was 2·4 mm. At this thickness, the results given in Fig. 4.9 show that the four-pole magnetic field configuration gives about twice as much birefringent phase change as that given by the simple transverse magnetic field. Using a roughly optimum ferrite tube thickness of 2·8 mm with a four-pole magnetic bias field, the variation of birefringent phase change with frequency is shown in Fig. 4.10. In both these sets of measurements, the parameter is the strength of the magnetic bias field which was measured at the centre of one pole of the electromagnet supplying the four-pole field where the field entered the ferrite.

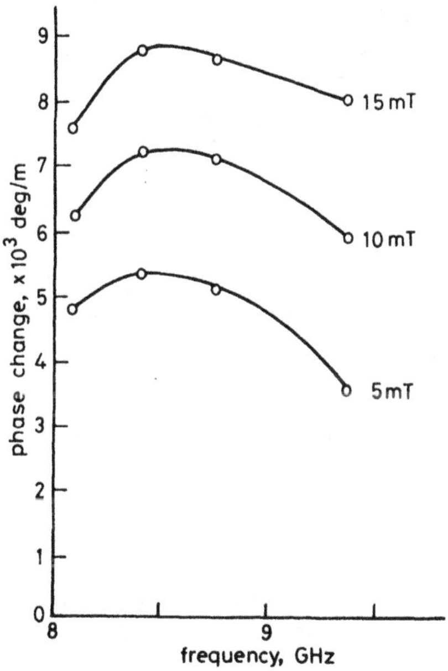

Fig. 4.10 *Birefringent phase change versus frequency caused by a ferrite tube of 20·3 mm inside diameter with a four-pole biasing magnetic field in circular waveguide of 26 mm diameter*

For both a ferrite tube with a simple transverse magnetic field and one with a four-pole magnetic field, the mathematical analysis is long and complicated and no easy solution is available. However, a statement of the governing differential equations is given in the next section.

4.7 Theoretical analysis

The theoretical analysis of the circumferentially magnetised ferrite tube in circular waveguide has already been given in Sections 4.2 and 4.3. In this section, the governing equations which describe the transversely magnetised ferrite tube are given. The biasing magnetic field in the ferrite is shown in Fig. 4.11 for the two configurations of a simple transverse magnetic field and the four-pole magnetic field. It is assumed that the ferrite is magnetised to saturation and that the direction of that magnetisation varies sinusoidally between the poles. If ϕ is the angle between the direction of the biasing magnetic field in the ferrite and an r-directed component of the field, $\phi = m\theta$ where $m = 1$ for the transverse magnetising field shown in Fig. 4.11a and $m = 2$ for the four-pole field shown in Fig. 4.11b. Then, in the cylindrical polar co-ordinate system r, θ, and z, the tensor permeability in the ferrite is given by an expression similar to eqn. 2.51

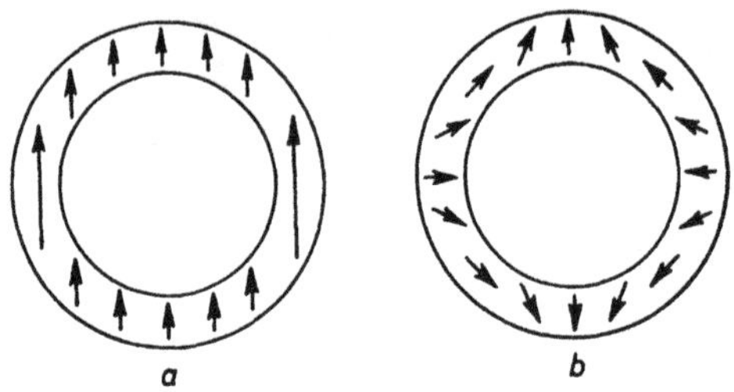

Fig. 4.11 *Theoretical orientation of the magnetic field in a ferrite tube*
a Simple transverse field
b Four-pole field

$$\boldsymbol{\mu} = \begin{vmatrix} (\mu \cos^2\phi + \mu_0 \sin^2\phi) & (\mu_0 - \mu) \sin\phi \cos\phi & j\kappa \sin\phi \\ (\mu_0 - \mu) \sin\phi \cos\phi & (\mu \sin^2\phi + \mu_0 \cos^2\phi) & j\kappa \cos\phi \\ -j\kappa \sin\phi & -j\kappa \cos\phi & \mu \end{vmatrix} \quad (4.31)$$

If it is assumed that the ferrite is in a uniform waveguide with fields propagating in the z-direction with a propagation constant, β, Maxwell's curl equations, eqns. 2.31 and 2.32, for the fields in the ferrite become

$$\frac{1}{r}\frac{\partial E_z}{\partial \theta} + j\beta E_\theta = -j\omega[(\mu \cos^2\phi + \mu_0 \sin^2\phi) H_r + (\mu_0 - \mu) \sin\phi \cos\phi\, H_\theta + j\kappa \sin\phi\, H_z] \quad (4.32)$$

$$-j\beta E_r - \frac{\partial E_z}{\partial r} = -j\omega[(\mu_0 - \mu) \sin\phi \cos\phi\, H_r + (\mu \cos^2\phi + \mu_0 \sin^2\phi) H_\theta + j\kappa \cos\phi\, H_z] \quad (4.33)$$

$$\frac{\partial E_\theta}{\partial r} + \frac{1}{r} E_\theta - \frac{1}{r}\frac{\partial E_r}{\partial \theta} = -j\omega[-j\kappa \sin\phi\, H_r - j\kappa \cos\phi\, H_\theta + \mu H_z] \quad (4.34)$$

and

$$\left.\begin{aligned} \frac{1}{r}\frac{\partial H_z}{\partial \theta} + \mathrm{j}\beta H_\theta &= \mathrm{j}\omega\varepsilon E_r \\ -\mathrm{j}\beta H_r - \frac{\partial H_z}{\partial r} &= \mathrm{j}\omega\varepsilon E_\theta \\ \frac{\partial H_\theta}{\partial r} + \frac{1}{r}H_\theta - \frac{1}{r}\frac{\partial H_r}{\partial \theta} &= \mathrm{j}\omega\varepsilon E_z \end{aligned}\right\} \qquad (4.35)$$

It can be seen that a full field analysis of the ferrite tube in circular waveguide is very difficult because of the angle dependent functions on the right-hand side of eqns. 4.32 to 4.34. For the four-pole field, Garbas and Makarov[18] have a perturbation solution from similar equations for a thin ferrite layer adjacent to the wall of circular waveguide. Their theoretical results give a phase change of 180° in a length of ferrite, R, for a ferrite thickness of 0·04R in a waveguide of diameter 2R. In terms of the waveguide diameter of 26 mm used in the experimental investigation quoted in the last section, this gives a phase change of 14 000°/m for a ferrite tube thickness of 0·5 mm.

An alternative approach to the theoretical analysis is given by Kotovas and Sugurovas.[19] They simplify the expression for the permeability tensor by taking $\mu = \mu_0$. However, they also make allowance for the possibility of an unsaturated magnetic field in the ferrite by letting κ in eqn. 4.32 be proportional to the radius from the axis of the circular waveguide. They seek solutions for both the saturated and unsaturated ferrite. They indicate a series which can be used to provide a solution of the differential equations governing the fields in the ferrite.

Expressed in our terminology, they[19] obtain the following expressions for the field components, which are also solutions to eqns. 4.32 to 4.35 when $\mu = \mu_0$ and $\phi = 2\theta$. The wave equation is given by the two coupled wave equations

$$\frac{\partial^2 H_z}{\partial r^2} + \frac{1}{r}\frac{\partial H_z}{\partial r} + \frac{1}{r^2}\frac{\partial^2 H_z}{\partial \theta^2} + \left[\frac{\omega^2\varepsilon(\mu_0^2 - \kappa^2)}{\mu_0} - \beta^2 - \frac{\kappa\beta \sin 2\theta}{\mu_0 r}\right]H_z = \frac{\omega\kappa\varepsilon}{\mu_0}\left[\cos 2\theta\,\frac{\partial E_z}{\partial r} - \frac{\sin 2\theta}{r}\frac{\partial E_z}{\partial \theta}\right] \qquad (4.36)$$

and

$$\frac{\partial^2 E_z}{\partial r^2} + \frac{1}{r}\frac{\partial E_z}{\partial r} + \frac{1}{r^2}\frac{\partial^2 E_z}{\partial \theta^2} + (\omega^2\varepsilon\mu_0 - \beta^2)E_z = \omega\kappa\left[\cos 2\theta\left(\frac{H_z}{r} - \frac{\partial H_z}{\partial r}\right) + \frac{\sin 2\theta}{r}\frac{\partial H_z}{\partial \theta}\right] \qquad (4.37)$$

and the other components of the fields are given by

$$E_r = \frac{\mathrm{j}}{k^2}\left[-\beta\frac{\partial E_z}{\partial r} - \omega\kappa\beta \cos 2\theta\, H_z - \omega\mu_0\frac{1}{r}\frac{\partial H_z}{\partial \theta}\right] \qquad (4.38)$$

$$E_\theta = \frac{\mathrm{j}}{k^2}\left[-\beta\frac{1}{r}\frac{\partial E_z}{\partial\theta} + \omega\kappa\beta\sin 2\theta\, H_z + \omega\mu_0\frac{\partial H_z}{\partial r}\right] \tag{4.39}$$

$$H_r = \frac{\mathrm{j}}{k^2}\left[\omega\varepsilon\frac{1}{r}\frac{\partial E_z}{\partial\theta} - \omega^2\kappa\varepsilon\sin 2\theta\, H_z - \beta\frac{\partial H_z}{\partial r}\right] \tag{4.40}$$

$$H_\theta = \frac{\mathrm{j}}{k^2}\left[-\omega\varepsilon\frac{\partial E_z}{\partial r} - \omega^2\kappa\varepsilon\cos 2\theta\, H_z - \beta\frac{1}{r}\frac{\partial H_z}{\partial\theta}\right] \tag{4.41}$$

When suitable functions are found for E_z and H_z that satisfy eqns. 4.36 and 4.37, these may be substituted into eqns. 4.38 to 4.41 and boundary conditions applied to solve the problem.

If it is desired to model a ferrite-filled circular waveguide biased with a four-pole transverse magnetic field, it is necessary to use an alternative simplification to eqns. 4.32 to 4.35, which allows the cross-diagonal component of the tensor permeability, κ, to vary with radius. The alternative forms of eqns. 4.36 and 4.37 under these conditions are also given by Kotovas and Sugurovas,[19] and are used in a subsequent paper by Kotov *et al.*[20] to derive the propagating conditions in ferrite-filled circular waveguide biased with a four-pole field. The static magnetic field distribution in the ferrite is expressed in terms of a potential in order to determine the effect of the non-magnetic gap between the ferrite rod and the biasing magnetic field. A characteristic equation is given which can be solved numerically and some results of propagation constant of the HE_{11}-mode for two perpendicular polarisations are given. The calculated variation of differential phase change with frequency shows argeement with experimental results.

4.8 References

1 See, for example, BADEN FULLER, A. J.: 'Microwaves', 2nd edn. (Pergamon, 1979), p. 122

2 SUHL, H. and WALKER, L. R.: 'Topics in guided wave propagation through gyromagnetic media. Pt. II', *Bell Syst. Tech. J.* 1954, **33,** pp. 939–986

3 IVANOV, K. P.: 'Propagation along azimuthally magnetised ferrite loaded circular guide', *Rad. Sci.*, 1984, **19,** pp. 1305–1310

4 BOLLE, D. M. and HELLER, G. S.: 'Theoretical considerations on the use of circularly symmetric TE modes for digital ferrite phase shifters', *IEEE Trans.*, 1965, **MTT-13,** pp. 421–426

5 FOX, A. G., MILLER, S. E. and WEISS, M. T.: 'Behaviour and applications of ferrites in the microwave region', *Bell Syst. Tech. J.*, 1955, **34,** pp. 5–103

6 BERNUES, F. J. and BOLLE, D. M.: 'The digital twin-ferrite-toroid circular waveguide phaser', *IEEE Trans.*, 1973, **MTT-21,** pp. 842–845

7 POTEKHIN, A. I. and YURGENSON, R. R.: 'Theory of electromagnetic wave propagation in a gyrotropic medium', *Rad. i Elek.*, 1970, **15,** pp. 307–315 (English translation in *Radio Engng. & Electron. Phys.*, **15,** pp. 266–273)

8 POTEKHIN, A. I. and YURGENSON, R. R.: 'Design of certain microwave lines with azimuthally magnetised ferrite', *ibid.*, pp. 456–464 (pp. 392–399)

9 INCE, W. J. and TSANDOULAS, G. N.: 'Modal inversion in circular waveguides – Pt. II: Application to latching nonreciprocal phasers', *IEEE Trans.*, 1971, **MTT-19**, pp. 393–400

10 TSANDOULAS, G. N. and INCE, W. J.: 'Modal inversion in circular waveguides – Pt. I: Theory and phenomenology', *ibid.*, pp. 386–392

11 See, for example, BADEN FULLER, A. J.: 'Microwaves', 2nd edn. (Pergamon, 1979), p. 140

12 WEISS, M. T. and FOX, A. G.: 'Magnetic double refraction at microwave frequencies', *Phys. Rev.*, 1952, **88**, pp. 146–147

13 TURNER, E. H.: 'A new non-reciprocal medium using ferrites', *Proc. IRE*, 1953, **41**, p. 937

14 CACHERIS, J.: 'Microwave single-sideband modulator using ferrites', *Proc. IRE*, 1954, **42**, pp. 1243–1247

15 KARAYIANIS, N. and CACHERIS, J.: 'Birefringence of ferrites in circular waveguide', *Proc. IRE*, 1956, **44**, pp. 1414–1421

16 CLARRICOATS, P. J. B.: 'Microwave ferrites', (Chapman & Hall, 1961) p. 144

17 LAX, B. and BUTTON, K. J.: 'Microwave ferrites and ferrimagnetics' (McGraw-Hill, 1962), p. 347

18 GARBAS, Ch. and MAKAROV, S.: 'Dispersive properties of a circular waveguide with a ferrite surface layer in the field of a magnetic quadrupole', *Liet. Fiz. Rink.*, 1982, **22**, 1, pp. 79–85 (English translation in *Sov. Phys. Collect*, **22**, 1, pp. 56–60)

19 KOTOVAS, M and SUGUROVAS, V.: 'Propagation of electromagnetic waves in a ferrite with quadrupole magnetisation', *Liet. Fiz. Rink.*, 1982, **22**, 4, pp. 46–54 (English translation in *Sov. Phys. Collect*, **22**, 4, pp. 34–40)

20 KOTOV, M. N., PUZAKOV, A. N., CHAPLINSKII, V. V. and SHUGUROV, V. K.: 'Analysis of a ferrite-filled circular waveguide with transverse quadrupole magnetisation', *Liet. Fiz. Rink.*, 1983, **23**, 2, pp. 33–40 (English translation in *Sov. Phys. Collect*, **23**, 2, pp. 26–31)

Chapter 5

Circular waveguide devices

5.1 Longitudinally magnetised rotator

If a rod or tube of ferrite inside circular waveguide is magnetised in a direction along the axis of the waveguide, its effective permeability to circularly polarised waves will be given by values similar to those shown in Fig. 5.1 (see also Fig. 2.5). Below saturation, the variation of permeability is approximately linear with change of biasing magnetic field. As explained in Section 2.8, the different effective permeability to the two hands of circular polarisation gives rise to rotation of the direction of polarisation of a linearly polarised plane wave. Similarly, in circular waveguide, the plane of polarisation of a linearly polarised dominant TE_{11}-mode is rotated by the longitudinally magnetised ferrite material. Below saturation, the variation in rotation is proportional to the strength of the biasing magnetic field. This can be seen from some of my experimental results reproduced in Fig. 5.2.[1] Rotation is also nonreciprocal in that, if in traversing a given ferrite device in the forward direction the wave is rotated clockwise, traversing in the reverse direction the wave will be rotated anticlockwise. Also the results given in Chapter 3 show that rotation of the dominant HE_{11}-mode (which is what the dominant TE_{11}-mode becomes in ferrite loaded waveguide) occurs for most configurations of ferrite rod or tube symmetrically mounted in circular waveguide.

In order to reduce reflections due to the transition from empty waveguide to ferrite loaded waveguide, some form of matching taper or step transition is needed. Almost every possible combination has been tried at some time. The commonest situation is where the ferrite is mounted in waveguide having the same diameter as the adjacent air-filled waveguide. Then the matching transition can consist of tapers in the cross-section of the ferrite rod or tube. Because ferrite has to be shaped by grinding, it is usually preferred if dielectric matching tapers can be used. Alternatively, matching may be provided by stepped dielectric transitions. When the ferrite is mounted in reduced diameter waveguide, a match may sometimes be provided using reduced diameter dielectric-filled waveguide or the steps or taper to reduced diameter waveguide may provide the match. Illustrations of the combination of dielectric tapers and change of waveguide diameter are given by Karayianis and Cacheris[2] but in the context of transversely magnetised ferrite rods and tubes.

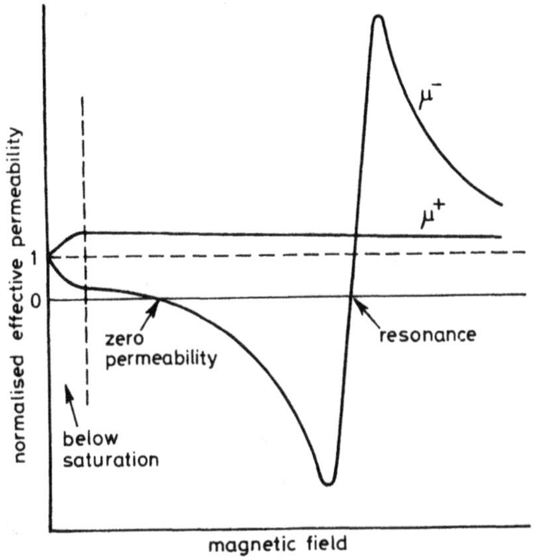

Fig. 5.1 *Variation of the microwave normalised effective permeability, μ/μ_0, with change of magnetic field of an infinite ferrite medium for two opposite hands of circularly polarised plane waves at a fixed frequency*

Referring to the theoretical results given in Chapter 3, rotation is given by the difference in propagation constants between the positive and negative circularly polarised HE_{11}-modes. Some relevant results are given in Figs. 3.4, 3.8 and 3.13. It is seen that rotation is influenced by the concentration of power in the ferrite for one hand of circular polarisation. This makes the rotation frequency dependent. However, for a plane wave in an infinite ferrite medium, eqn. 2.58 shows that rotation is not then frequency dependent. It is the scale factor of ferrite diameter to free-space wavelength that tends to concentrate the microwave fields into the ferrite. Ohm[3] has shown that surrounding a ferrite rod with a tube of dielectric material controls the frequency dependence of rotation. It appears that the dielectric material reduces the concentration of the microwave power in the ferrite. However, it also reduces the rotation available from a given length of ferrite rod. Both the increased length required and the dielectric material increase the power absorbed by the rotation device. An alternative system for controlling the frequency dependence of rotation due to a ferrite rod is suggested by Chait and Sakiotis.[4] They mount the ferrite rod at the centre of a ridged circular waveguide having four symmetrical ridges.

For a ferrite rod Faraday rotator, there is the problem of how to mount the ferrite at the centre of the circular waveguide. It may be mounted in a dielectric sleeve; the dielectric can control the variation of rotation with frequency but it also reduces the rotation available from a given length of ferrite, or the ferrite can be mounted in a sleeve of expanded polystyrene whose permittivity is almost the same as that of air. Alternatively, the ferrite rod may be supported by disks of dielectric material. Dielectric matching elements at each end of the rod can

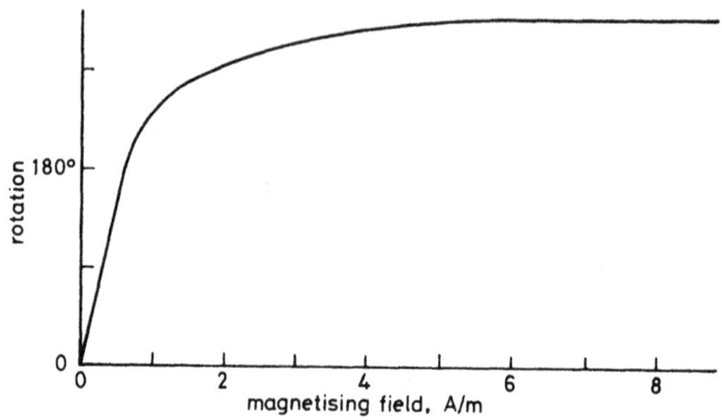

Fig. 5.2 *Measured change of rotation of a linearly polarised TE_{11}-mode in 22.8 mm diameter circular waveguide due to a 6.25 mm diameter ferrite rod 50 mm long versus magnetic field at 9.37 GHz. The ferrite is described in Reference 1.*

span the waveguide diameter and also provide support for the rod. If the rotator is to be used in a high power system, it will be necessary to cool the ferrite, which may be difficult if the ferrite is mounted in a dielectric sleeve. An alternative that solves both the problem of mounting and cooling is to use a ferrite tube adjacent to the wall of the waveguide. This has been investigated by Rizzi,[5] who shows that there is a much higher loss due to the ferrite tube than with a ferrite rod. However, ease of mounting and cooling may still make the ferrite tube preferable. Some calculated results of rotation for a particular size of ferrite tube, taken from the results given in Fig. 3.13, are shown in Fig. 5.3. The variation of rotation with frequency is less than with a ferrite rod but is not as good as the results quoted by Ohm[3] and Chait and Sakiotis.[4]

The theoretical design of ferrite rod or tube rotator elements is given in Chapter 3. However, the theory considers only an infinite length of ferrite loaded waveguide and no consideration is given to any end effects. One of the end effects is the matching of the electromagnetic fields to the empty waveguide. Obviously a poor match will lead to multiple reflections from the ends of the ferrite which introduces variations into the rotation obtained. The error due to the reflected waves may increase or decrease the rotation depending on the phase length of the ferrite. As can be seen from the results given in Chapter 3, the propagation constant in the ferrite loaded waveguide depends on the state of magnetisation, so that it is necessary to obtain a match at the operating condition. In most states of magnetisation, the propagation constant is different for the two hands of circular polarisation, but it is still possible to obtain a match to the combination of both hands of circular polarisation for a linearly polarised wave. If the rotator is being used in a control situation, it is impossible to match the ferrite control element for all conditions of magnetisation and a compromise must be accepted unless the matching is obtained by a taper at each end of the ferrite element. Even then, being able to attain a good match for all conditions of magnetisation is unlikely. Another end effect is the demagnetisation of the ends of the ferrite element. Experimental results I have quoted[1] show that for a rod of ferrite the rotation is proportional to the length of the rod minus some small dimension

$$\theta = \psi(l - c) \tag{5.1}$$

where θ is the rotation, ψ is the rotation per unit length, l is the length of the rod and c is an experimental constant. For the same experimental conditions as those given in Fig. 5.2, the experimental constant is 2.8 mm.

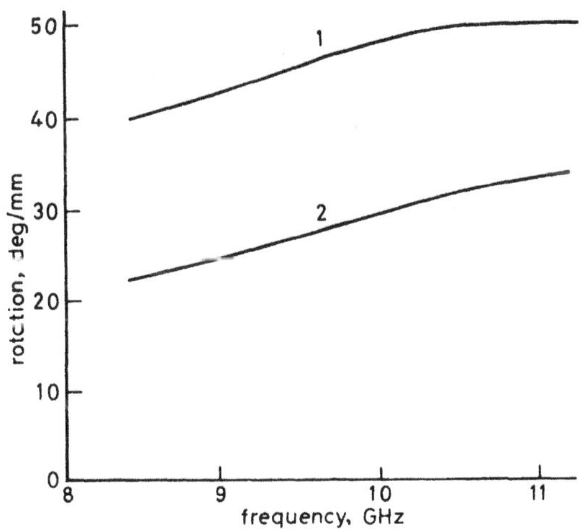

Fig. 5.3 *Calculated rotation due to a ferrite tube in 14 mm diameter circular waveguide derived from the results given in Fig. 3.13*
Ferrite tube 1: i.d. = 8·0 mm, o.d. = 14 mm, 2: i.d. = 9·2 mm, o.d. = 14 mm. The ferrite properties are $\mu_r = 0{\cdot}98$, $\kappa_r = 0{\cdot}63$ and $\varepsilon_r = 11{\cdot}2$

Except possibly in an antenna feed, a rotator is not used directly to produce a wave with a variable angle of polarisation of a linearly polarised wave. It is used in conjunction with a waveguide system to provide circulation, isolation, attenuation, phase change etc. and these devices are described in the following sections.

5.2 Faraday rotation circulator

The circuit symbol for a four-port circulator is shown in Fig. 5.4. If power is incident on port 1, it will come out of port 2 and there will be no power coupled to the other ports. Similarly, power incident on port 2 will come out of port 3 etc. The diagram shows a four-port circulator because the Faraday rotation circulator is a four-port circulator, but there is no restriction in principle to the number of ports in a circulator. A three-port circulator is shown in Fig. 1.1 and in Fig. 8.1. Circulators can be made using a number of different ferrite devices suitably interconnected; each will be described with the device providing the nonreciprocity. The important property of the circulator is its nonreciprocal characteristics. The ferrite circulator is one of the most important components in any microwave system, although now it is hardly ever seen in its Faraday rotation configuration. Faraday rotation devices are hardly ever used now because there are simpler alternatives, but they will all be described for completeness.

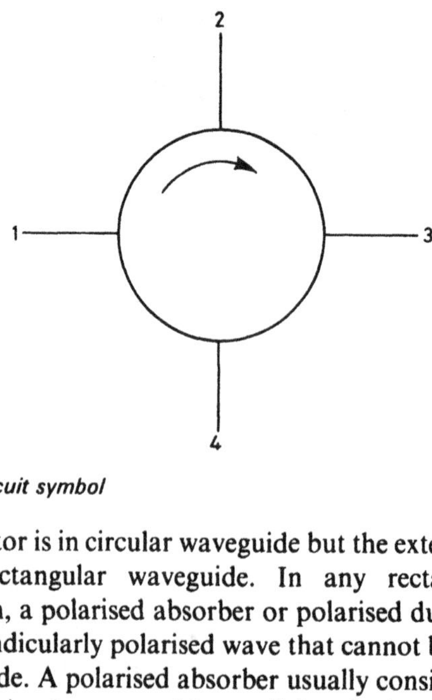

Fig. 5.4 *Circulator circuit symbol*

The Faraday rotator is in circular waveguide but the external connections are in conventional rectangular waveguide. In any rectangular to circular waveguide transition, a polarised absorber or polarised duplexer is required to eliminate any perpendicularly polarised wave that cannot be transmitted by the rectangular waveguide. A polarised absorber usually consists of an attenuating vane situated in the circular waveguide parallel to the broad face of the rectangular waveguide so as to absorb any perpendicular wave, as shown in Fig. 5.5. A polarised duplexer has metal vanes in the circular waveguide so as to deflect any perpendicular wave into the side arm rectangular waveguide, as shown in Fig. 5.6. The Faraday rotation circulator is shown in Fig. 5.7. It works as a circulator because a wave entering port 1 will be rotated 45° and will exit by port 2 with no power going to ports 3 and 4, and a wave entering port 2 will be rotated *in the same direction* 45° and will be polarised in a plane perpendicular to port 1, thus it will exit by port 3. It is easy to continue this argument to show complete circulator action between other ports.

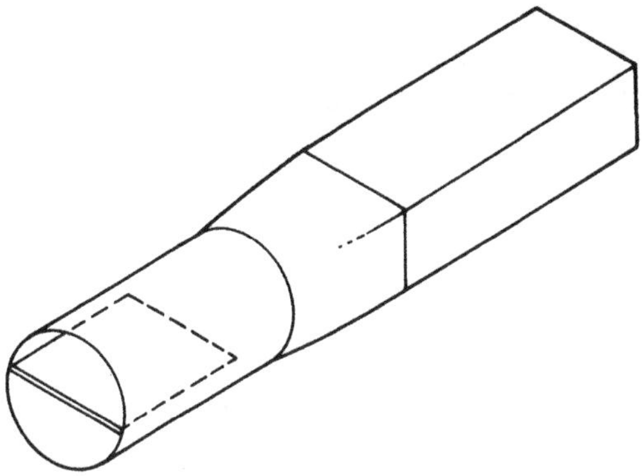

Fig. 5.5 *Polarised absorber*

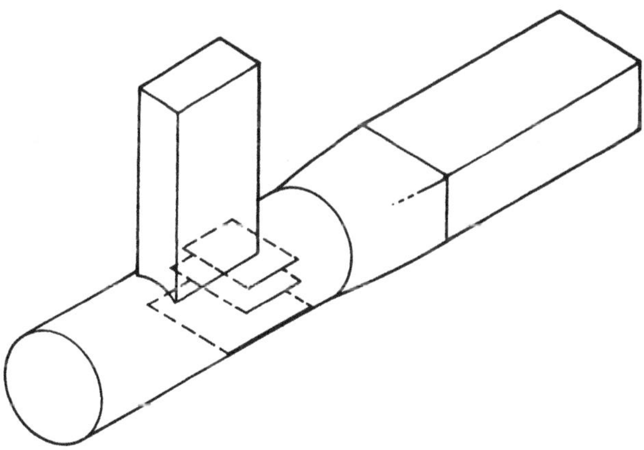

Fig. 5.6 *Polarised duplexer*

Isolator: If the Faraday rotation circulator is furnished with matched terminations on the two side arms, ports 3 and 4, it becomes an isolator because the forward wave passes from ports 1 to 2 unattenuated, whereas the reverse wave is absorbed in the matched termination. As an alternative, for low power use, the polarisation duplexers in the circulator may be replaced by polarisation absorbers to make an isolator.

Switch: If the direction of the 45° rotation in the circulator is reversed, the circulator will act as a switch, thus the input to port 1 can be switched between ports 2 or 4 as required. An input to any other port will similarly be switched between the two ports at the opposite end of the device.

In the 1950s much effort went into the design of Faraday rotation circulators. Papers such as those by Ohm,[3] Chait and Sakiotis[4] and Rizzi[5] give good examples of Faraday rotation circulator designs. Subsequently, for most applications, the Y-junction circulator, described in Chapter 8, and the phase change circulator, described in Section 7.3, were found to be smaller, simpler and easier to manufacture. However, Faraday rotation devices are being investigated for use at millimetre wave frequencies, as discussed in Section 10.4.

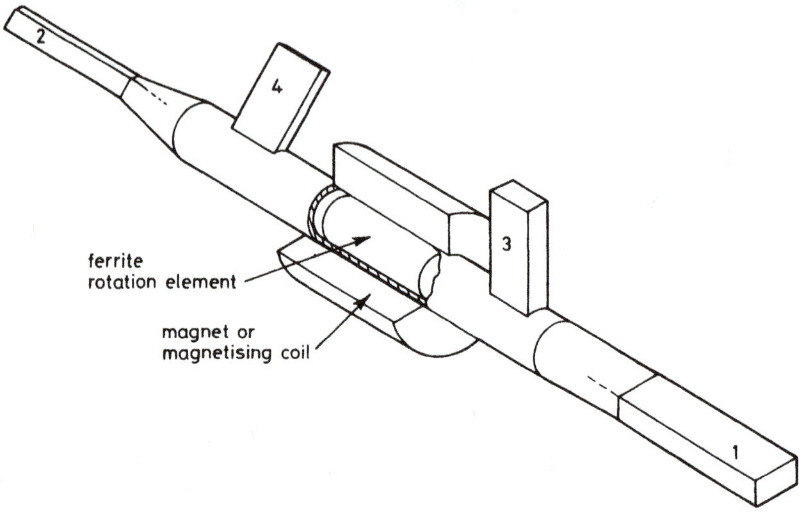

Fig. 5.7 *Faraday rotation circulator*

5.3 Variable attenuation and phase

If the ferrite element in the Faraday rotation circulator is made to give a variable angle of rotation, the power connected to the two output ports 2 and 4 becomes variable, so that a variable attenuator can be constructed. Since it is no longer necessary to have 45° rotation for the straight-through connection, a ferrite rotator variable attenuator is usually constructed with the input and output rectangular waveguides aligned as shown in Fig. 5.8. This is a reciprocal variable attenuator. The input TE_{10}-mode in the rectangular waveguide is transformed into the TE_{11}-linearly polarised mode in the circular waveguide with vertical electric field which is unaffected by the polarisation absorber. With no rotation, this wave is transformed back into the TE_{10}-mode in the output rectangular waveguide without loss. If the wave is rotated through an angle θ, $\sin\theta$ of the wave is absorbed in the output polarised absorber and the attenuation through the device is

$$\text{attenuation} = 20\log(\sec\theta)\ \text{dB} \tag{5.2}$$

The device shown in Fig. 5.8 is suitable for low-power applications only, since the attentuated power is absorbed in the polarisation absorber. For high-power applications the polarisation absorbers are replaced with polarisation duplexers attached to suitable matched terminations. This device is a reciprocal attenuator since the attenuation is independent of the direction of rotation.

Rotation is controlled by varying the current in a coil round the outside of the waveguide. If the rate of variation is fast, the waveguide acts as a shorted turn. Cutting a suitable slot in the waveguide wall will eliminate the shorted turn. An alternative is to use thin-wall waveguide; the thin wall needs to be

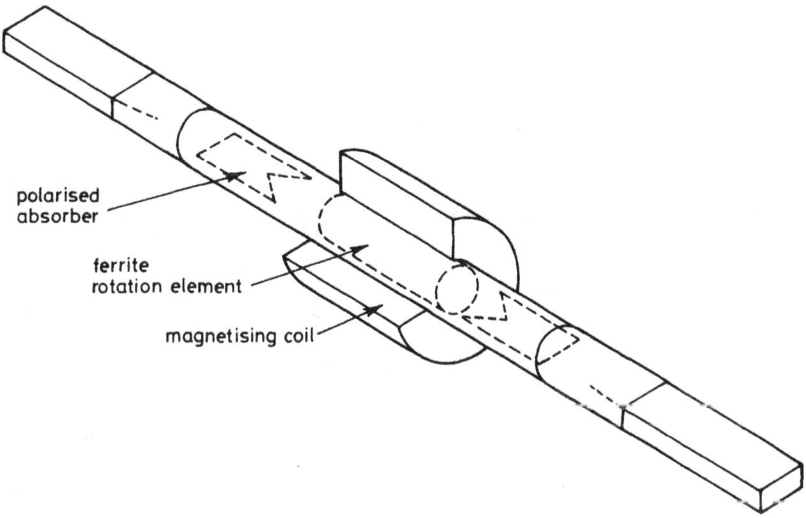

Fig. 5.8 *Variable attenuator*

thicker than the skin depth at the microwave frequency and thinner than the skin depth at the control frequency. For the slotted waveguide, a longitudinal slot will radiate depending on its position relative to the electromagnetic field pattern in the waveguide. Providing a choke around the slot with a second slotted waveguide outside the first helps to prevent radiation. The slotted section needs to be separated from the rest of the waveguide by a circumferential slot so that the end of the waveguide cannot provide a shorted turn. The radiation from the circumferential slot is usually negligtible. Another system uses a helical slot in the waveguide wall.

A longitudinally magnetised ferrite rotator element will also provide variable phase to a circularly polarised wave in the circular waveguide. The amount of phase change depends on the hand of circular polarisation so that the rotator is a nonreciprocal variable phase changer. Construction of the variable phase changer is similar to that of the variable attenuator shown in Fig. 5.8, except that the polarised absorbers are replaced by circular polarisers or quarter-wave plates. It has also been suggested that by continuously advancing the phase of the signal, its frequency may be changed.[6] However, this is probably better provided by the transverse field birefringent device described in Section 5.7. For some applications, particularly in antenna feeds, nonreciprocity is undesirable, and the dual mode longitudinally magnetised device has been devised. The longitudinally magnetised phase change element is combined with nonreciprocal circular polarisers to counteract the nonreciprocal phase change effect. One method is to precede the quarter-wave plate circulator polarisers with nonreciprocal 45° rotators, as shown in Fig. 5.9, and as proposed by Sakiotis, Simmons and Chait[7] in 1952.

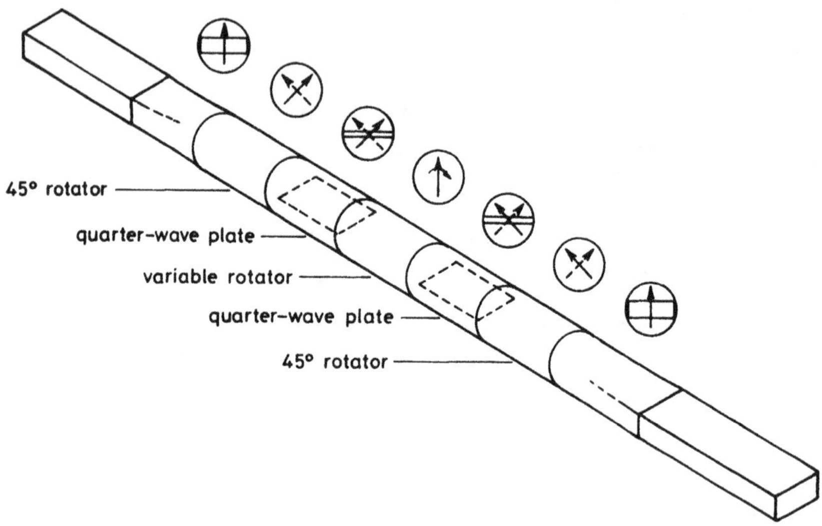

Fig. 5.9 *Reciprocal variable phase changer. The rotators are nonreciprocal*

With reference to Fig. 5.9, the rotators are all longitudinally magnetised ferrite elements which are nonreciprocal. A wave travelling from right to left is represented by the dotted arrows showing the direction of the maximum electric field in the waveguide section between the 45° rotator and the quarter-wave plate circular polariser. The 45° rotation is clockwise and the wave in the centre section is left-hand circularly polarised. A wave travelling in the reverse direction from left to right is represented by the solid line arrows. The 45° rotation is counter-clockwise and the wave is right-hand circularly polarised. In that way, the electromagnetic field in the variable rotator is always rotating in the same relative direction as the precession of the electrons in the ferrite. To make a more compact device, nonreciprocal circular polarisers may replace the 45° rotator and quarter-wave plate. They are the quadrupole birefringent device described in Section 5.6.

One problem with ferrite control devices that depend on the variation of the magnetic field in the ferrite is hysteresis. Some typical results for the rotation due to a longitudinally magnetised ferrite tube 12.2 mm outside diameter and

8.4 mm inside diameter at 14 GHz are shown in Fig. 5.10. From the diagram it can be seen that some ferrite control components need to adopt special techniques to overcome hysteresis. Another problem with the use of coil driven ferrite control devices is the electrical power needed to drive the magnetising current through the magnetising coil. An alternative makes use of the latching properties of the magnetisation in the ferrite.

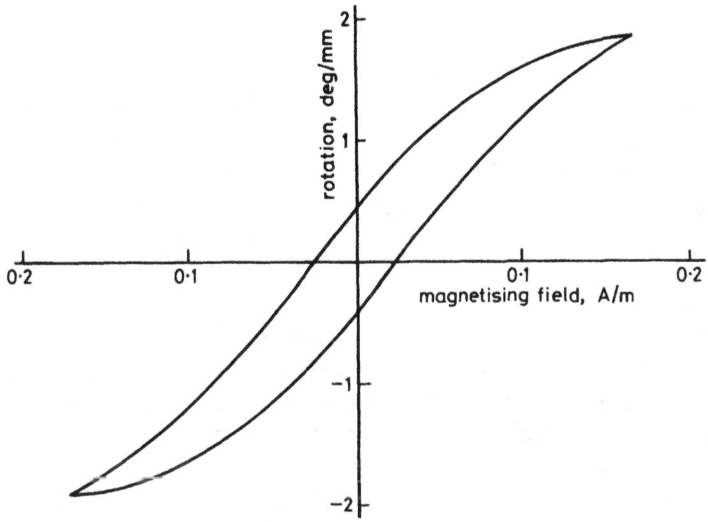

Fig. 5.10 *Rotation versus magnetising field for a longitudinally magnetised ferrite tube at 14 GHz*

Many microwave ferrite materials have the square-loop hysteresis curve, as shown in Fig. 7.2. This means that, if a loop of ferrite material is magnetised to saturation and the magnetising field removed, the ferrite will remain in a state of remanent magnetisation which is close to its saturation magnetisation. If a magnetising coil is wound round one arm of the ferrite loop or if a current carrying wire threads the loop, a current pulse of sufficient intensity will serve to magnetise the ferrite loop to its remanent state and the ferrite will remain magnetised in that state until another current pulse of sufficient intensity serves to reverse the direction of the remanent magnetisation. Boyd[8] has used such a system to produce a latching reciprocal phase changer. The remanent state provides two fixed values, positive and negative, of phase change. He calls it a 'dual-mode' device because the forward wave through the ferrite is a right-hand circularly polarised mode whereas the reverse wave is a left-hand circularly polarised mode. His device consists of a centre portion of longitudinally magnetised ferrite to provide a given amount of phase change. On either side of this central section are short ferrite sections which are transversely magnetised with a fixed quadrupole field to act as nonreciprocal circular polarisers. At each end are thin resistive film elements which act as polarised absorbers. Impedance matching elements are added to transform to standard rectangular waveguides. The whole combination of ferrite elements is metallised to form a waveguide structure and a ferrite yoke is added over the outside to provide latching operation of the centre phase change element.

5.4 Isolator and circular polariser

Resonance isolation

If a ferrite rod at the centre of circular waveguide is magnetised to the resonance condition shown in Fig. 5.1, the energy of one of the circularly polarised waves will be absorbed in the ferrite. This is because the field at the centre of a circular waveguide supporting the TE_{11}-mode approximates to that of a plane wave, and resonance absorption, as described in Section 2.1, occurs. Such a device becomes an isolator if the ferrite section of waveguide is located between two reciprocal quarter-wave plate circular polarisers. If the forward wave magnetic field is rotating in the direction which in unaffected by the ferrite, the reverse wave field will be rotating in the opposite direction and will be absorbed in the ferrite owing to its resonant state. If the ferrite rod is very thin, to ensure that the field is truely circularly polarised, it may need to be excessively long to provide adequate isolation or attentuation of the reverse wave. If the ferrite rod is fatter, there will be regions in the rod where the transverse magnetic field is not circularly polarised, which leads to a small amount of unwanted resonance absorption of the forward wave. Except at low microwave frequencies, the magnetic fields required for resonance are large and are particularly difficult to supply as a longitudinal field at the centre of circular waveguide. For this reason, resonance isolation is hardly ever used in a longitudinally magnetised device. It is used in the transversely magnetised device in rectangular waveguide as described in Sections 6.9 and 7.5. Resonance absorption in circular waveguide could be used to remove one hand of circular polarisation of a linearly polarised wave to provide a circular polariser which also removed half the signal power, but such an application has never come to my notice.

Field-displacement isolation

Referring again to Fig. 5.1, if the ferrite is magnetised to the zero permeability condition for the negative circularly polarised wave, the wave will be rejected from the ferrite. Duncan and Swern[9] investigated the effect in a number of longitudinally magnetised ferrite rods inside circular waveguide. If the whole ferrite rod appears to have zero permeability, the electromagnetic fields will be rejected from the rod and the wave will propagate in the TE_{11} coaxial line mode in the air space surrounding the ferrite rod. In order to encourage the coaxial mode of propagation, they used conical metal tapers at each end of the ferrite rod to provide a match between the rod and the empty waveguide. However, the transverse magnetic field in the ferrite will not be circularly polarised throughout and the ellipticity becomes appreciable near to the outside of the ferrite rod. Waldron[10] gives theoretical results of the magnetic field strength variation across the waveguide cross-section for a ferrite magnetised below saturation. These results have been used to plot the ellipticity inside the ferrite in Fig. 5.11, which show that the ellipticity is large near to the surface of the rod. The effect may be expected to be more pronounced for the ferrite magnetised to the zero permeability condition. This means that some penetration of the electromagnetic field occurs even for the ferrite magnetised to the zero permeability condition.

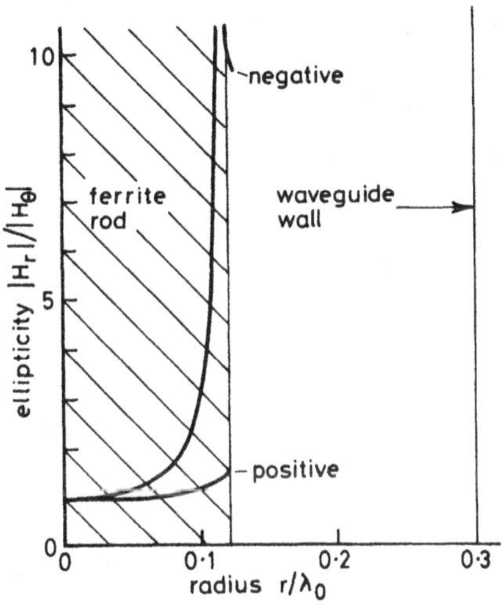

Fig. 5.11 *Ellipticity of the transverse electromagnetic field components for a longitudinally magnetised ferrite rod in circular waveguide*
The ferrite properties are $\mu_r = 1$, $\kappa_r = 0{\cdot}5$ and $\varepsilon_r = 10$. The waveguide diameter is $0{\cdot}6\ \lambda_0$ and the ferrite rod diameter is $0{\cdot}24\ \lambda_0$. The data is taken from Waldron[10]

Duncan and Swern[9] construct a ferrite absorption isolator using either a lossy ferrite material for the rod or a ferrite tube filled with lossy material. If the wave is circularly polarised in the circular waveguide, one hand of circular polarisation will be attenuated while the opposite hand of circular polarisation will be unaffected. The overall system is similar to that described in the previous paragraph for a resonance isolator. It has the advantage that a smaller magnetic field is required than for resonance isolation and that the field-displacement effect occurs for a wide variation of magnetic field and thus operates over a wide bandwidth. As with the resonance isolator, the system can also be used as an absorption mode circular polariser. However, this longitudinally magnetised field-displacement isolator suffers from most of the disadvantages of the longitudinally magnetised resonance isolator described in the last paragraph and is not normally used. The field-displacement effect may be used much more conveniently with transversely magnetised ferrite in rectangular waveguide as described in Sections 6.6 and 7.6.

5.5 Circumferentially magnetised ferrite tube – latching phase changer

The mode of operation of the circumferentially magnetised ferrite tube phase changer has been described in Section 4.2. A summary will be repeated here. A single ferrite tube circumferentially magnetised, as shown in Fig. 5.12, acts as a variable phase changer to the TE_{01}-mode in circular waveguide. There are two optimum positions for the ferrite tube given by the solution to eqn. 4.2, and the phase change in any given length of ferrite tube will be approximately doubled by using two ferrite tubes magnetised, as shown in Fig. 4.2. One ferrite tube may be magnetised by a current carrying wire along the axis of the waveguide. There is no longitudinal or radial component of the electric field of the TE_{01}-mode in the waveguide, thus such a wire threading the ferrite tube will have negligible effect on the electromagnetic field in the waveguide. A suitable pulse of current in the wire will magnetise the ferrite tube to saturation circumferentially and will leave the ferrite tube in its remanent magnetised state. Using a ferrite material having a square hysteresis loop means that there is not much difference in magnetisation between the remanent and saturation magnetisations.

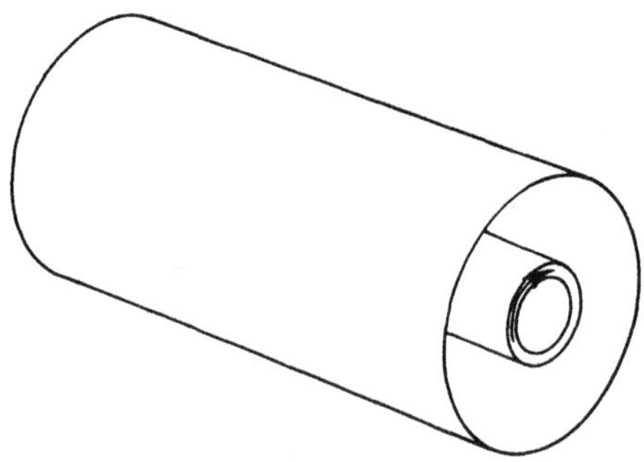

Fig. 5.12 *Circumferentially magnetised ferrite tube phase changer*

The axial current carrying wire may also be used to magnetise the ferrites in the two tube device as described by Bernues and Bolle.[11] As shown in Fig. 4.2, there are two concentric ferrite tubes magnetised in opposite directions. They may be switched to their remanent conditions by using two different strengths of current pulse in the axial wire. The first pulse is of sufficient strength and direction to magnetise the outer tube to its remanent condition. Because of the $1/r$ variation of field owing to electric current the second pulse of current of opposite polarity is of just sufficient strength to magnetise the inner tube but is of insufficient strength to disturb the magnetisation of the outer tube. Then the two ferrite tubes are magnetised as shown in Fig. 4.2. Design calculations for such a device are given by Bernues and Bolle.[11] They do not assess the problem of transforming from a commonly used rectangular waveguide mode into the TE_{01}-mode is circular waveguide but waveguide transformers have been designed in connection with low-loss TE_{01}-mode waveguide transmisstion.[12] In the presence of the ferrite, mode conversion can become a problem, but dielectric loading of the waveguide causes the TE_{01}-mode to become the dominant mode, as shown by Ince and Tsandoulas.[13] Bernues and Bolle's design incorporates dielectric loading[11] in the space between the two ferrite tubes and in the centre of the waveguide, as shown in Fig. 5.13. The dielectric material has a

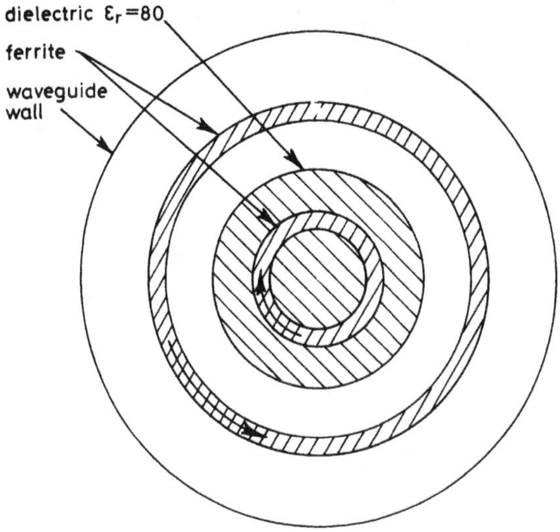

Fig. 5.13 *Twin ferrite tube dielectric loaded phase changer*

permittivity $\varepsilon_r = 80$, which could be rutile. The differential phase change is maximised when the inner ferrite tube is positioned at the point where the magnetic field in the unloaded waveguide is circularly polarised. However, the outer tube needs to be of a smaller diameter than that given by the other position of the circularly polarised magnetic field in empty waveguide. It size is largely governed by the need to magnetise the tube by two different pulses of current in the magnetising wire. Increasing the thickness of the tubes increases the differential phase change but it reduces the frequency bandwidth over which the TE_{01}-mode is the dominant mode.

5.6 Birefringence – quarter-wave plate

In Section 4.5, it has been shown how a transversely magnetised ferrite rod or tube in circular waveguide or a ferrite tube with a four-pole magnetic bias field is birefringent. It causes a nonreciprocal differential phase change between two perpendicular linearly polarised TE_{11}-mode waves in circular waveguide. Depending on the angle between the direction of the linear polarisation in the waveguide and the direction of the static biasing magnetic field on the ferrite, the wave experiences a larger or smaller phase change in passng through the device. Fig. 5.14 shows the electric field vector for the x- and y-directed modes having the maximum differential phase difference for the four-pole field device. The variation in transmission phase delay with variation in the angle between the axis of the four-pole field and the direction of polarisation of the linearly

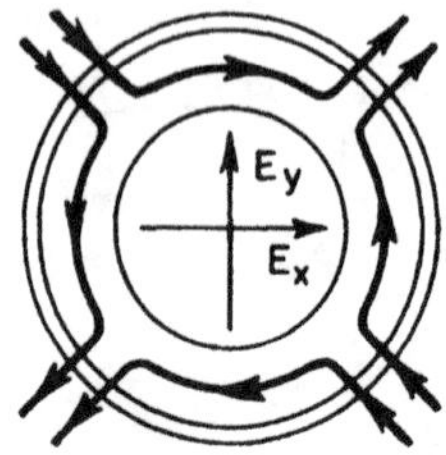

Fig. 5.14 *Ferrite tube loaded waveguide with a four-pole transverse field acting as a birefringent device*

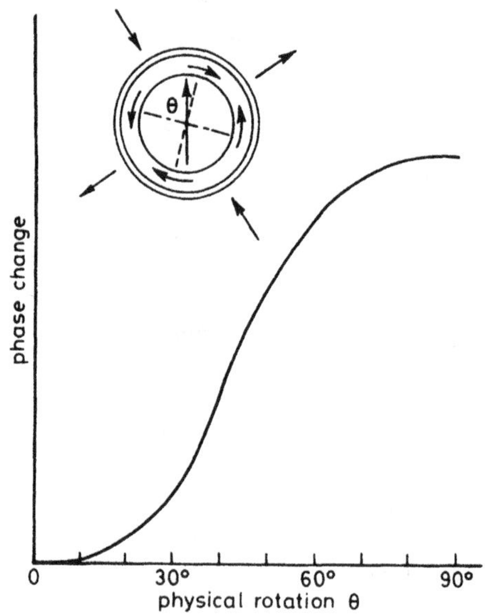

Fig. 5.15 *Variation of phase of a linearly polarised TE_{11}-mode in circular waveguide due to a ferrite tube birefringent device as the angular direction of the transverse four-pole magnetising field is varied*

polarised mode in the circular waveguide is shown in Fig. 5.15, so that varying the angular position of the biasing four-pole field varies the phase delay due to the device on a linearly polarised wave. The device could be used as a variable phase changer, although I have never seen a report of such a use. This is not to be confused with varying the angle of a ferrite half-wave plate as described by Boyd[14] and discussed in Section 5.7.

If the maximum phase difference that two perpendicular modes experience in traversing a birefringent device is 90°, the device is a quarter-wave plate and it may be used as a circular polariser.[15] If the birefringent device is a transversely magnetised ferrite element, as shown in Fig. 5.16*a*, then it is also a reciprocal device and a reverse wave will experience the same relationship between the electromagnetic field direction and the biasing magnetic field as the forward wave.[2] However, for the ferrite tube biased by a four-pole field, as shown in Fig. 5.16*b*, the reverse wave will experience the field shown in Fig. 5.16*c* and the differential phase change is the negative of that of the forward wave, so that the

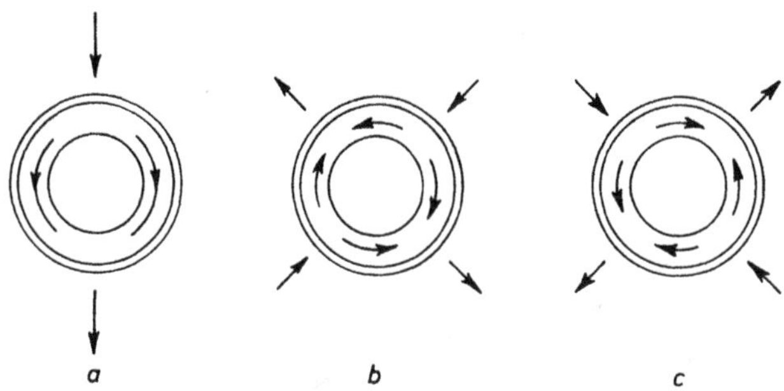

Fig. 5.16 *Ferrite tube birefringent devices*
a Transverse magnetisation, *b* Four-pole field, *c* Four-pole field of opposite hand

four-pole field device is nonreciprocal. Such a four-pole transversely-magnetised ferrite nonreciprocal circular polariser is used in the dual-mode phase changer described by Boyd[8] and in Section 5.3. Sultan[16] shows that, for a four-pole magnetic field device, the phase change is linear with magnetising current and suggests that such a device could be used as a variable phase changer.

5.7 Birefringence – half-wave plate

In Section 5.6 we have discussed how a transversely magnetised ferrite rod or tube in circular waveguide with either a simple transverse field, as shown by Fig. 5.16*a*, or a four-pole field, as shown by Fig. 5.16*b* or *c*, causes birefringent phase change for the TE_{11}-mode in circular waveguide. If the maximum phase difference that two perpendicular modes experience in traversing the device is 180°, then the device is a half-wave plate and it will act as a variable phase changer on circularly polarised waves or as a rotator for linearly polarised waves. The use of a dielectric half-wave plate as a variable phase changer was first reported by Fox[17] and is the basis of the rotary phase changer. The construction of the Fox or rotary phase changer is shown in Fig. 5.17. The input transformer to circular waveguide is followed by a polarised absorber to absorb any cross-polarised mode that may be inadvertently generated in the circular waveguide.

Next comes a circular polariser which is a dielectric slab quarter-wave plate at 45° to the plane of the linearly polaried wave at the input. The phase changer is a dielectric slab half-wave plate or a birefringent ferrite half-wave plate whose angular position may be varied. The centre section of the phase changer containing the half-wave plate rotates with respect to the rest of the waveguide system. If the angle of rotation is θ, the phase change is given by

$$\phi = 2\theta \tag{5.3}$$

The output then consists of a circular polariser, a polarised absorber and a transformer to rectangular waveguide as at the input.

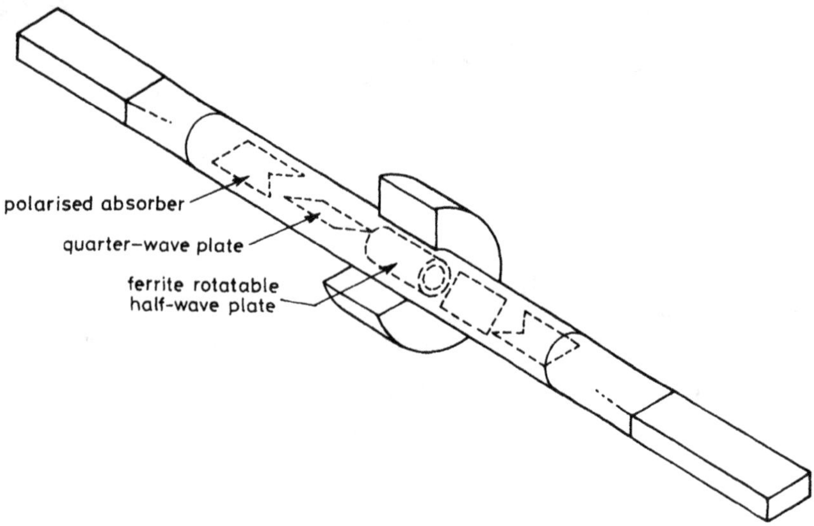

Fig. 5.17 *Rotary phase changer*

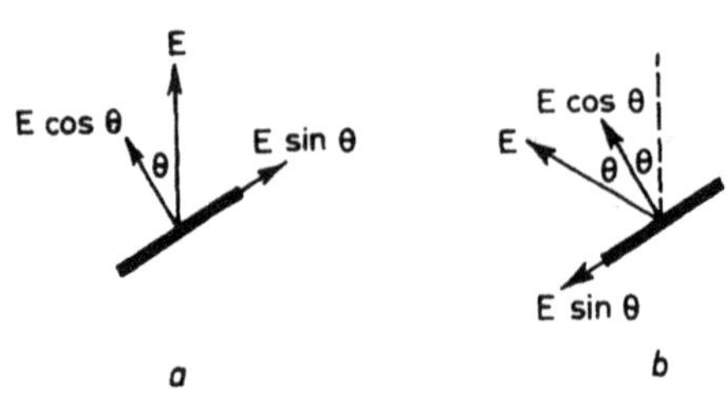

Fig. 5.18 *Half-wave plate*
a Input linearly polarised wave, *b* Output linearly polarised wave rotated through 2θ

The effect of a variable-angle half-wave plate on a linearly polarised wave is shown in Fig. 5.18. It is seen that the linearly polarised wave is rotated through the angle 2θ. It can easily be shown by a similar argument that a circularly polarised wave is advanced in phase by the angle 2θ and has its direction of circular polarisation reversed. The half-wave plate may be provided by a ferrite birefringement device, and, if the transverse field is provided by a magnetic yoke similar to a machine stator, the angle of the half-wave plate may be varied electrically. If the angle of the half-wave plate can be advanced continuously, the frequency of the microwave signal is changed and the device becomes a single-sideband modulator. In the late 1950s much effort was expended in attempting to develop such a device. A discussion of the publications on magnetic birefringence in ferrites has been given in Section 4.5. Fox, Miller and Weiss[15] give some of the applications for a ferrite birefringent section, including showing how it may be used with a continuously rotating external magnetic field as a continuous phase changer or frequency changer of circularly polarised waves. Cacheris[18] describes a single-sideband modulator operating with a modulation frequency of 20 kHz at 9.375 GHz. Karayianis and Cacheris[2] develop this further with experimental results on ferrite rods and tubes in circular waveguide at X-band. Glass[19] gives details of the construction of a transversely-magnetised ferrite-tube single-sideband modulator in reduced diameter waveguide. At one magnetic field setting the device provided 180° differential phase change over a 4% frequency band at X-band. No results are given of the operation of the device as a frequency changer.

As shown in Fig. 5.18, the half-wave plate may be used as a rotator, and such a device is used by Mohr.[20] He describes an electronically rotating ferrite half-wave plate for use in a conical scan conversion system. The ferrite half-wave plate consists of a ferrite tube adjacent to the wall of thin-walled waveguide and the magnetic field is provided by a two-phase four-pole induction motor winding. Operation was possible up to a maximum drive frequency of 1 kHz. Sultan[16] gives an analysis of the complete rotary phase changer system. He allows for angular position errors of each of the differential phase change elements in the device and for errors in the quarter-wave and half-wave differential phase changers. For example, for a variable angle variable differential phase change section between two perfect circular polarisers, where θ is the angle of rotation of the differential phase change section and β is the differential phase change, the phase change is given by

$$\phi = 2\theta + \text{constant} \tag{5.4}$$

It is independent of the value of the differential phase change β except for providing a constant error. The error also provides a constant attenuation that is independent of the angle θ. When $\beta = 180° \pm 30°$, the additional attenuation is 0.3 dB. If the output polarised absorber is omitted, it is found that the output is linearly polarised at an angle

$$\psi = \tfrac{1}{2}\beta - 90° \tag{5.5}$$

Thus, as β departs from 180°, the output polarisation rotates away from the vertical. Fox, Miller and Weiss[15] were the first to show that a variable differential phase change section between two circular polarisers is equivalent to a Faraday rotator. However, we have already shown that a single half-wave plate on its own acts as a rotator to a linearly polarised wave. Sultan[16] presents some experimental results based on a ferrite half-wave plate consisting of a ferrite tube with a four-pole magnetic biasing field.

In order to improve the bandwidth performance of a ferrite-tube quadrupole-field half-wave plate, Nakahara *et al.*[21] used a metal rod at the centre of the circular waveguide in the ferrite section. They obtained a 0.6 dB insertion loss over an 8% frequency band at 6.4 GHz. They used a 5 mm diameter metal rod inside 30 mm and 26 mm diameter waveguides with a 3.8 mm thick ferrite tube adjacent to the waveguide wall. Boyd[22,14] reports the design and manufacture of a ferrite-filled circular waveguide half-wave plate being used as a variable phase changer. The ferrite-filled waveguide is magnetised with a four-pole field. Even though the ferrite at the centre of the waveguide cannot be contributing to the differential phase change, such devices appear to be commercially practical with a total insertion loss over a 9% bandwidth of less than 0.6 dB. Solid ferrite rods are certainly easier to manufacture than tubes. Boyd[22] gives perturbation design formulas which agree with experiment.

5.8 References

1 BADEN FULLER, A. J.: 'Microwave propagation through round waveguide partially filled with ferrite', *Proc. IEE*, 1961, **108C**, pp. 339–348

2 KARAYIANIS, N. and CACHERIS, J.: 'Birefringence of ferrites in circular waveguide', *Proc. IRE*, 1956, **44**, pp. 1414–1421

3 OHM, E. A.: 'A broad-band microwave circulator', *IRE Trans.*, 1956, **MMT-4**, pp. 210–217

4 CHAIT, H. N. and SAKIOTIS, N. G.: 'Broad-band ferrite rotators using quadruply-ridged circular waveguide', *IRE Trans.*, 1959, **MTT-7**, pp. 38–41

5 RIZZI, P. A.: 'High-power ferrite circulators', *IRE Trans.*, 1957, **MTT-5**, pp. 230–237

6 O'HARA, F. J. and SCHARFMAN, H.: 'A ferrite serrodyne for microwave frequency translation', *IRE Trans.*, 1959, **MTT-7**, pp. 32–37

7 SAKIOTIS, N. G., SIMMONS, A. J. and CHAIT, H. N.: 'Applications of ferrites to the microwave antenna problem', Report of N.R.L. Progress, Jan. 1952, pp. 1–5. (reprinted in WHICKER, L. R.: 'Ferrite control components. Vol. 1' (Artech House, 1974), pp. 17–31)

8 BOYD, C. R.: 'A dual-mode latching reciprocal ferrite phase shifter', *IEEE Trans.*, 1970, **MTT-18**, pp. 1119–1124 (reprinted in WHICKER, L. R.: 'Ferrite control components. Vol. 2' (Artech House, 1974), pp. 142–147)

9 DUNCAN, B. J. and SWERN, L.: 'Effects of zero ferrite permeability on circularly polarised waves', *Proc. IRE.*, 1957, **45**, pp. 647–655

10 WALDRON, R. A.: 'Features of cylindrical waveguides containing gyromagnetic media', *J. Brit. IRE.*, 1960, **20**, pp. 695–706, see Fig. 2

11 BERNUES, F. J. and BOLLE, D. M.: 'The digital twin-ferrite-toroid circular waveguide phaser', *IEEE Trans.*, 1973, **MTT-21**, pp. 842–845

12 KARBOWIAK, A. E.: 'Trunk waveguide communication' (Chapman & Hall, 1965)

13 INCE, W. J. and TSANDOULAS, G. N.: 'Modal inversion in circular waveguides Pt. II: Application to latching nonreciprocal phasers', *IEEE Trans.*, 1971, **MTT-19**, pp. 393–400

14 BOYD, C. R.: 'Analog rotary-field ferrite phase shifters', *Microwave J.*, 1977, **20**, 12, pp. 41–43

15 FOX, A. G., MILLER, S. E. and WEISS, M. T.: 'Behaviour and applications of ferrites in the microwave region', *Bell Syst. Tech. J.*, 1955, **34**, pp. 5–103

16 SULTAN, N. B.: 'Generalized theory of waveguide differential phase sections and application to novel ferrite devices', *IEEE Trans.*, 1971, **MTT-19**, pp. 348–357

17 FOX, A. G.: 'An adjustable wave-guide phase changer', *Proc. IRE*, 1947, **35**, pp. 1489–1498

18 CACHERIS, J.:'Microwave single-sideband modulator using ferrites', *Proc. IRE*, 1954, **42**, pp. 1243–1247

19 GLASS, H. I.: ' A short rugged ferrite half-wave plate as a single-sideband modulator', *IRE Trans.*, 1959, **MTT-7**, p. 295

20 MOHR, M. C.: 'New and unique ferrite devices', *Microwave J.*, 1970, **13**, 1, pp. 58–64

21 NAKAHARA, S., KUREBAYSHI, H. and MIZOBUCHI, A.: 'A 6GHz rotating field phaser', *IEEE Trans.*, 1972, **MAG-8**, pp. 544–547

22 BOYD, C. R.: 'Design of ferrite differential phase shift sections', IEEE–MTTS Int. Symp. Digest, 1975, pp. 240–242

Chapter 6

Transversely magnetised ferrite in rectangular waveguide

6.1 Introduction

The effect of a transversely magnetised ferrite medium on a plane wave is discussed in Section 2.9. However, for the dominant TE_{10}-mode in rectangular waveguide, the field approximates to a plane wave only at the centre of the waveguide. To get a maximum interaction between the ferrite material and a

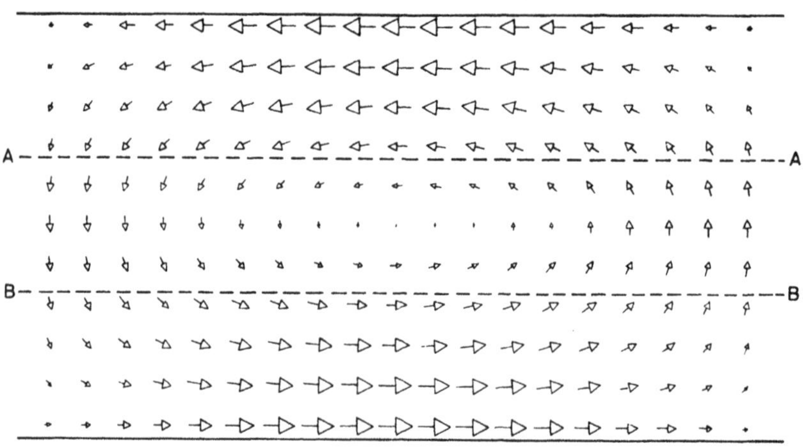

Fig. 6.1 *Magnetic field intensity of the TE_{10}-mode in a plane parallel to the broad face of rectangular waveguide*

microwave field the magnetic component of the microwave field needs to be circularly polarised in a plane perpendicular to the direction of magnetisation in the ferrite. For the TE_{10}-mode there is a position in the cross-section of the waveguide where the magnetic field is circularly polarised in the plane of the broad face of the waveguide, as shown by the plot of the magnetic field in the plane of the broad face of the waveguide given in Fig. 6.1.[1] The size of the arrow denotes the strength of the field. It is seen that, for the particular frequency chosen for this plot, the magnetic field is circularly polarised in the plane of the plot along the line AA and is circularly polarised with the opposite hand along the line BB. Then, if a thin slab of ferrite is mounted in the waveguide as shown in Fig. 6.2, its effective permeability will be μ^+ or μ^-, as shown in Fig. 5.1. If the wave is propagating in the opposite direction, the magnetic field will be circularly polarised in the opposite direction and the effective permeability seen by the wave will be different to the effective permeability seen by the forward wave. Such a device is nonreciprocal. Depending on the strength of the magnetisation, it may be used as a phase changer, a field displacement isolator or as a resonance isolator.

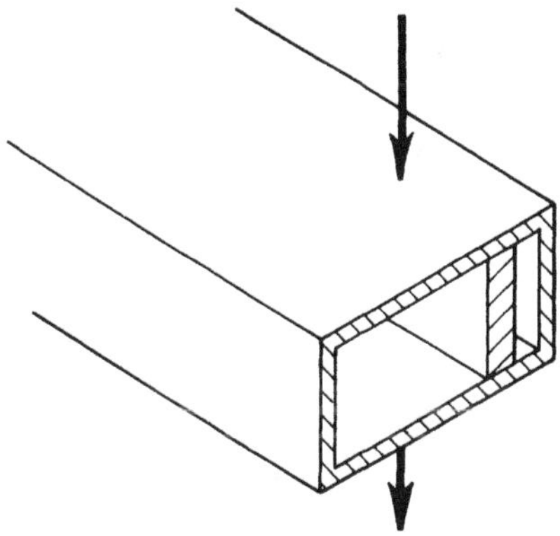

Fig. 6.2 *Ferrite slab in rectangular waveguide*

The position of the circularly polarised magnetic field in the waveguide may be obtained from a study of the expressions for the field in empty rectangular waveguide. Such expressions are given in many textbooks.[2] From the field quantities, the ellipticity is given by

$$\frac{H_x}{H_z} = \frac{\beta}{k_c^2}\frac{\pi}{a}\tan\frac{\pi x}{a} = \sqrt{\left[\left(\frac{\lambda_c}{\lambda_0}\right)^2 - 1\right]}\left(\frac{\lambda_c}{2a}\right)\tan\frac{\pi x}{a} \tag{6.1}$$

where x is measured from one wall of the waveguide, a is the broad dimension of the waveguide and λ_0 is the free-space wavelength in a space of whatever dielectric is filling the waveguide. For the TE_{10}-mode, $\lambda_c = 2a$, and the expression for the ellipticity reduces to

$$\frac{H_x}{H_z} = \sqrt{\left(\frac{4a^2}{\lambda_0^2} - 1\right)}\tan\frac{\pi x}{a} \tag{6.2}$$

To find the position of circular polarisation, the ellipticity is unity and eqn. 6.2 gives

$$\frac{x}{a} = \frac{1}{\pi}\cot^{-1}\left[\sqrt{\left(\frac{4a^2}{\lambda_0^2} - 1\right)}\right] \tag{6.3}$$

The relationship given by eqn. 6.3 is plotted in Fig. 6.3, which shows that, at any particular frequency, there is only one position (or two symmetrical positions) where a thin slither of ferrite material will have the effective permeability given in Fig. 5.1. When the ferrite slab has a finite thickness, as shown in Fig. 6.2, the dielectric properties of the ferrite tend to concentrate the microwave fields into the ferrite and the variation of the position of circular polarisation with variation of frequency is much reduced. A full mathematical analysis of the ferrite slab in rectangular waveguide as shown in Fig. 6.2 is given in the next sections.

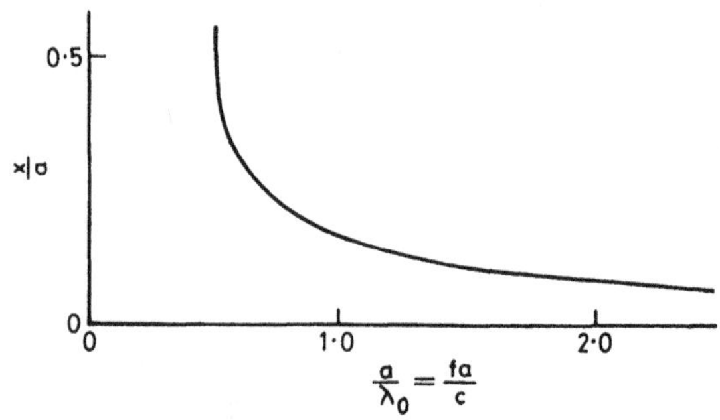

Fig. 6.3 *Position of the circularly polarised magnetic field in the cross-section of rectangular waveguide as the frequency is changed*

6.2 Field analysis in the ferrite

Provided that a ferrite slab fills the full height of the waveguide, as shown in Fig. 6.4, theoretical analysis of the propagating conditions in the waveguide is relatively easy. The geometry of the waveguide system is shown in Fig. 6.4. Regions 1 and 3 are filled with air or some other dielectric material whose material properties are ε_0 and μ_0. Region 2 is the ferrite, magnetised perpendicular to the broad face of the waveguide as shown in Fig. 6.4, having the permittivity ε and a tensor permeability which is given by eqn. 2.59

$$\boldsymbol{\mu} = \begin{vmatrix} \mu & 0 & j\kappa \\ 0 & \mu_0 & 0 \\ -j\kappa & 0 & \mu \end{vmatrix} \tag{6.4}$$

For the fields in the ferrite, the fundamental relationships are given by Maxwell's equations, eqns. 2.31 and 2.32, combined with material properties given by eqn. 6.4. Therefore

$$\frac{\partial E_z}{\partial y} + j\beta E_y = -j\omega(\mu H_x + j\kappa H_z) \tag{6.5}$$

$$-j\beta E_x - \frac{\partial E_z}{\partial x} = -j\omega\mu_0 H_y \tag{6.6}$$

$$\frac{\partial E_y}{\partial x} - \frac{\partial E_x}{\partial y} = -j\omega(-j\kappa H_x + \mu H_z) \tag{6.7}$$

$$\frac{\partial H_z}{\partial y} + j\beta H_y = j\omega\varepsilon E_x \tag{6.8}$$

$$-j\beta H_x - \frac{\partial H_z}{\partial x} = j\omega\varepsilon E_y \tag{6.9}$$

$$\frac{\partial H_y}{\partial x} - \frac{\partial H_x}{\partial y} = j\omega\varepsilon E_x \tag{6.10}$$

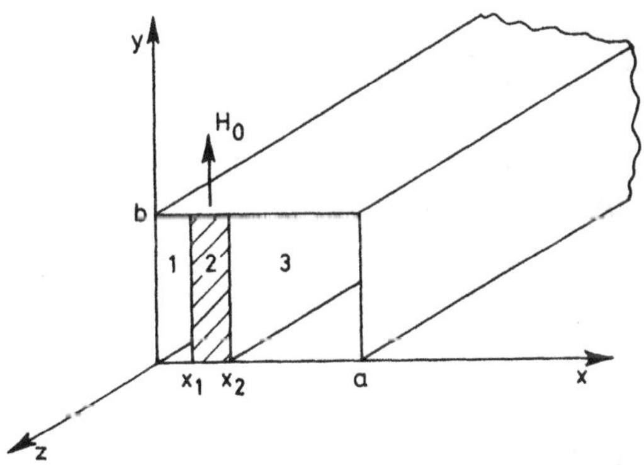

Fig. 6.4 *Transversely magnetised ferrite slab in rectangular waveguide*
Regions 1 and 3 are air, region 2 is ferrite

where β is the propagation constant for propagation along the length of the waveguide. For the higher order modes in the waveguide it is noticed that there is no separation into TE- and TM-modes, but for the dominant mode these equations simplify. For the TE_{10}-mode, there is no variation of the field quantities in the y-direction so that $\partial/\partial y = 0$. Also in the rectangular waveguide, the fields appropriate to a TM_{10}-mode cannot exist so that $E_z = 0$ and $E_x = 0$. Then eqns. 6.5 to 6.10 simplify to

$$\beta E_y = -\omega(\mu H_x + j\kappa H_z) \tag{6.11}$$

$$\frac{\partial E_y}{\partial x} = -j\omega(-j\kappa H_x + \mu H_z) \tag{6.12}$$

$$j\omega\varepsilon E_y = -j\beta H_x - \frac{\partial H_z}{\partial x} \tag{6.13}$$

Eliminating the magnetic field components from these equations gives

$$\frac{\partial^2 E_y}{\partial x^2} + \left[\omega^2 \varepsilon \left(\frac{\mu^2 - \kappa^2}{\mu}\right) - \beta^2\right] E_y = 0 \tag{6.14}$$

Let

$$\mu_e = (\mu^2 - \kappa^2)/\mu \tag{6.15}$$

and

$$k_f^2 = \omega^2 \varepsilon \mu_e - \beta^2 \tag{6.16}$$

Then the solution to eqn. 6.14 is

$$E_y = [A \exp(jk_f x) + B \exp(-jk_f x)] \exp j(\omega t - \beta z) \tag{6.17}$$

where A and B are arbitrary constants specified by the boundary conditions. Eqns. 6.11 and 6.12 can be solved to give expressions for the magnetic field components

$$H_x = \frac{1}{\omega \mu_e \mu}\left(-\mu \beta E_y + \frac{\partial E_y}{\partial x}\right)$$

$$H_z = \frac{j}{\omega \mu_e \mu}\left(\kappa \beta E_y - \frac{\partial E_y}{x}\right)$$

6.3 Single full-height ferrite slab

Having found an expression for the fields in the ferrite, it is now necessary to apply boundary conditions to eliminate the arbitrary constants in the expressions for the fields. The simplest geometry to consider is the single ferrite slab in rectangular waveguide shown in Fig. 6.4. The first published results of a theoretical analysis for this geometry were by Kales, Chait and Sakiotis.[3] The results of a much fuller analysis together with expressions for the fields in the ferrite and dielectric regions have been given by Lax, Button and Roth,[4] and they also give the characteristic equation for the symmetrical twin-slab ferrite loaded waveguide considered in the next section. A very full analysis containing these and many more results for full-height ferrite slabs in rectangular waveguide is given by Lax and Button.[5] The mathematical results from Lax, Button and Roth[4] are also given by Helszajn.[6] Clarricoats[7] gives an extended discussion of the modes propagating in ferrite loaded rectangular waveguide although he gives less practical results than Lax and Button.[5] Waldron[8] gives a more general mathematical analysis which is not confined to the TE_{m0}-modes in the rectangular waveguide, although he only gives results for the TE_{10}-mode. We will now proceed to derive the characteristic equation for the TE_{m0}-modes in the ferrite loaded rectangular waveguide.

For the single ferrite slab in rectangular waveguide shown in Fig. 6.4, eqn. 6.17 gives an expression for the field in the ferrite, region 2. It is necessary to match the fields across the boundaries of the ferrite slab in order to eliminate the arbitrary constants A and B, but first we need to specify the fields in the other two regions of the waveguide. They can be postulated from the normal TE_{10}-mode fields in empty waveguide to be[2]

$$E_{y1} = C \sin kx \exp \mathrm{j}(\omega t - \beta z) \tag{6.18}$$

$$E_{y3} = D \sin k(a - x) \exp \mathrm{j}(\omega t - \beta z) \tag{6.19}$$

where

$$k^2 = \omega^2 \varepsilon_0 \mu_0 - \beta^2 \tag{6.20}$$

The boundary conditions at the wall of the waveguide are implicit in eqns. 6.18 and 6.19. C and D are two further arbitrary constants. Exp $\mathrm{j}(\omega t - \beta z)$ occurs in all the expressions for the field quantities and will be omitted in all subsequent equations. There are four arbitrary constants in the expressions for the electric field and these may be eliminated by matching the components of the fields parallel to the narrow face of the waveguide at each face of the ferrite slab. First we need expressions for the fields in the three regions of the waveguide. Expressions for the electric field are given in eqns. 6.17 to 6.19. Then the expressions for the magnetic fields are

$$H_{z1} = \frac{\mathrm{j}Ck}{\omega\mu_0} \cos kx \tag{6.21}$$

$$H_{z2} = -\frac{1}{\omega\mu_e}\left[k_f \{A \exp(\mathrm{j}k_f x) - B \exp(-\mathrm{j}k_f x)\} + \frac{\mathrm{j}\kappa\beta}{\mu}\{A \exp(\mathrm{j}k_f x) + B \exp(-\mathrm{j}k_f x)\}\right] \tag{6.22}$$

$$H_{z3} = -\frac{\mathrm{j}Dk}{\omega\mu_0} \cos k(a - x) \tag{6.23}$$

$$H_{x1} = -\frac{C\beta}{\omega\mu_0} \sin kx \tag{6.24}$$

$$H_{x2} = \frac{1}{\omega\mu_e}\left[\frac{\mathrm{j}\kappa k_f}{\mu}\{A \exp(\mathrm{j}k_f x) - B \exp(-\mathrm{j}k_f x)\} - \beta\{A \exp(\mathrm{j}k_f x) + B \exp(-\mathrm{j}k_f x)\}\right] \tag{6.25}$$

$$H_{x3} = -\frac{D\beta}{\omega\mu_0} \sin k(a - x) \tag{6.26}$$

Equating the values of the electric field each side of the boundaries at $x = x_1$ and $x = x_2$ gives

$$\left.\begin{aligned} C \sin kx_1 &= A \exp(jk_f x_1) + B \exp(-jk_f x_1) \\ D \sin k(a - x_2) &= A \exp(jk_f x_2) + B \exp(-jk_f x_2) \end{aligned}\right\} \tag{6.27}$$

Two similar equations may be obtained from suitable substitution into eqns. 6.21 to 6.23 for the longitudinal magnetic field.

In order to find the characteristic equation for propagation in the ferrite loaded waveguide, it is necessary to eliminate the four arbitrary constants from the four equations given by equating the values of the fields at each side of the boundaries. Then we obtain the determinant

$$\begin{vmatrix} F_1 \exp(jk_f x_1) & F_2 \exp(-jk_f x_1) & \dfrac{jk}{\mu_0}\cos kx_1 & 0 \\ F_1 \exp(jk_f x_2) & F_2 \exp(-jk_f x_2) & 0 & -\dfrac{jk}{\mu_0}\cos k(a - x_2) \\ \exp(jk_f x_1) & \exp(-jk_f x_1) & -\sin kx_1 & 0 \\ \exp(jk_f x_2) & \exp(-jk_f x_2) & 0 & -\sin k(a - x_2) \end{vmatrix} = 0 \tag{6.28}$$

where

$$F_1 = -\frac{k_f}{\mu_e} - j\frac{\kappa\beta}{\mu\mu_e}$$

and

$$F_2 = \frac{k_f}{\mu_e} - j\frac{\kappa\beta}{\mu\mu_e}$$

When put equal to zero, this determinant gives the characteristic equation describing propagation through the ferrite loaded waveguide. On multiplying out the determinant we obtain

$$\begin{aligned} &\frac{kk_f}{\mu_0\mu_e}\{\tan kx_1 + \tan k(a - x_2)\} \cot k_f(x_2 - x_1) \\ &\quad - \frac{k\kappa\beta}{\mu_0\mu\mu_e}\{\tan kx_1 - \tan k(a - x_2)\} \\ &\quad - \left[\left(\frac{k_f}{\mu_e}\right)^2 + \left(\frac{\kappa\beta}{\mu\mu_e}\right)^2\right] \tan kx_1 \tan k(a - x_2) \\ &\quad + \left(\frac{k}{\mu_0}\right)^2 = 0 \end{aligned} \tag{6.29}$$

This expression may easily be evaluated using a computer to give values for the propagation constant β. Some typical results are given in Fig. 6.5. It will be seen from eqn. 6.29 that the second term on the left-hand side of the equation contains the product $\kappa\beta$ whereas all the other terms only contain κ^2 or β^2. The term in $\kappa\beta$ represents the gyromagnetic effect of the ferrite giving nonreciprocal propagation. Altering the sign of κ or of β alters the propagation constant but altering the sign of both together keeps the propagation conditions the same. Eqn. 6.29 is presented in the form it is because it gives a minimum number of intermediate expressions for computational purposes. The same expression is given by Lax, Button and Roth[4] and others[5–8] in a slightly different form. However, the relationship was first reported by Kales, Chait and Sakiotis[3] in a form closely resembling eqn. 6.29.

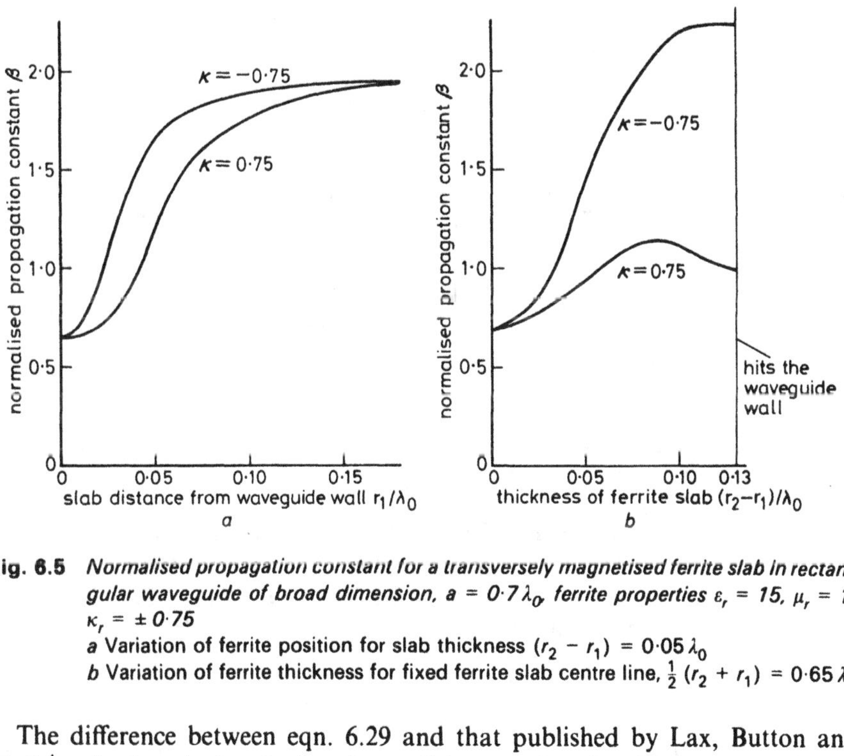

Fig. 6.5 *Normalised propagation constant for a transversely magnetised ferrite slab in rectangular waveguide of broad dimension, $a = 0{\cdot}7\lambda_0$, ferrite properties $\varepsilon_r = 15$, $\mu_r = 1$, $\kappa_r = \pm 0{\cdot}75$*
a Variation of ferrite position for slab thickness $(r_2 - r_1) = 0{\cdot}05\,\lambda_0$
b Variation of ferrite thickness for fixed ferrite slab centre line, $\frac{1}{2}(r_2 + r_1) = 0{\cdot}65\,\lambda_0$

The difference between eqn. 6.29 and that published by Lax, Button and Roth[4] and others is that the Lax form of the equation simplifies computation of the propagation constant when the thickness of the ferrite slab is constant and its position is variable. However, with modern computational power, such simplification is not necessary and the more general form of eqn. 6.29 is equally useful. Having found the propagation constant, it is then possible to substitute back into the expressions for the fields in the ferrite and dielectric regions to determine the relative strengths of the fields in different positions in the waveguide. Expressions for evaluating the arbitrary constants were given by

Lax, Button and Roth[4] and by Helszajn.[6] This enables them to give a number of plots of the variation of field strength across the waveguide. Lax and Button[5] give the largest number of field strength plots from these results.

6.4 Two symmetrical full-height ferrite slabs

Another geometry which is similar to that considered in the last section is that shown in Fig. 6.6. This is the configuration used to make a latching phase changer as described in Section 7.2. Because the ferrite slabs on the opposite sides of the waveguide are magnetised in the opposite direction, both ferrite slabs present the same permeability to the wave in the waveguide and the fields are symmetrical. Therefore the characteristic equation for the propagation in

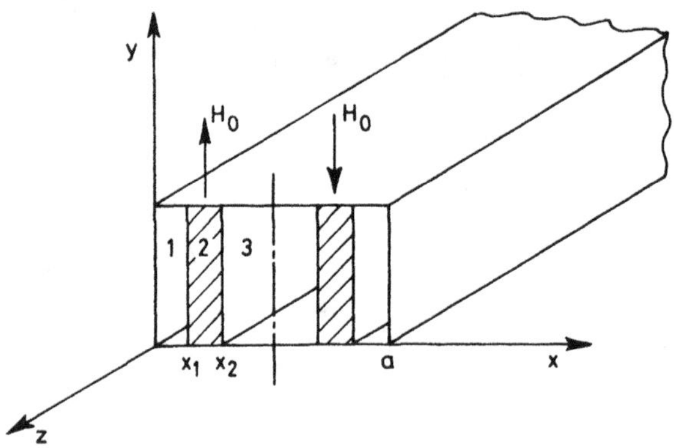

Fig. 6.6 *Two ferrite slabs symmetrically mounted in rectangular waveguide which have a symmetrical field pattern*

the waveguide is similar to eqn. 6.29. The boundary condition for region 3 is now a magnetic wall along the centre line of the waveguide. The expressions for the fields in regions 1 and 2 are still given by eqns. 6.17, 6.18, 6.21, 6.22, 6.24 and 6.25. The fields in region 3 are given by

$$E_{y3} = D \cos k(\tfrac{1}{2}a - x) \tag{6.30}$$

$$H_{z3} = \frac{jDk}{\omega\mu_0} \sin k(\tfrac{1}{2}a - x) \tag{6.31}$$

$$H_{x3} = -\frac{D\beta}{\omega\mu_0} \cos k(\tfrac{1}{2}a - x) \tag{6.32}$$

Elimination of the arbitrary constants from the equations obtained by equating the values of the fields on each boundary of the ferrite slab gives the characteristic equation

$$\frac{kk_f}{\mu_0\mu_e}\{\tan kx_1 \tan k(\tfrac{1}{2}a - x_2) - 1\} \cot k_f(x_2 - x_1)$$
$$+ \frac{k\kappa\beta}{\mu_0\mu\mu_e}\{1 + \tan kx_1 \tan k(\tfrac{1}{2}a - x_2)\}$$
$$+ \left[\left(\frac{k_f}{\mu_e}\right)^2 + \left(\frac{\kappa\beta}{\mu\mu_e}\right)^2\right] \tan kx_1$$
$$+ \left(\frac{k}{\mu_0}\right)^2 \tan k(\tfrac{1}{2}a - x_2) = 0 \qquad (6.33)$$

This expression was first given by Lax, Button and Roth[4] in a slightly different form and subsequently by others.[5-7] Lax, Button and Roth[4] and Lax and Button[5] also give some results for the variation of the strength of the field components across the cross-section of the waveguide. Other more complicated geometries are considered in the next section.

In a further consideration of the twin-slab ferrite phase changer, Ince and Stern[9] investigate the effect of the centre section of the waveguide being filled with a dielectric material. They[9] give a very full theoretical analysis of both propagation constant and attenuation constant for the configuration of Fig. 6.6 with region 3 filled with a dielectric material. The analysis assumes that the ferrite is at its remanent magnetisation condition, as it would be in a latching phase changer. They show that the differential phase change is an almost linear function of the remanent magnetisation. They also show that differential phase change increases approximately linearly with ferrite thickness until a large proportion of the microwave energy is concentrated in the ferrite. If the dielectric region is made very thin, the ferrite material at the centre of the waveguide is not contributing to the differential phase change because the microwave field is not circularly polarised at the centre of the waveguide. Then the ferrite at the centre of the waveguide may be replaced by dielectric material of the same permittivity. The ratio of the differential phase change to the insertion phase is a maximum for a ferrite slab width of between $0{\cdot}04\,\lambda_0$ and $0{\cdot}06\,\lambda_0$ for a wide range of dielectric permittivities and sizes. The frequency sensitivity of the differential phase change may be controlled by varying the width of the waveguide. For wide waveguide, the phase change increases with change of frequency, whereas it decreases for narrow waveguide. At an intermediate size of waveguide, the variation of differential phase change with change of frequency may be very small.

An earlier paper by Schlomann[10] also analysed the geometry of Fig. 6.6 for latching phase changer applications. He also gives eqn. 6.33 as the characteristic equation. He gives the results of considering the attenuation of the ferrite so that a figure of merit may be calculated. He assumes that the magnetic loss factor μ'' is dominant and that κ'' and ε'' may be taken to be zero. He also considers the effect of additional high-power loss on the power handling capabilities of the phase changer.

6.5 Full-height dielectric and ferrite slabs

Both the geometries shown in Figs. 6.4 and 6.6 give nonreciprocal phase change or isolation for the TE_{10}-mode in the waveguide. However, for two symmetrically positioned ferrite slabs in the waveguide, if the slabs are magnetised in the same direction, as shown in Fig. 6.7, the microwave field pattern in the waveguide is not symmetrical because each ferrite slab has a different effective permeability. Since a reverse wave will see the same combination of effective permeability, the ferrite loaded waveguide shown in Fig. 6.7 is a reciprocal device rather than a nonreciprocal one.

A theoretical analysis of the reciprocal two-slab ferrite phase changer shown in Fig. 6.7 with the ferrite slabs adjacent to the waveguide wall has been given by Pokusin.[11] He investigates the various modes that are able to propagate in the ferrite loaded waveguide. However, a full analysis of the structure shown in Fig. 6.7 is also possible.

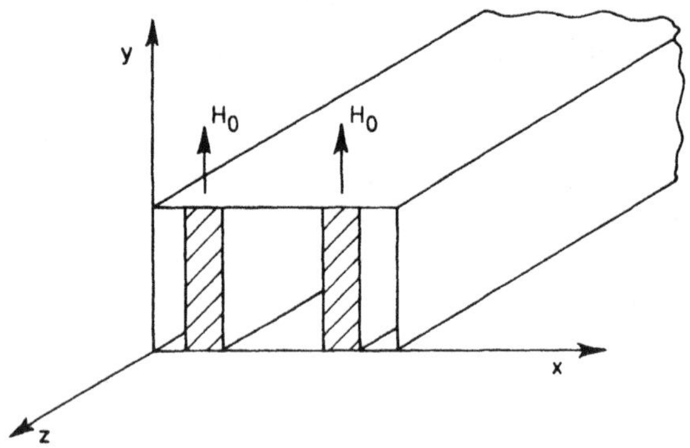

Fig. 6.7 *Two ferrite slabs symmetrically mounted in rectangular waveguide which provide reciprocal phase change*

Following the techniques used to analyse propagation in a waveguide loaded by single ferrite slab, it is possible to specify expressions for the fields in each of the regions shown in Fig. 6.7 to match the values for the fields across each of the boundaries and finally to produce an 8 by 8 matrix similar to eqn. 6.28. Any combination of full-height ferrite and dielectric slabs in rectangular waveguide may be similarly analysed. However, it is only the ratio of E_y and H_z that needs to be matched at each boundary. Seidel,[12] and later Gardiol,[13] have shown how these fields at each boundary of a uniform material are related by a simple 2 by 2 matrix. Then the characteristic equation is obtained from the product of the matrices representing each of the regions in the cross-section of the waveguide. Gardiol and Vander Vorst[14] have applied this technique to the analysis of resonance isolators in rectangular waveguide, some incorporating dielectric slabs next to the ferrite slab. Their program allows for the loss components of the permeability and permittivity and calculates the attenuation due to the ferrite loaded waveguide. However, for a variety of reasons which are given in Sections 6.9 and 7.5, the resonance isolator usually consists of partial-height ferrite slabs and the analysis given here is of only limited use for the design of resonance isolators.

Referring to the geometry of Fig. 6.4, Gardiol[13] relates the values of E_y and H_x in the planes $x = x_1$ and $x = x_2$ by the matrix

$$\begin{vmatrix} E_y \\ H_z \end{vmatrix}_{x=x_2} = \begin{vmatrix} \cos k_f d + \dfrac{\beta\kappa}{\mu k_f} \sin k_f d & -\mathrm{j}\dfrac{\omega\mu_e}{k_f} \sin k_f d \\ -\mathrm{j}\dfrac{(\omega^2\varepsilon\mu - \kappa^2)}{\omega\mu k_f} \sin k_f d & \cos k_f d - \dfrac{\beta\kappa}{\mu k_f} \sin k_f d \end{vmatrix} \begin{vmatrix} E_y \\ H_z \end{vmatrix}_{x=x_1} \tag{6.34}$$

where $d = x_2 - x_1$ is the thickness of the ferrite slab, and all the other terms have already been defined. Similar terms may be defined for the air spaces so that for the complete cross-section of the waveguide

$$\begin{vmatrix} E_y \\ H_z \end{vmatrix}_{x=a} = |T_3|\,|T_2|\,|T_1| \begin{vmatrix} E_y \\ H_z \end{vmatrix}_{x=0} = |T| \begin{vmatrix} E_y \\ H_z \end{vmatrix}_{x=0} = \begin{vmatrix} t_{11} & t_{12} \\ t_{21} & t_{22} \end{vmatrix} \begin{vmatrix} E_y \\ H_z \end{vmatrix}_{x=0} \tag{6.35}$$

where $|T_2|$ is the matrix defined in eqn. 6.34 and $|T_1|$ and $|T_3|$ are the matrices representing regions 1 and 3, respectively.

The electric field at the waveguide walls must be zero so the two values of E_y in eqn. 6.35 must be zero. However, for a non-trivial solution to eqn. 6.35, H_z must not be zero, which can only be true if

$$t_{12} = 0 \tag{6.36}$$

This gives the characteristic equation for the ferrite loaded waveguide. Substitution of the appropriate expressions into eqns. 6.35 and 6.36 leads to eqn. 6.29. Eqn. 6.35 may easily be extended to represent any combination of ferrite and dielectric slabs in rectangular waveguide. Since programs for the multiplication of matrices are easily available, programming eqn. 6.36 becomes particularly simple.

Some simplification in the mathematics appears to be possible if region 1 is eliminated and the ferrite slab is placed adjacent to the sidewall of the waveguide. However, additional modes appear to propagate under these conditions. Although the work of Pokusin[11] was referred to earlier because his mathematical analysis was similar to that here, he was not the first to investigate the modes propagating in a ferrite loaded waveguide with the ferrite slab or slabs adjacent to the waveguide wall. In particular, it was discovered that a surface wave type of mode could propagate at the ferrite/air interface when the other surface of the slab abutted on to a perfectly conducting surface. Bresler[15] showed that, in rectangular waveguide with a transversely magnetised ferrite slab adjacent to the wall of the waveguide, there are two kinds of waveguide modes, a *surface wave mode* bound to the ferrite/air interface and *ordinary waveguide modes*. For $\mu_e < 0$, with increasing ferrite slab thickness, the waveguide modes become cut off and only the surface wave type of mode may propagate. In certain conditions only one mode can propagate supporting propagation in one direction only. This started a discussion about the thermodynamic paradox whereby a section of ferrite loaded waveguide would transmit microwave power into a closed cavity but not allow any microwave power to flow out. A solution has been given by Barzilai and Gerosa[16] who showed both experimentally and theoretically that there is also a lossy surface wave supported at the ferrite metal interface when the metal has a finite conductivity. Power will flow out of the cavity in this mode and heat up the ferrite waveguide wall interface. Lax and Button[5] give an extended discussion of these anomalous modes in ferrite loaded rectangular waveguide.

6.6 Field displacement

If the ferrite slab in rectangular waveguide, as shown in Fig. 6.2, is magnetised to the position of zero permeability shown on Fig. 5.1, then the microwave field tends to be rejected from the ferrite for one direction of propagation. However, it is concentrated in the ferrite for propagation in the opposite direction, because that wave sees the normal positive permeability of the ferrite. If an attenuating vane is placed adjacent to the ferrite slab at a position of minimum electric field, a field-displacement isolator is produced. The forward wave is rejected from the ferrite, the attenuating vane is at the position of minimum field and the wave is unattenuated. The backward wave, however, is concentrated in the ferrite slab with a peak of the electric field at the position of the attenuating vane and it is attenuated. Such a device was first suggested by Fox, Miller and Weiss.[17] The design of a practical field-displacement isolator, together with some analysis based on earlier theory,[3,4] was given by Weisbaum and Seidel[18] and Weisbaum and Boyet.[19] Such a device may be expected to have two advantages over the resonance isolator. It is more likely to have broadband isolation than the resonant isolator, because, as seen from Fig. 5.1, the permeability is changing slowly with frequency around the zero permeability condition, whereas resonance is a direct function of frequency. Secondly, the field-displacement isolator will require a much smaller magnetic field than a resonance isolator for the same frequency of operation.

Using the theoretical analysis given by Lax, Button and Roth,[4] Button[20] has analysed the operation of the field-displacement isolator. He showed that the thickness of the slab is not critical provided that it is about $0{\cdot}1a$ or greater. The propagation constant is still given by the solution to eqn. 6.29. The forward wave is characterised by the condition $\beta_n < 1$, so that k is real and there is a sinusoidal variation of the field strength in regions 1 and 3. However, for the reverse wave $\beta_n > 1$, k is imaginary and there is a hyperbolic sine variation of the fields in regions 1 and 3. A sketch of the probable electric field strength variation across the cross-section of the waveguide is given in Fig. 6.8. It is seen that a resistive film on one face of the ferrite slab will absorb the reverse wave which has a peak electric field strength in the plane of the film, whereas the forward wave is unattenuated because it has zero electric field strength in the plane of the film. An ideal isolator design will require:

1 a hyperbolic sine dependence of the fields in the air space for the reverse wave
2 a sinusoidal dependence for the forward wave which is zero at the face of the ferrite slab where the resistive film is located.

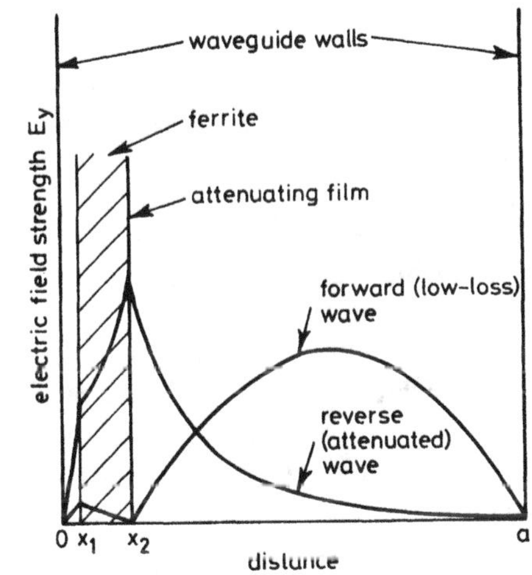

Fig. 6.8 *Electric field strength variation across the cross-section of the waveguide for the field-displacement isolator*

Weisbaum and Seidel[18] have shown that the electric field strength at the face of the ferrite slab can only be brought to zero if $\mu_e < 0$. Then the position of the ferrite slab may be found by letting

$$\sin k(a - x_2) = 0 \tag{6.37}$$

and substituting into eqn. 6.29. The first condition for the reverse wave is satisfied by using a thick ferrite slab slightly spaced from the waveguide wall. Comstock and Fay[21] have taken the analysis further by including the effect of the resistive film absorber. They show how the field distribution in the waveguide is dependent on the resistivity of the absorbing sheet.

The transversely magnetised field-displacement isolator in rectangular waveguide is not quite the same as the field-displacement effect in circular waveguide described in Section 5.4. With the longitudinally magnetised ferrite rod in circular waveguide, the whole of the ferrite rod is in a circularly polarised magnetic field and there is little penetration of the microwave fields into the ferrite. The ferrite material itself may be lossy and provide isolation by absorbing the microwave power for propagation in one direction. For the transversely magnetised field-displacement isolator, however, the microwave fields are not completely rejected from the ferrite in the forward direction of propagation and a low-loss ferrite needs to be used for good isolator performance.

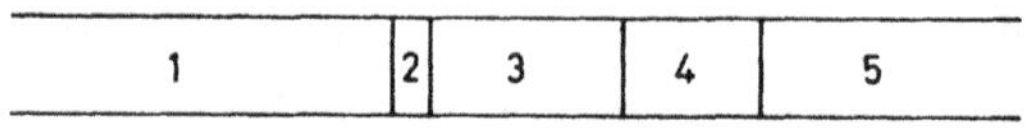

Fig. 6.9 *Ferrite slab in parallel plate waveguide which will provide field displacement effects*
Region 1 is air, region 2 is absorber, region 3 is ferrite, region 4 is dielectric and region 5 is air

6.7 Parallel plate waveguide

The theoretical analysis of transversely magnetised ferrite in parallel plate waveguide has been used as a simplification of ferrite loaded stripline or microstrip. The method of analysis is similar to that used in this chapter and the results show a field-displacement effect. Consider the general parallel plate waveguide, shown in Fig. 6.9, which is loaded with five different regions in the waveguide cross-section. Typically these might be as follows: regions 1 and 5 are air and theoretically extend to infinity; region 2 is an absorbing material; region 3 is the ferrite magnetised perpendicular to the plane of the waveguide, and region 4 is a dielectric material. This geometry may be analysed by matching the fields at each boundary with a technique similar to that used in Sections 6.3 and 6.4 or alternatively by using the transverse matrix of Gardiol[13] given in eqns. 6.34 and 6.35. Such an analysis was first given by Hines[22] in investigating the stripline field-displacement phase changer. He proposed magnetic walls at the outside boundaries of regions 2 and 4 as the outside boundaries of his waveguide. Region 2 was air and regions 3 and 4 were ferrite and dielectric, respectively. Talisa and Bolle[23] analyse a structure more like Fig. 6.9, except that region 2, the absorbing material, extends to infinity. In another paper,[24] they analyse the exact structure given in Fig. 6.9, and in yet another[25] they analyse the field-displacement phase changer where region 2 of Fig. 6.9 is omitted. A discussion of their results is given in Chapter 9 about stripline and microstrip devices. Helszajn and Downs[26] take the analysis further and give some plots of the electromagnetic field variation across the cross-section of the waveguide which clearly show the field-displacement effect. They are also able to identify regions of circularly polarised magnetic field in the ferrite and show that the power flow is displaced towards that edge for which the scalar permeability is $(\mu + \kappa)$ and away from that for which the permeability is $(\mu - \kappa)$.

6.8 Perturbation theory

For certain conditions of ferrite control devices in rectangular waveguide carrying high power, the ferrite will absorb power and the full-height ferrite slab of Figs. 6.2 or 6.6 or 6.7 becomes difficult to cool. Therefore many ferrite devices are designed using thin ferrite slabs fixed to the broad walls of the waveguide, as shown in Fig. 6.10.

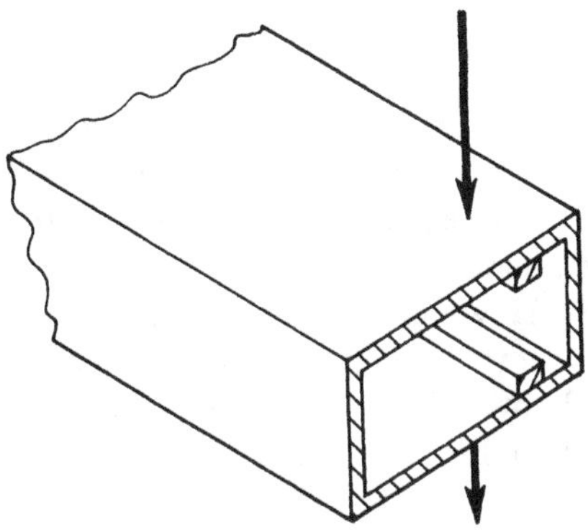

Fig. 6.10 *Partial-height ferrite slabs in rectangular waveguide which will make an isolator or phase changer*

Analysis of the propagating conditions in the waveguide becomes much more difficult. There is no full analysis available. A numerical technique using finite elements[27] has been used to analyse the effect of a dielectric slab in rectangular waveguide and ought to be also applicable to the ferrite problem. For many situations, the cross-section of the ferrite slab is small compared to the cross-sectional area of the rectangular waveguide and it is satisfactory to use perturbation techniques to analyse the effect of the ferrite on the propagation constant. This analysis was given by Lax[28] and with more detail by Schlomann.[29] It is also given in the two books by Lax and Button[5] and Helszajn.[6]

Perturbation theory assumes that the ferrite makes only a small change in the propagation constant compared with the propagation constant in empty waveguide, and that, except in the ferrite, the fields are also those of the empty waveguide. Perturbation theory is developed in Chapter 12 mainly in the context of small ferrite samples perturbing the resonant frequency of a cavity. Here the application is slightly different, but Section 12.7 does give the derivation of the perturbation formula. If the propagation constant of the empty waveguide is γ_0, then the propagation constant of the perturbed waveguide is given by eqn. 12.73

$$\gamma + \gamma_0^* = \frac{j\omega \int_{\Delta S} [(\boldsymbol{\Delta\mu} \cdot \boldsymbol{H}) \cdot \boldsymbol{H}_0^* + (\Delta\varepsilon \boldsymbol{E}) \cdot \boldsymbol{E}_0^*] \mathrm{d}S}{\int_S \boldsymbol{z} \cdot (\boldsymbol{E} \times \boldsymbol{H}_0^* + \boldsymbol{E}_0^* \times \boldsymbol{H}) \mathrm{d}S} \tag{6.38}$$

where $\boldsymbol{E}$ and $\boldsymbol{H}$ are the fields in the perturbed system and $\boldsymbol{E}_0^*$ and $\boldsymbol{H}_0^*$ are the complex conjugate of the fields in the empty waveguide, $\boldsymbol{\Delta\mu}$ is the difference in the tensor permeability and $\Delta\varepsilon$ is the difference in the scalar permittivity between the perturbed and unperturbed conditions. When the unperturbed condition is empty waveguide, the difference permeability and permittivity in eqn. 6.38 leads to the use of magnetic and electric susceptibilities as in eqn. 12.73. Provided that the perturbation is small, the use of unperturbed values for the field quantities in the denominator of eqn. 6.38 is valid and the expression in the denominator becomes

$$2\int_S \boldsymbol{z}\cdot(\boldsymbol{E}_0 \times \boldsymbol{H}_0^*)\,\mathrm{d}S \tag{6.39}$$

However, a similar substitution for the unperturbed values of the field quantities for the actual values of the field quantities in the ferrite will lead to errors. There is a large discontinuity of the permeability and permittivity at the boundary of the ferrite and the field components inside the ferrite are appreciably different from the unperturbed values. However, it is easy to calculate the internal fields in terms of the external field components if the demagnetising factors due to the shape of the ferrite slab are known. Then the terms in the numerator which are products of the susceptibility and the internal field are equal to the product of an external susceptibility and the unperturbed field

$$\Delta\boldsymbol{\mu}\cdot\boldsymbol{H} = \boldsymbol{\chi}\cdot\boldsymbol{H} = \boldsymbol{\chi}^{\mathrm{e}}\cdot\boldsymbol{H}_0 \tag{6.40}$$

where $\boldsymbol{\chi}^{\mathrm{e}}$ is the external susceptibility, and a similar equation relates $\Delta\varepsilon$ to the external dielectric susceptibility $\chi_{\mathrm{e}}^{\mathrm{e}}$.

The form of the external susceptibility may be derived from a consideration of the internal and external fields of the ferrite in terms of the demagnetising factors N_x, N_y and N_z of the ferrite shape. A similar analysis of the demagnetising factors for a small ferrite sample in a cavity is given in Section 12.3. Then the components of the internal fields $H_{\mathrm{i}x}$ and $H_{\mathrm{i}z}$ are given by

$$H_{\mathrm{i}x} = H_{\mathrm{o}x} - \frac{N_x M_x}{\mu_0} = H_{\mathrm{o}x} - \frac{N_x}{\mu_0}(\chi H_{\mathrm{i}x} + \mathrm{j}\kappa H_{\mathrm{i}z}) \tag{6.41}$$

$$H_{\mathrm{i}z} = H_{\mathrm{o}z} - \frac{N_z M_z}{\mu_0} = H_{\mathrm{o}z} - \frac{N_z}{\mu_0}(-\mathrm{j}\kappa H_{\mathrm{i}x} + \chi H_{\mathrm{i}z}) \tag{6.42}$$

Rearranging these equations gives expressions for the internal field components

$$H_{\mathrm{i}x} = \frac{(\mu_0 + N_z\chi)H_{\mathrm{o}x} - \mathrm{j}N_x\kappa H_{\mathrm{o}z}}{D} \tag{6.43}$$

$$H_{\mathrm{i}z} = \frac{(\mu_0 + N_x\chi)H_{\mathrm{o}z} + \mathrm{j}N_z\kappa H_{\mathrm{o}x}}{D} \tag{6.44}$$

where

$$D = \mu_0 + (N_x + N_z)\chi + N_x N_z(\chi^2 - \kappa^2)/\mu_0$$

Substituting into eqn. 6.40 enables us to obtain expressions for the external permeability and permittivity

$$\chi^e_{xx} = \frac{\mu_0\chi + N_z(\chi^2 - \kappa^2)}{D} \tag{6.45}$$

$$\chi^e_{zz} = \frac{\mu_0\chi + N_x(\chi^2 - \kappa^2)}{D} \tag{6.46}$$

$$\chi^e_{xz} = -\chi^e_{zx} = \frac{j\kappa\mu_0}{D} \tag{6.47}$$

Waldron[30] has shown that the demagnetising factors are independent of the direction of the static magnetisation in the ferrite and are functions of shape alone. By definition they obey the relationship

$$N_x + N_y + N_z = 1 \tag{6.48}$$

Demagnetisation factors can only be calculated exactly for bodies of ellipsoidal shape and are given in Fig. 12.1; other shapes such as rods and disks are approximated by a suitable ellipsoidal shape. The demagnetisation factors of the three common shapes shown in Fig. 6.11 are

$$\left.\begin{array}{llll} \text{sphere} & N_x = N_y = N_z = \frac{1}{3} & & \\ \text{rod} & N_x = N_y = \frac{1}{2}, & N_z = 0 & \\ \text{disk} & N_x = N_y = 0, & N_z = 1 & \end{array}\right\} \tag{6.49}$$

For the particular condition of the ferrite strip perturbing the dominant TE_{10}-mode in rectangular waveguide, the external fields are given by[2]

$$\left.\begin{aligned} E_{oy} &= -\frac{j\omega\mu_0 H_0}{k}\sin kx \\ H_{ox} &= \frac{j\beta_0}{k}H_0\sin kx \\ H_{oz} &= H_0\cos kx \end{aligned}\right\} \tag{6.50}$$

These expressions may be substituted into eqn. 6.38 to give[31]

$$\begin{aligned} \gamma_\pm + \gamma_0^* = {} & j\frac{\Delta S}{S\beta_0}(\omega^2\varepsilon_0\mu_0\chi^e_e\sin^2 kx + \beta_0^2\chi^e_{xx}\sin^2 kx \\ & + k^2\chi^e_{zz}\cos^2 kx \pm j\chi^e_{xz}k\beta_0\sin 2kx) \end{aligned} \tag{6.51}$$

The $\pm$ signs show that normally the propagation is nonreciprocal. Taking the real part of eqn 6.51 will give the attenuation component of the propagation constant and will involve the imaginary components of the complex external susceptibility.

Taking the imaginary part of eqn. 6.51 gives an equation for the propagation constant in the ferrite loaded waveguide. Such an expression is given by Helszajn,[31] but only consists of eqn. 6.51 with the real components of the complex susceptibility. However, in considering an expression for the differential phase change, only one term in eqn. 6.51 is different for the two directions of propagation, and we have, for the differential phase change

$$\beta_+ - \beta_- = 2\frac{\Delta S}{S}\mathrm{j}\chi^{e'}_{xz}\, k \sin 2kx \qquad (6.52)$$

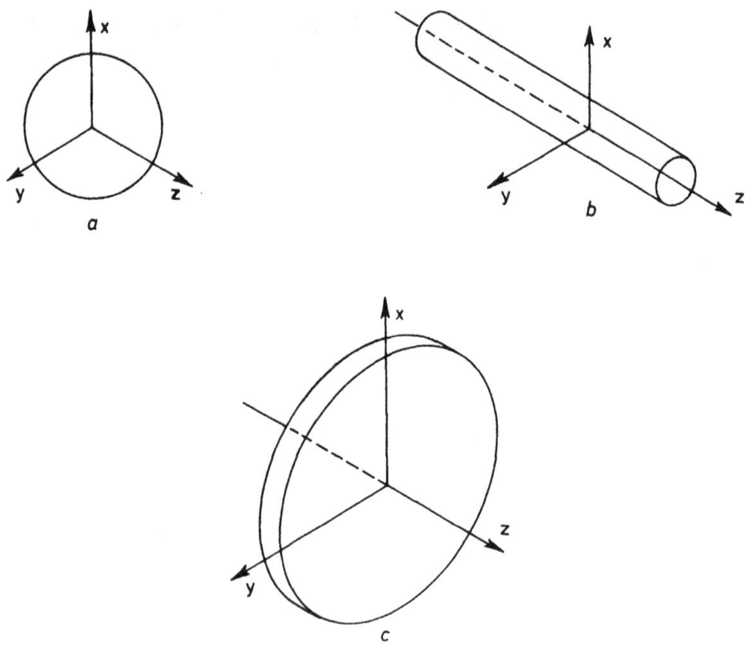

Fig. 6.11 *Some simple shapes whose demagnetisation factors are given in eqn. 6.49*
a Sphere
b Rod
c Disk

For a thin, wide ferrite slab parallel to the broad wall of the waveguide, which may be assumed to have the demagnetisation factors of a ferrite disk, eqn. 6.52 reduces to

$$\beta_+ - \beta_- = 2\frac{\Delta S}{S}\kappa' k \sin 2kx \qquad (6.53)$$

6.9 Resonance isolation

Consider the single slab of ferrite in rectangular waveguide shown in Fig. 6.2. If the ferrite is magnetised to its resonance condition, it will absorb power from the microwave field propagating in the waveguide. Because the magnetic field of a wave propagating in the opposite direction is circularly polarised in the opposite sense, resonance absorption only occurs for a wave propagating in one direction along the waveguide. Such a device becomes a resonance absorption isolator. The forward wave is propagated through the device with a minimum loss and the reverse wave is attenuated by resonance absorption in the ferrite. The design of full-height ferrite slab resonance isolators follows from the same theory as that used to design full-height ferrite slab phase changers, given earlier in this chapter. It becomes necessary to consider the loss components of the permeability tensor. Such a design is given by Gardiol and Vander Vorst.[14]

However, if the resonance isolator is to be used in a microwave system using an appreciable quantity of microwave power, it becomes difficult to cool a ferrite slab which has absorbed the microwave power. Therefore most resonance isolators are designed using thin ferrite slabs fixed to the waveguide walls, as shown in Fig. 6.10, so that perturbation theory is also used in the design of resonant isolators. The attenuation constant is given by taking the real part of eqn. 6.51. For the resonance isolator, the loss components of the permeability of the ferrite will dominate and the imaginary component of the electric susceptibility will be very small and can be neglected. Expressions for the loss have been given by Lax,[28] Lax and Button[5] and Helszajn[31]

$$\alpha_{\pm} = \frac{\Delta S}{S}\left(\beta^2 \chi^{e''}_{xx} \sin^2 kx + k^2 \chi^{e''}_{zz} \cos^2 kx \pm j\chi^{e''}_{xz}\, k\beta \sin 2kx\right) \tag{6.54}$$

The reverse-to-forward attenuation ratio is given by

$$R = \alpha_+/\alpha_- \tag{6.55}$$

where expressions for α_+ and α_- are obtained from eqn. 6.54. With variation of x, the maximum value of R occurs when

$$\tan^2 kx = \frac{k^2 \chi^{e''}_{zz}}{\beta^2 \chi^{e''}_{xx}} \tag{6.56}$$

Then the maximum value of R is obtained by substituting the value of x given by eqn. 6.56 into eqns. 6.55 and 6.54

$$R_{max} = \frac{(\chi^{e''}_{xx}\chi^{e''}_{zz})^{1/2} + \chi^{e''}_{xz}}{(\chi^{e''}_{xx}\chi^{e''}_{zz})^{1/2} - \chi^{e''}_{xz}} \tag{6.57}$$

Therefore the conditions for resonance and the possible performance of a resonant isolator consisting of ferrite loaded rectangular waveguide similar to that shown in Fig. 6.10, are dependent on the shape of the ferrite as well as on the properties of the ferrite material. Schlomann[29] has shown that generally the position of the ferrite for maximum reverse attenuation is not the same as the positon giving minimum forward attenuation, although the position for maximum reverse-to-forward ratio, R_{max}, is nearly coincident with the position for minimum forward attenuation.

Further information may be obtained from eqns. 6.56 and 6.57 when some simplifying assumptions are made. If it is assumed that the ferrite consists of thin slabs in the H-plane, $N_x = N_z = 0$, and $\chi^e_{xx} = \chi^e_{zz} = \chi$, and $\chi^e_{xz} = -j\kappa$. Then the position for R_{max} is given by

$$\tan kx = k/\beta \tag{6.58}$$

which is the position for circular polarisation in empty guide and is the same as eqn. 6.3. Then

$$R_{max} = \frac{\chi'' + \kappa''}{\chi'' - \kappa''} \tag{6.59}$$

Substituting from eqns. 2.24 and 2.26 into eqn. 6.59 and using eqn. 2.27 gives

$$R_{max} = \left(\frac{4\omega_0}{\gamma \Delta H}\right)^2 \tag{6.60}$$

so that the maximum reverse-to-forward attenuation ratio is inversely proportional to the square of the resonance linewidth of the ferrite material. The simplifications leading to eqns. 6.58 and 6.60 assume that the magnetic field inside the ferrite is circularly polarised. For less simple geometries, substitution from eqn. 6.50 into eqns. 6.43 and 6.44 gives an expression for the ratio of these internal fields

$$\frac{H_{ix}}{H_{iz}} = \frac{j(\mu_0 + N_z\chi)\beta_0 \sin kx + jN_x\kappa k \cos kx}{(\mu_0 + N_x\chi)k \cos kx + N_z\kappa\beta_0 \sin kx} \tag{6.61}$$

For there to be circular polarisation, this ratio must equal $\pm$ j, and the position of circular polarisation is given by

$$\tan kx = \pm \frac{k}{\beta_0}\left[\frac{\mu_0 + N_x(\chi \mp \kappa)}{\mu_0 + N_z(\chi \mp \kappa)}\right] \tag{6.62}$$

For long thin strips of ferrite in the waveguide, $N_z = 0$ and eqn. 6.62 simplifies to give real solutions to the equation. Also, when the ferrite is being used for phase change a long way away from resonance, solutions to eqn. 6.62 are possible because the imaginary components of the complex susceptibility are small. However, for the design of resonance isolators, the imaginary components of the complex susceptibility cannot be ignored and eqn. 6.62 may not have real solutions. In fact the internal field is not circularly polarised when $N_z \neq 0$ and, as we have already seen, the ferrite strip should not be placed at the point of circular polarisation in the empty waveguide.

One problem in resonance isolators is that when the ferrite absorbs the microwave power, it heats up. Because the saturation magnetisation and hence the field required for ferrimagnetic resonance is temperature sensitive, the isolator performance will deteriorate as it warms up. In an attempt to overcome this temperature sensitivity, Haken and Haza-Radlitz[32] have investigated the optimum shape of the ferrite slab in order to reduce the temperature sensitivity. By calculations involving demagnetisation factors, they show that the temperature dependence of the resonance frequency or the nonreciprocal phase change or Faraday rotation can be made very small for certain shapes of the ferrite body.

6.10 References

1 BADEN FULLER, A. J. and DOS SANTOS, M. L. X.: 'New method for the display of three-dimensional vector fields', *Proc. IEE*, 1980, **127A**, pp. 435–442

2 See, for example, BADEN FULLER, A. J.: 'Microwaves' 2nd edn. (Pergamon Press, 1979), p. 88

3 KALES, M. L., CHAIT, H. N. and SAKIOTIS, N. G.: 'A nonreciprocal microwave component', *J. Appl. Phys.*, 1953, **24**, pp. 816–817

4 LAX, B., BUTTON, K. J. and ROTH, L. M.: 'Ferrite phase shifters in rectangular wave guide', *J. Appl. Phys.*, 1954, **25**, pp. 1413–1421

5 LAX, B. and BUTTON, K. J.: 'Microwave ferrites and ferrimagnetics' (McGraw-Hill, 1962)

6 HELSZAJN, J.: 'Principles of microwave ferrite engineering' (Wiley, 1969), p. 139

7 CLARRICOATS, P. J. B.: 'Microwave ferrites' (Chapman & Hall, 1961)

8 WALDRON, R. A.: 'Ferrites' (Van Nostrand, 1961)

9 INCE, W. J. and STERN, E.: 'Nonreciprocal remanence phase shifters in rectangular waveguide', *IEEE Trans.*, 1967, **MTT-15**, pp. 87–95

10 SCHLOMANN, E.: 'Theoretical analysis of twin-slab phase shifters in rectangular waveguide', *IEEE Trans.*, 1966, **MTT-14**, pp. 15–23

11 POKUSIN, D. N.: 'Electromagnetic waves in a rectangular waveguide with two ferrite slabs', *Rad. i Elek.*, 1963, **8**, pp. 73–83 (English translation in *Rad. Eng. & Elect. Phys.*, **8**, pp. 66–75)

12 SEIDEL, H.: 'Ferrite slabs in transverse electric mode wave guide', *J. Appl. Phys.*, 1957, **28**, pp. 218–226

13 GARDIOL, F. E.: 'Anisotropic slabs in rectangular waveguides', *IEEE Trans.*, 1970, **MTT-18**, pp. 461–467

14 GARDIOL, F. E. and VANDER VORST, A. S.: 'Computer analysis of E-plane resonance isolators', *IEEE Trans.*, 1971, **MTT-19**, pp. 315–322

15 BRESLER, A. D.: 'On the TE_{n0} modes of a ferrite slab loaded rectangular waveguide and the associated thermodynamic paradox', *IRE Trans.*, 1960, **MTT-8**, pp. 81–95

16 BARZILAI, G. and GEROSA, G.: 'Rectangular waveguides loaded with magnetised ferrite and the so-called thermodynamic paradox', *Proc. IEE*, 1966, **113**, pp. 285–288

17 FOX, A. G., MILLER, S. E. and WEISS, M. T.: 'Behaviour and applications of ferrites in the microwave region', *Bell Syst. Tech. J.*, 1955, **34**, pp. 5–103

18 WEISBAUM, S. and SEIDEL, H.: 'The field displacement isolator', *Bell Syst. Tech. J.*, 1956, **35**, pp. 877–898

19 WEISBAUM, S. and BOYET, H.: 'Field displacement isolators at 4, 6, 11 and 24 KMC', *IRE Trans.*, 1957, **MTT-5**, pp. 194–198

20 BUTTON, K. J.: 'Theoretical analysis of the operation of the field displacement ferrite isolator', *IRE Trans.*, 1958, **MTT-6**, pp. 303–308

21 COMSTOCK, R. L. and FAY, C. E.: 'Operation of the field displacement isolator in rectangular waveguide', *IRE Trans.*, 1960, **MTT-8**, pp. 605–611

22 HINES, M. E.: 'Reciprocal and nonreciprocal modes of propagation in ferrite stripline and microstrip devices', *IEEE Trans.*, 1971, **MTT-19**, pp. 442–451

23 TALISA, S. H. and BOLLE, D. M.: 'On the modeling of the edge-guided mode stripline isolators', *IEEE Trans.*, 1979, **MTT-27**, pp. 584–591

24 TALISA, S. H. and BOLLE, D. M.: 'A five-region model for the edge-guided mode ferrite-loaded stripline isolator', *IEEE Trans.*, 1979, **MAG-15**, pp. 1001–1003

25 BOLLE, D. M. and TALISA, S. H.: 'The edge-guided mode nonreciprocal phase shifter', *IEEE Trans.*, 1979, **MTT-27**, pp. 878–882

26 HELSZAJN, J. and DOWNS, H.: 'Field displacement, circular polarisation, scalar permeabilities and differential phase shift in edge mode ferrite devices', IEE Coll. Ferrite materials devices and techniques, Digest No. 1985/15, 1985, pp. 4.1–4.5

27 WEBB, J. P., MAILE, G. L. and FERRARI, R. L.: 'Finite-element solution of three-dimensional electromagnetic problems', *Proc. IEE*, 1983, **130H**, pp. 153–159

28 LAX, B.: 'Frequency and loss characteristics of microwave ferrite devices', *Proc. IRE*, 1956, **44**, pp. 1368–1386

29 SCHLOMANN, E.: 'On the theory of the ferrite resonance isolator', *IRE Trans.*, 1960, **MTT-8**, pp. 199–206

30 WALDRON, R. A.: 'Electromagnetic fields in ferrite ellipsoids', *Brit. J. Appl. Phys.*, 1959, **10**, pp. 20–22

31 Reference 6, p. 148

32 HAKEN, W. and HAZA-RADLITZ, C.: 'Ferritkorper mit temperaturunabhangigen gyromagnetischen Eigenschaften (Ferrite bodies with temperature independent gyromagnetic properties)', *Archiv der Elektrischen Ubertragung.*, 1959, **13**, pp. 157–160

Chapter 7

Rectangular Waveguide Devices

7.1 Introduction

The general theory of transversely magnetised ferrite in rectangular waveguide is given in Chapter 6. In particular, Fig. 6.1 shows the planes where the magnetic field in the rectangular waveguide is circularly polarised in a plane parallel to the broad face of the waveguide. If a slab of ferrite is placed at one of the these positions of circular polarisation, and magnetised perpendicular to the broad face of the waveguide, the microwave fields in the waveguide see the effective permeability of the ferrite μ^+ and μ^- as shown by Fig. 5.1. Changing the biasing magnetic field on the ferrite will provide variable phase change, resonance absorbtion or field displacement. A possible position for the ferrite in the waveguide is shown in Fig. 6.2. For devices used with high power, or where power is deliberately absorbed in the ferrite, the full-height ferrite slab is difficult to cool and two small ferrite slabs are mounted on the waveguide wall, as shown in Fig. 6.10. The theory of full-height ferrite slab or slabs on their own or in conjunction with full-height dielectric slabs is known and is given in Sections 6.2 to 6.6. No exact theory of the partial height ferrite slabs is known, but since the slabs are usually small, perturbation theory is applicable and is given in Sections 6.8 and 6.9. In the rest of this chapter, it will be assumed that the necessary design theory is known and applied. Generally devices will be described phenomenologically by means of values for μ^+ and μ^- taken from Fig. 5.1.

However, transversely magnetised ferrite devices are not the only ones to give satisfactory performance in rectangular waveguide. It is also possible to provide variable phase change by using a longitudinally magnetised ferrite rod of appropriate diameter at the centre of rectangular waveguide. It appears that some kind of circular waveguide mode is generated in the ferrite which ought to experience Faraday rotation. With the constraint of the rectangular waveguide walls, rotation cannot occur and a large phase change takes place instead. The longitudinally magnetised phase changer is described in Section 7.7.

7.2 Variable nonreciprocal phase changer

The use of nonreciprocal phase changers in phased array antenna systems is widespread. Ince and Temme,[1] as well as describing both ferrite and diode phase changers, describe the various antenna feeds and the application of the phase changers. Three different ferrite devices are used to provide phase change:

1 a transversely magnetised ferrite slab or slabs analysed in Sections 6.2 to 6.5;
2 a longitudinally magnetised ferrite rod in rectangular waveguide described in Section 7.7,
3 a longitudinally magnetised ferrite in circular waveguide described in Section 5.3.

In this section we discuss the transversely magnetised ferrite in rectangular waveguide. In Section 6.1 it has already been shown that the magnetic field in rectangular waveguide is circularly polarised in the plane of the broad face of the waveguide. If a slab of ferrite is placed at the position of circular polarisation

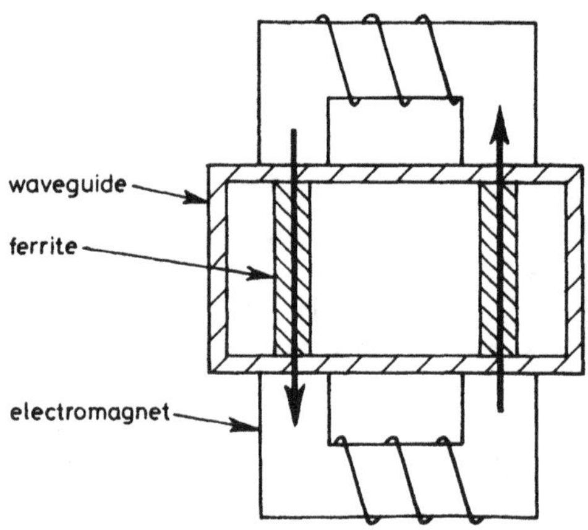

Fig. 7.1 *Ferrite nonreciprocal phase changer*

and magnetised perpendicular to the plane of the circular polarisation, the effective permeability of the ferrite may be controlled by varying the strength of the biasing magnetic field. Because the hand of rotation of the circular polarisation is opposite on opposite sides of the waveguide, two ferrite slabs magnetised in opposite directions on opposite sides of the waveguide give the most efficient device. The arrangement, together with an external biasing control magnet, is shown in Fig. 7.1. Such a device will provide both variable and nonreciprocal phase change.

The theory for the design of the ferrite loaded waveguide shown in Fig. 7.1 is given in Section 6.4, and the most efficient combination of ferrite in waveguide may be chosen. In particular, the theory gives a relationship between the dimensions of the ferrite and waveguide and the propagation constant for specified values of ferrite parameters. Ince and Stern[2] have shown that enhanced performance is possible if the centre region of the waveguide between the two ferrite slabs is filled with a dielectric material, whereas the outer region between the ferrite and the waveguide wall is filled with air. For many control situations the phase delay will need to be switched between particular values and remain at a particular value for some time rather than to be continuously variable. Since Fig. 7.1 depicts a closed magnetic circuit, use may be made of the *square* hysteresis loop of microwave ferrite material, with the ferrite operated in its remanent condition. A typical hysteresis loop is shown in Fig. 7.2. The mode of magnetisation of the phase changer is:

> The ferrite is magnetised to saturation, the point *a* on the hysteresis loop, by a pulse of current in the magnetising coil. When the pulse has ended, the magnetisation will relax to the remanent condition, the point *b* on the hysteresis loop, and remain in that condition.

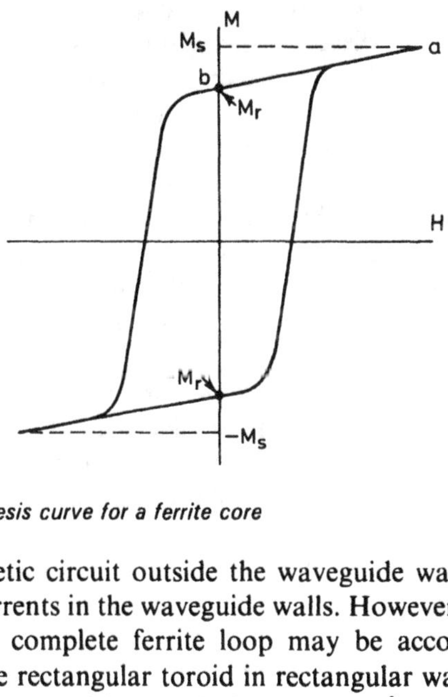

Fig. 7.2 *Typical hysteresis curve for a ferrite core*

With half the magnetic circuit outside the waveguide walls, fast switching is inhibited by eddy currents in the waveguide walls. However, if a large drive coil is not required, the complete ferrite loop may be accomodated inside the waveguide walls. The rectangular toroid in rectangular waveguide is shown in Fig. 7.3. It was first reported by Treuhaft and Silber.[3] The switching current is carried by a single wire along the centre-line of the waveguide which at all times is perpendicular to the electric field in the waveguide and has very little effect on the propagating conditions.

A good description of the latching nonreciprocal phase changer is given in the reprint book by Whicker.[4] A section of waveguide loaded with a ferrite toroid, as shown in Fig. 7.3, will provide two possible values of phase change depending on the direction of the remanent magnetisation in the ferrite. It may be analysed as approximately equal to the twin slab ferrite phase changer shown in Fig. 7.1 which is governed by eqn. 6.33. It is assumed that the section of the ferrite toroid that is magnetised parallel to the broad face of the waveguide does not contribute to the phase change. In the context of designing remanent phase changers,

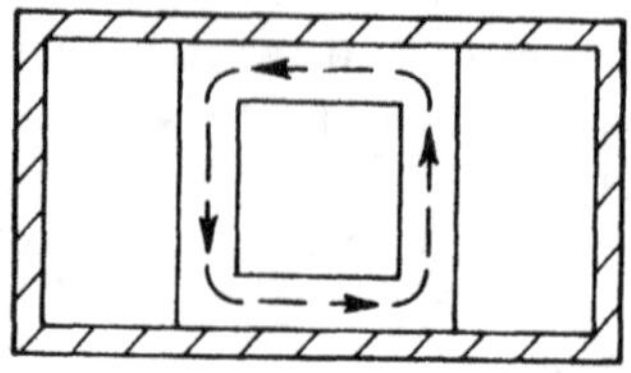

Fig. 7.3 *Latching ferrite phase changer*

Schlomann[5] has applied the theory of eqn. 6.33 to the design of ferrite loaded waveguide devices similar to that shown in Fig. 7.3. However, such a design makes no allowance for the effect of the sections of ferrite adjacent to the broad wall of the waveguide. As they provide no gyromagnetic effect, they may be considered as if they were dielectric material. If the whole centre of the ferrite toroid is filled with dielectric, we approximate to the simpler geometry of two ferrite slabs with a dielectric material between them. The dielectric in the centre is useful in supporting the wire which carries the biasing electric current pulses.

The practical geometry of a remanence or latching phase changer is shown in Fig. 7.4*a*. The wire carrying the biasing electric current pulses is supported by the dielectric material and is everywhere perpendicular to the electric field in the waveguide so that it has negligible effect on the microwave performance of the device. It is analysed by using the theory of Chapter 6 on the simplified geometry of Fig. 7.4*b*. Ince and Stern[2] have analysed such a geometry in the design of latching remanence phase changers. However, differences in the measured differential phase change of 20% or more have been observed compared with that predicted by the simple theory given here. Gardiol[6] has provided a more accurate analysis by applying a variational principle to convert the geometry of Fig. 7.4*a* into that of Fig. 7.4*b*. He obtained good agreement between his theoretical results and experimental results for many designs of practical interest. However, he assumes the magnetic field is entirely confined to the ferrite at the corners and at the top and bottom of the toroid, so that his theory does not give such good results for narrow toroids or toroids with chamfered corners.

For the ferrite toroid having square corners, the region near the outer corner will carry very little magnetic biasing field and will be contributing nothing to the microwave performance of the device. In microwave terms it is only contributing to the loss. As a consequence, Clark[7] investigated the effect of chamfering the corners of the ferrite toroid, as shown in Fig. 7.5. He showed that chamfering the corners not only reduced the microwave attenuation but also increased the differential phase change. The best condition was with $d = w$, which

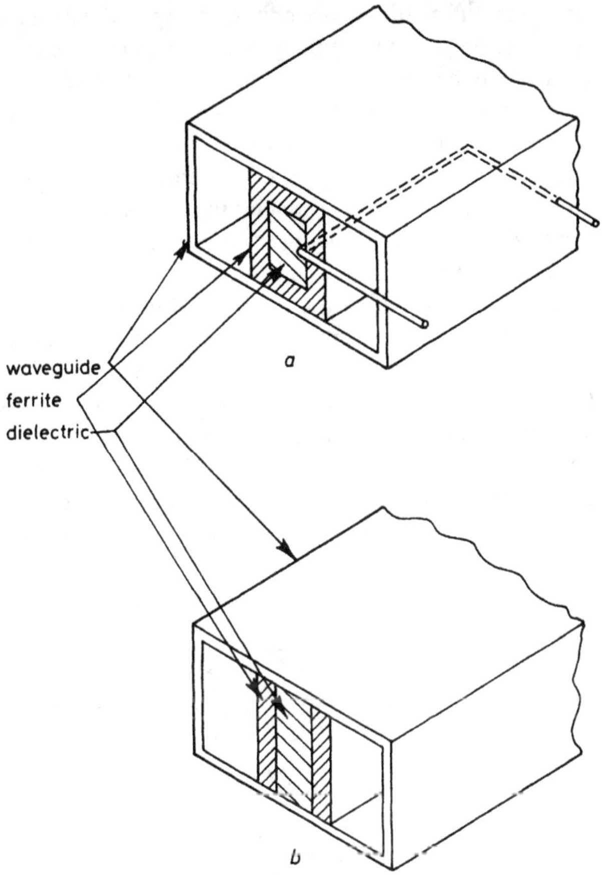

Fig. 7.4 *a* Latching ferrite phase changer
b Simplified for calculation

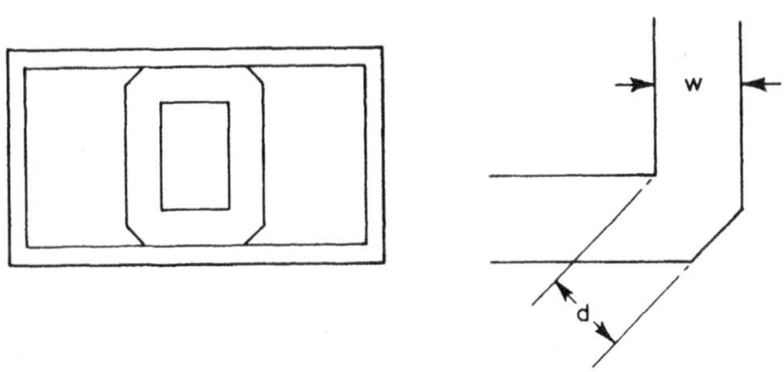

Fig. 7.5 *Ferrite toroid with chamfered corners in a latching ferrite phase changer*

gave a 20% increase in differential phase change compared with a similar device without chamfered corners. This improvement in performance is obtained without any degradation in any other properties of the phase changer. However, Ince, Temme and Willwerth[8] show that his increase of 20% is not normally obtained and was possibly due to the mechanical stresses in the ferrite distorting the remanent magnetic field in the region of the corners. Using fully annealed ferrite toroids, in order to relieve any machining stresses, they show that the increase in differential phase change owing to chamfering varies with the permittivity of the dielectric material filling the centre of the toroid and is not more than 10%. An electrostatic model of the field distribution, computed by Weiss and Bryant,[9] suggests that chamfering the corner redistributes the microwave fields in the region of the corner, making the electric field bend parallel to the remanent field lines. Then the microwave magnetic field will be more transverse to the remanent magnetisation than when the corner is not chamfered.

An alternative approach to the problem of the ferrite at the top and bottom of the toroid contributing nothing to the phase change but only increasing the microwave loss, is given by mounting the ferrite in grooved waveguide, as shown in Fig. 7.6. Then the ferrite which is not contributing to phase change is in a region of reduced microwave field strength, and the figure of merit, being the

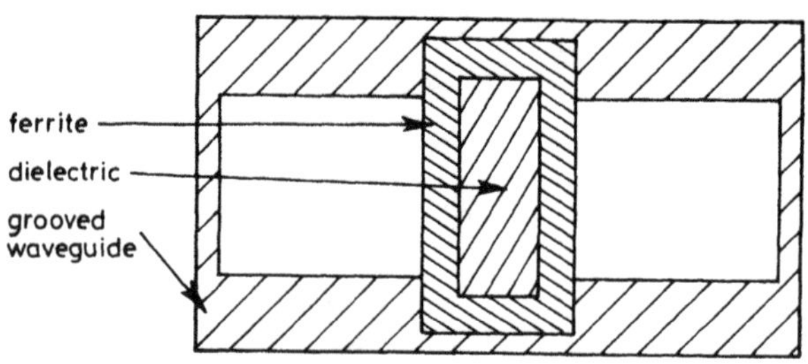

Fig. 7.6 *Ferrite toroid in grooved waveguide*

phase change divided by the attenuation, is increased. Mizobuchi and Kurebayashi[10] have investigated a phase changer in grooved waveguide, similar to that shown in Fig. 7.6, using the theory of Ince and Stern[2] and obtained good agreement between theory and experimental measurements. They used a ferrite toroid of the usual waveguide height in reduced height grooved waveguide and obtain a figure of merit improvement of 20% compared with the phase changer in rectangular waveguide.

The construction of a six-bit latching remanent phase changer is shown in Fig. 7.7. Each section is pulsed to its saturation magnetisation condition and then left in its remanent condition. The length of each section is chosen to provide a given quantity of differential phase change which is then switched on or off. The binary sequence of phase elements is ideally compatible with computer generated control sequences. In order to avoid magnetic interaction, the sections are usually separated by dielectric spacers which also provide an easy path to accomodate the drive wires. Dielectric matching transformers are used at each end. Normally each core is driven with a current pulse generating a magnetising field at least five times the coercive force. However, such a device has limited accuracy because differential phase change errors can occur owing to temperature sensitivity of the remanent magnetisation, variations of remanent magnetisation between different pieces of ferrite, and mechanical tolerances.

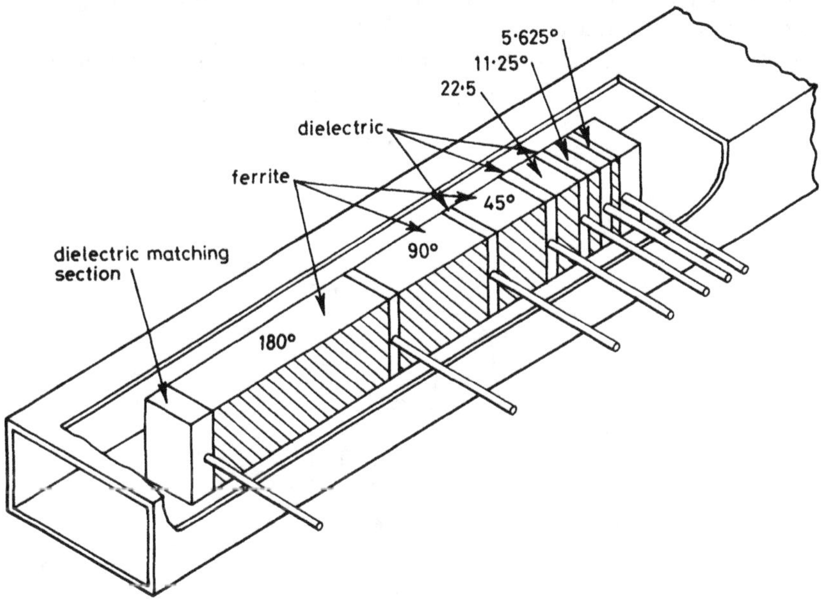

Fig. 7.7 *Six-bit latching remanent phase changer*

An alternative drive system, described by DiBartolo, Ince and Temme,[11] uses one long ferrite toroid threaded by one drive wire in the phase changer. The ferrite is sufficiently long to provide more than the maximum required differential phase change. Then the current pulses in the drive circuit are controlled so as to take the magnetisation only partially round the hysteresis loop and to leave the ferrite magnetised at less than its full remanent magnetisation. Such a system may be adjusted to compensate for variations in the remanent magnetisation owing to temperature variations or manufacturing tolerances. The ferrite core is first reset to its remanent condition with a large positive or negative reset pulse and then set to an intermediate condition by a set pulse of opposite polarity, as shown by Fig. 7.8. The reset pulse sets the magnetisation at the point a. Then a second pulse of the opposite polarity brings the toroid to the point b, and at the end of the pulse the magnetisation will relax to the point c. The magnetisation path b–c is effectively reversible, thus a second pulse of the same polarity but larger would take the magnetisation along the path c to b before continuing up the main hysteresis loop. If $v(t)$ is the instantaneous value of the voltage of the driver pulse, energy conservation shows that

$$\text{Total incremental flux} = \Delta\psi = \int_0^t v(t)\,dt \tag{7.1}$$

ignoring any losses in the electrical drive circuit, and assuming that the leakage inductance is negligible and that the generator impedence is low. The driver circuits are arranged to provide a constant voltage over a variable time to give the required phase delay setting of the phase changer.

In the theoretical design of the latching phase changer by the theory given in Chapter 6, values for μ and κ may be obtained by putting appropriate values for H_0 and M_0 into eqn. 2.15 according to the remanent condition. However, for normal high frequency use, $\omega \gg \gamma H_0$ and eqn. 2.15 simplifies to

$$\mu = \mu_0; \qquad \kappa = \frac{\gamma M_r}{\omega} \tag{7.2}$$

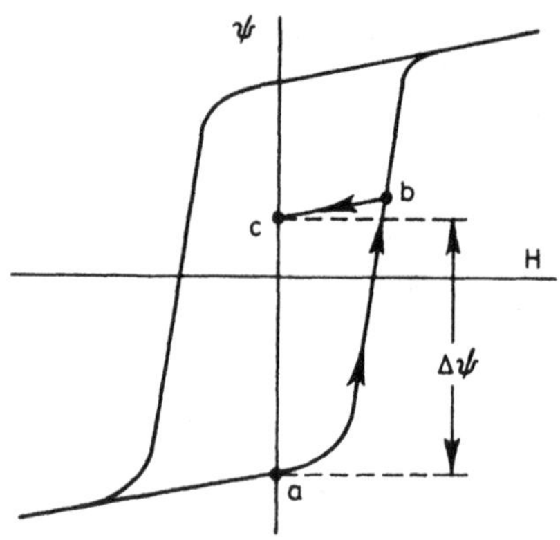

Fig. 7.8 *Hysteresis loop for a partially magnetised ferrite core*

These values are used by Schlomann[5] in his theoretical analysis of the twin-slab ferrite phase changer in rectangular waveguide. However, later he[12] goes on to discuss the microwave behaviour of partially magnetised ferrite material. His results are complicated and a good estimate is obtained by using eqn. 7.2 and a low frequency or static measurement of the remanent magnetisation.

7.3 Circulator

A circulator may be constructed by using a differential phase changer in conjunction with two 3dB directional couplers or hybrid T-junctions. Such a circulator is much more bulky than the equivalent Y-junction circulator, described in Chapter 8. However, because only half the power through the device goes through the ferrite phase changer, and because a rectangular waveguide phase changer can handle more power than a Y-junction circulator, high-power circualtors are constructed using ferrite differential phase change sections in rectangular waveguide. A phase changer having 180° differential phase change is called a *gyrator*. The principle of operation of a circulator using a gyrator was first given by Hogan,[13] and is shown in Fig. 7.9. The signals from the *H*-plane arm of hybrid 1 will arrive in phase at hybrid 2 and will be transmitted to the *H*-plane arm. However, in the reverse direction, signals from the *H*-plane arm

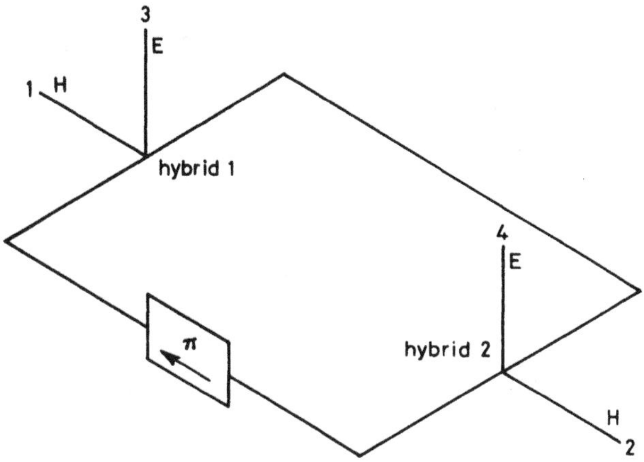

Fig. 7.9 *Circulator using two hybrid T-junctions and a differential phase changer*

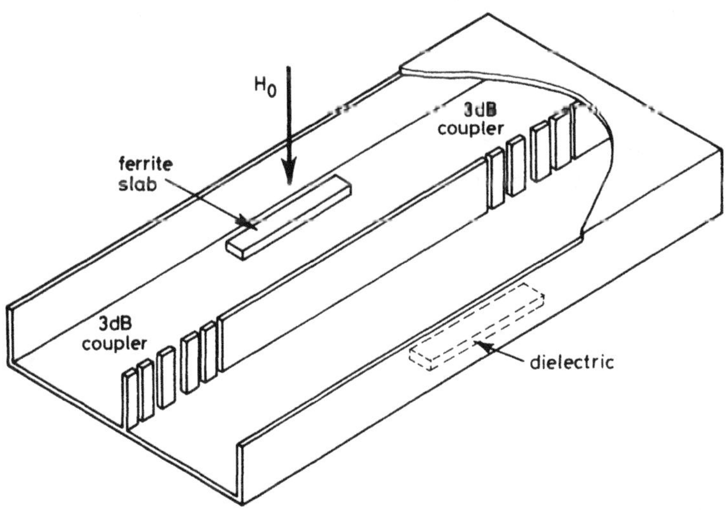

Fig. 7.10 *Circulator using two 3 dB couplers and a differential phase changer*

of hybrid 2 will arrive out of phase at hybrid 1 and will be transmitted to the *E*-plane arm. A compact device in rectangular waveguide using two sidewall 3 dB couplers was described by Fox, Miller and Weiss[14] and is shown in Fig.

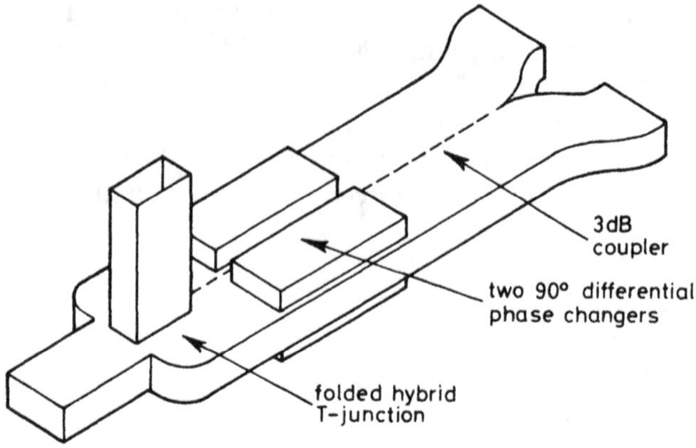

a

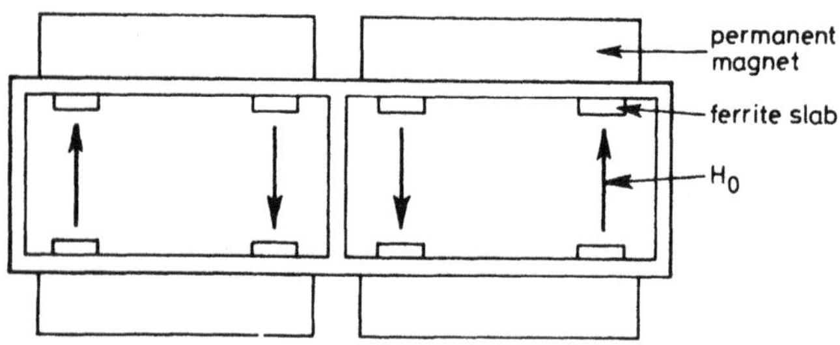

b

Fig. 7.11 *a* Circulator using a folded hybrid T-junction, a 3 dB coupler and two differential phase changers
b Possible construction of the two differential phase changers

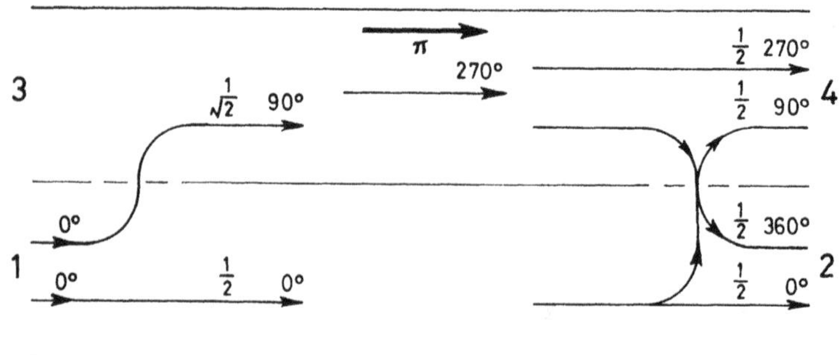

Fig. 7.12 *Mode of operation of the circulator shown in Fig. 7.10*

7.10. The most compact arrangement is attributed to Fox and is described by Lax and Button.[15] It consists of a folded hybrid T-junction, two 90° differential phase changers and a sidewall 3 dB coupler, and is shown in Fig. 7.11.

The mode of operation of the sidewall coupler device, shown in Fig. 7.10, is given in Fig. 7.12. Each 3 dB sidewall coupler introduces a 90° phase change in the signal coupled to the side arm. A forward signal experiences an additional 180° phase change in the differential phase changer and it is seen that the outputs add in the original main line to give an output at port 2. A reverse wave does not experience the 180° phase change in the differential phase changer and comes out of the side arm in output 3. The mode of operation of the folded hybrid T-junction device, shown in Fig. 7.11, is given in Fig. 7.13. An input to the *H*-plane arm of the folded hybrid T-junction will be equally split between the two arms of the circulator in phase. The signal in one arm will experience a 90° differential phase change which will then combine with the 90° phase change in the sidewall coupler to provide an output in that arm and no output in the other arm where the two signals are out of phase. A reverse wave will experience the 90° differential phase change in the other arm of the device, the two signals will arrive at the hybrid T-junction out of phase and be transmitted to the *E*-plane arm.

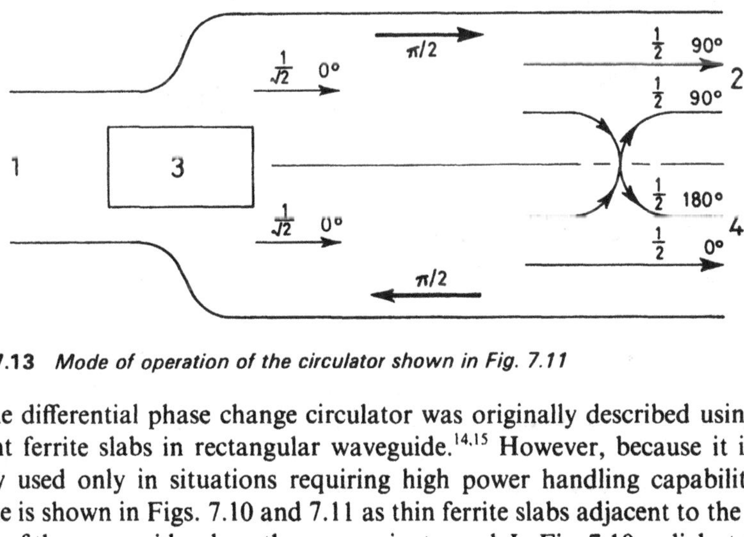

Fig. 7.13 *Mode of operation of the circulator shown in Fig. 7.11*

The differential phase change circulator was originally described using full-height ferrite slabs in rectangular waveguide.[14,15] However, because it is normally used only in situations requiring high power handling capability, the ferrite is shown in Figs. 7.10 and 7.11 as thin ferrite slabs adjacent to the broad walls of the waveguide where they are easier to cool. In Fig. 7.10, a dielectric slab is shown which is used to compensate for the insertion phase change of the differential phase changer. In the circulator shown in Fig. 7.11, the insertion phase of the two differential phase changers are equal. The performance of any of the circulators described in this section is dependent on the performance of the folded hybrid T-junction and the 3 dB couplers as well as on the performance of the differential phase changers.

The theoretical design of the ferrite slab differential phase changer has been discussed in Section 6.8, where perturbation theory is used to analyse a similar geometry. However, the volume of the ferrite is not really small, and the theoretical analysis no longer adequately represents the actual fields in the waveguide. Gardiol[16] has given an improved analysis of a slightly different geometry, shown in Fig. 7.14*a*.

The figure of merit, defined as the ratio of the differential phase change to the attenuation, increases when dielectric loading is added between the ferrite slabs as shown in Fig. 7.14*a*. Gardiol[16] has analysed the geometry of Fig. 7.14*a* as a perturbation of the dielectric loaded waveguide shown in Fig. 7.14*b*, for which an exact solution is available.[17] He[16] gives the mathematical expressions from which the differential phase change and the attenuation may be computed. The analysis will be most accurate when the dielectric material has the same permittivity as the ferrite, but even for other situations it gives useful results.

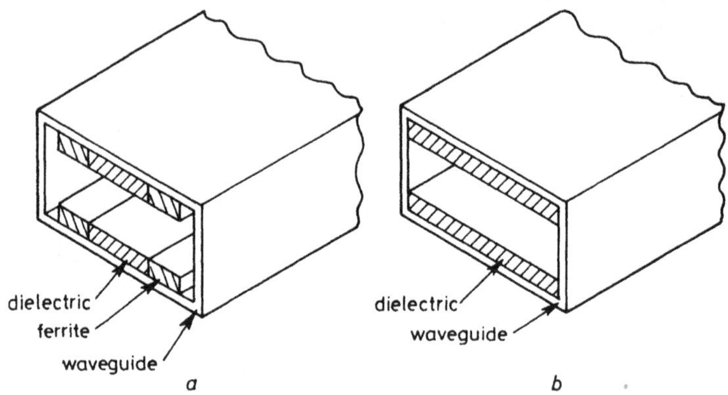

Fig. 7.14 *a* Dielectric loaded phase changer analysed by Gardiol[26]
b Dielectric loaded waveguide

The differential phase changer described in this section can also be used to construct a waveguide switch. If the two differential phase change sections in Fig. 7.11 are 90° latching phase changers, the channel to which the input signal is transmitted is dependent on the setting of the two differential phase changers. In this situation, the phase changers are probably not designed specially to handle high power but are remanent latching toroids, as shown in Figs. 7.3 to 7.5. Such a device has been described by Landry and Passaro.[18]

7.4 Variable reciprocal phase changer

Although the nonreciprocal phase changer shown in Figs. 6.1, 6.4, 6.6, 6.10, 7.1 and 7.11*b* is one of the most convenient methods of providing variable phase change in rectangular waveguide, nonreciprocity may not be required and may be a distinct disadvantage. Then a form of ferrite loaded waveguide is required which will provide variable reciprocal phase change. One such device is the longitudinally magnetised ferrite rod in rectangular waveguide described in Section 7.7. However, there is also a configuration of transversely magnetised ferrite slabs in rectangular waveguide which provides reciprocal phase change. It is shown in Fig. 6.7. A comparison between the reciprocal and the non-reciprocal phase changer is shown in Fig. 7.15. The ferrite on the left-hand side of the waveguide is magnetised in the same direction in both devices and will have the same effective permeability to the wave in the waveguide. However, the ferrite slab on the right-hand side of the waveguide is magnetised in the opposite direction in the two systems. For the reciprocal device, Fig. 7.15*a*, the ferrite on the right-hand side interacts with a magnetic field which is rotating in the opposite direction to that on the left, and its effective permeability will be different to that on the left. The effective permeability of the two ferrite slabs will be μ^+ and μ^-, respectively, as given in Fig. 5.1. A reverse wave will experience exactly the same effective permeability as a forward wave and the device will be variable but reciprocal. For the nonreciprocal device shown in Fig. 7.15*b*, on the other hand, both ferrite slabs will present an effective permeability of μ^+ to the forward wave and an effective permeability of μ^- to the reverse wave. No application of the transversely magnetised reciprocal phase changer is known because the longitudinally magnetised ferrite rod in rectangular waveguide, described in Section 7.7, gives a simpler and more compact device.

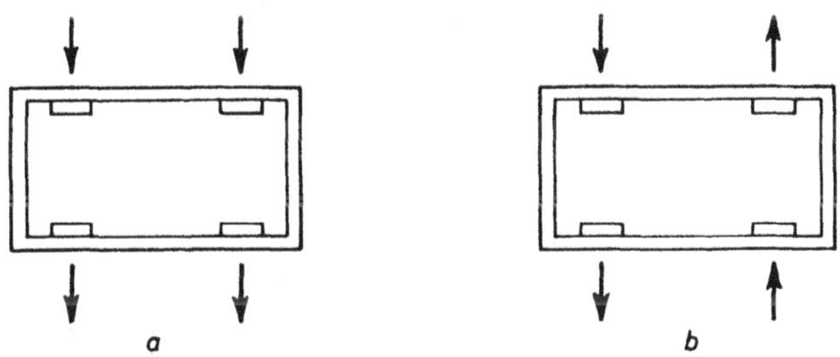

Fig. 7.15 *Phase changers in rectangular waveguide*
a Reciprocal
b Nonreciprocal

7.5 Resonance isolator

For the resonance isolator in rectangular waveguide the ferrite is positioned in the circularly polarised magnetic field and magnetised perpendicular to the broad face of the waveguide to its resonance condition. Then, for a wave travelling in one direction, the circularly polarised magnetic field interacts with the precessing electrons in the ferrite to excite them to resonance, whereas, for a wave travelling in the opposite direction, there is very little interaction between the precessing electrons and the wave in the waveguide. The wave travelling in one direction is attenuated as its power is absorbed by the resonance in the ferrite whereas the wave travelling in the opposite direction is very little affected. Because the magnetic field for resonance is much larger than that required for a transversely magnetised phase changer, the resonance isolator requires a large magnet and its construction is often as shown in Fig. 7.16. The thin ferrite slab adjacent to the broad wall of the waveguide, as well as aiding cooling of the ferrite when it is heated by absorbing power from the wave in the waveguide, also gives a better broadband performance and less deterioration of performance with variation of temperature compared to a full-height slab in the waveguide. However, the thin ferrite slab shown in Fig. 7.16 does require a much larger external magnetic field to bias it to resonance than the full-height ferrite slab shown in Fig. 6.2.

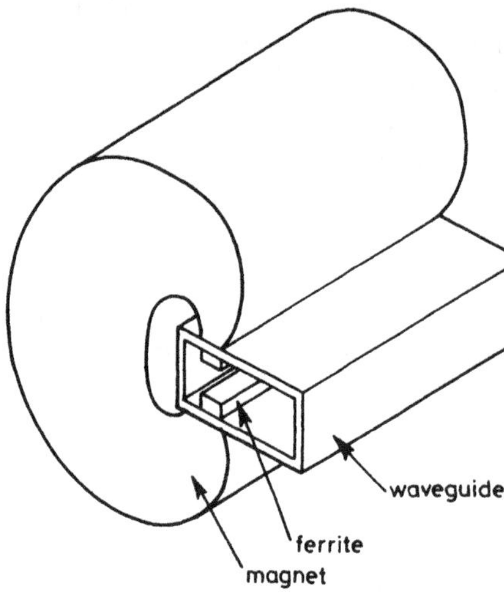

Fig. 7.16 *Resonance isolator*

The theory for the design of the resonance isolator is given in Section 6.9 and in the paper by Schlomann,[19] and the factors affecting the performance of such isolators are discussed in detail in the books by Clarricoats[20] and Lax and Button.[21] Resonance in ferrite materials is essentially a narrowband effect, and therefore most ingenuity has been spent in broadbanding resonance isolators. There are two approaches which may be used in broadbanding ferrite isolators: one is to use dielectric inserts in the waveguide so that the position of circular polarisation does not alter with change in frequency, and the other is to taper the field acting on the ferrite so that different parts of the ferrite are resonant at different frequencies. An example of the first technique is given by Grimes *et al.*,[22] who describe an experimental isolator in ridge waveguide having an octave bandwidth. Its construction is shown in Fig. 7.17, where the ridge waveguide keeps a very similar field pattern in the region between the ridges over a wide frequency range. Unfortunately, the microwave magnetic field of the dominant mode in emtpy ridge waveguide is not circularly polarised in a horizontal plane near to the walls of the reduced height section, but a dielectric insert is used, as shown in Fig. 7.17, which ensures that the field is circularly polarised in the ferrite. The dielectric used was alumina, $\varepsilon_r \sim 10$, and its optimum thickness was about half the width of the waveguide ridge. Using double ridge waveguide has the added advantage that the external biasing magnetic field is applied across only a small airgap.

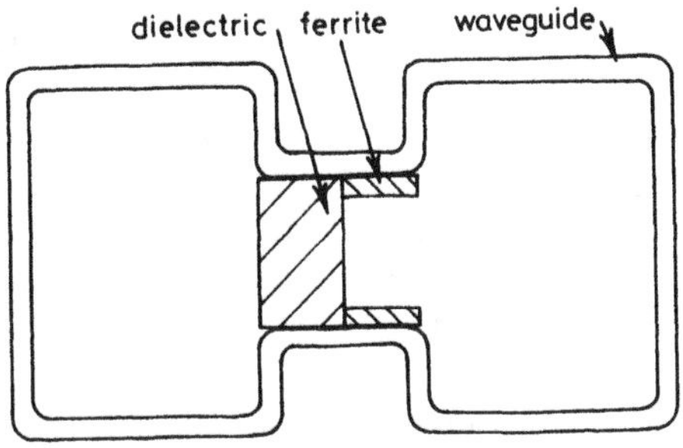

Fig. 7.17 *Resonance isolator in ridge waveguide*

Providing a different magnetic field on different parts of the ferrite slab is another way of achieving a broadband performance because different parts of the ferrite are resonant at different frequencies. As well as using a magnet with a tapered magnetic field along its length, the strength of the magnetic field in the ferrite may be altered by varying the thickness of the ferrite slab along its length or by stepping its thickness along its length. A further alternative is to use a ferrite slab with a varying thickness in its cross-section. Boronski[23] reported broadband isolator performance using ferrite strips of triangular cross-section.

7.6 Field-displacement isolator

The theory of field displacement due to a transversely magnetised ferrite slab in rectangular waveguide has been given in Section 6.6. The field distribution in the cross-section of the waveguide is given in Fig. 6.8. If an absorbing vane or an absorbing film is placed on one face of the ferrite slab, the wave travelling in one direction has a peak field strength at the position of the absorber and is attenuated, while the wave travelling in the reverse direction has a minimum field strength at the position of the absorber and is unattenuated. The field-displacement isolator has two advantages over the resonance isolator: the magnetic bias field used is much less than that needed for resonance, and the field-displacement effect is much less frequency sensitive than resonance.

The original design of field-displacement isolator by Weisbaum and Seidel[24] and Weisbaum and Boyet[25] gave a very good performance of 0·2 dB forward loss and better than 30 dB isolation over a 10% band. They used a less than full-height ferrite slab mounted away from the wall of the rectangular waveguide, as shown in FIg. 7.18, and shaped the resistive film so that it did not extend the full distance across the height of the ferrite slab. It was found that reducing the height of the ferrite slab from full height decreased the reflection coefficient owing to the ferrite in the waveguide and enabled a device with VSWR ~ 1·05 to be produced. The optimum design was very little different from that given by the theory in Section 6.6 and eqn. 6.29. Comstock and Fay[26] have analysed the full-height ferrite field-displacement isolator including the effect of the resistive sheet in their analysis. They studied both propagation constant and the variation of the field in the cross-section of the waveguide and obtained reasonable agreement with experimental measurements at 6 GHz.

Usually the field-displacement isolator is only suitable for low power applications with the ferrite completely removed from any cooling effect of the waveguide walls. For most applications now, the field-displacement isolator has probably been displaced by low power isolators based on variations of the Y-junction circulator described in Sections 8.1 and 8.5.

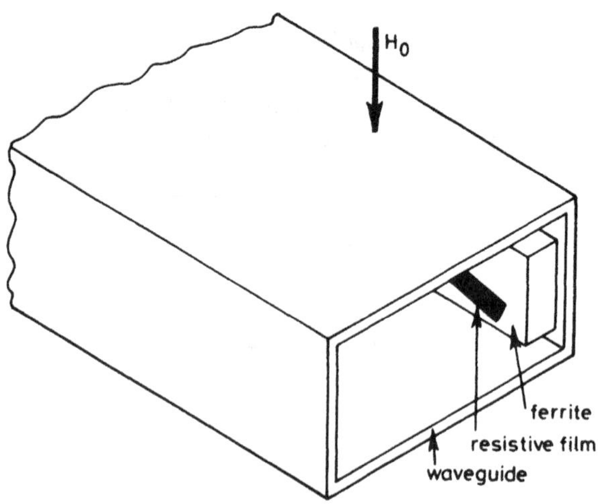

Fig. 7.18 *Field-displacement isolator*

7.7 Longitudinally magnetised ferrite – reciprocal phase changer

If a ferrite rod or slab of sufficient size is placed along the centre-line of rectangular waveguide and magnetised longitudinally, a large reciprocal phase change occurs. The effect was first reported by Bush[27] and extensive experimental results are given by Reggia and Spencer,[28] from whence the name Reggia–Spencer phase shifter[29] was adopted. Reggia and Spencer[28] showed that at 9·1 GHz with ferrite rods of various sizes in R 100 waveguide (0·9 × 0·4 inch inside dimensions) the phase change varied with size of ferrite rod for a fixed magnetic field as shown in Fig. 7.19. Up to a ferrite rod diameter of 7 mm the attenuation was small and largely attributable to the mismatch at the ends of the device as monitored by the input VSWR. Larger ferrite rods gave attenuation peaks as the magnetic field was varied, indicating that higher-order modes are being excited. A device similar to that shown in Fig. 7.20 with tapers on each end of the ferrite rod had a 5% bandwidth. Lax and Button[30] show how perturbation theory may be applied to this geometry and obtain good correlation between the experimental results of Bush[27] and their theoretical results.

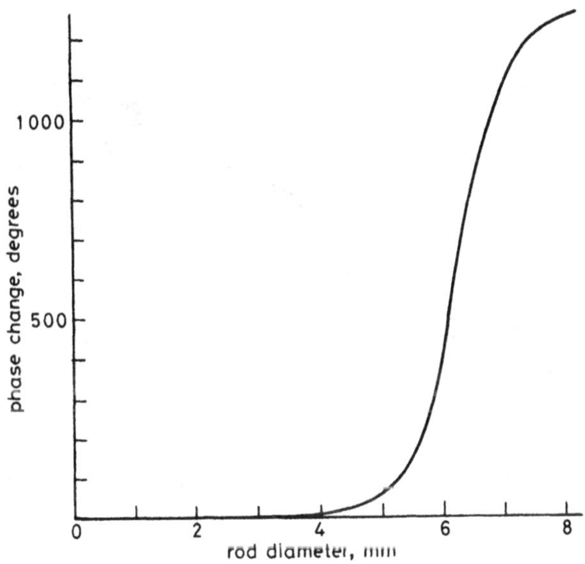

Fig. 7.19 *Longitudinally magnetised ferrite rod phase changer in rectangular waveguide – phase change plotted against diameter of the rod for 100 mm long rods magnetised at 2 A/m. The results are plotted from Reggla and Spencer*[28]

Rizzi and Gatlin[31] infer from the results in Fig. 7.19 that the wave is concentrated in the ferrite rod for rod diameters greater than 5 mm. They further suggest that the large phase changes obtained are due to circularly polarised modes excited in the ferrite, which are converted into large phase changes by the geometrical constraints of the rectangular waveguide. In particular, they set up a longitudinally magnetised phase changer having a discontinuity in the rectangular waveguide half way along the ferrite section. When the ferrite was magnetised they varied the orientation of the output waveguide with respect to the input with negligible loss of output power. In particular, when the two sections of waveguide were oriented with their *E*-planes perpendicular, so that in the absence of the magnetised ferrite rod there would be no propagation through the device, there was satisfactory propagation with low loss provided that the biasing magnetic field was above a certain minimum value. Weiss[32] shows how such a junction between two rectangular waveguides having their E-planes mutually perpendicular with a longitudinally magnetised ferrite rod in the centre of the junction can be used as a waveguide switch having an isolation ~ 60 dB in the 'off' state.

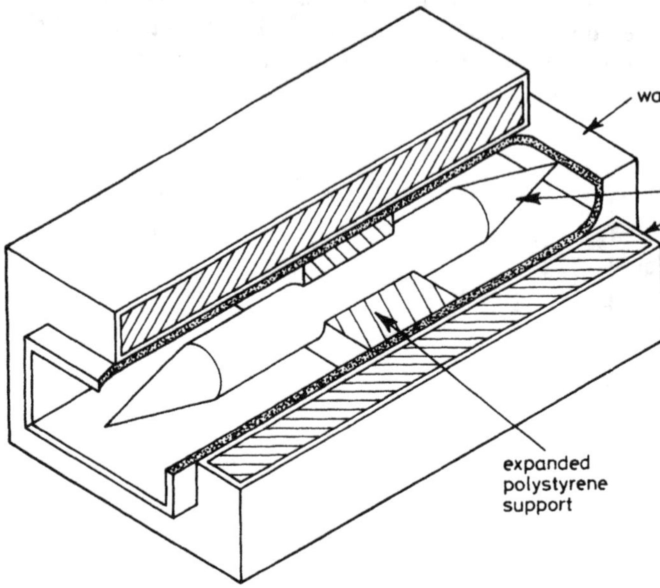

Fig. 7.20 *Longitudinally magnetised ferrite rod phase changer in rectangular waveguide*

Another switch has been described by Reggia[33] which also uses the longitudinally magnetised ferrite rod phase changer shown in Fig. 7.20. If the ferrite rod in the waveguide of Fig. 7.20 is split longitudinally and an absorbing film is inserted all along one diameter of the rod, and the rod is positioned so that

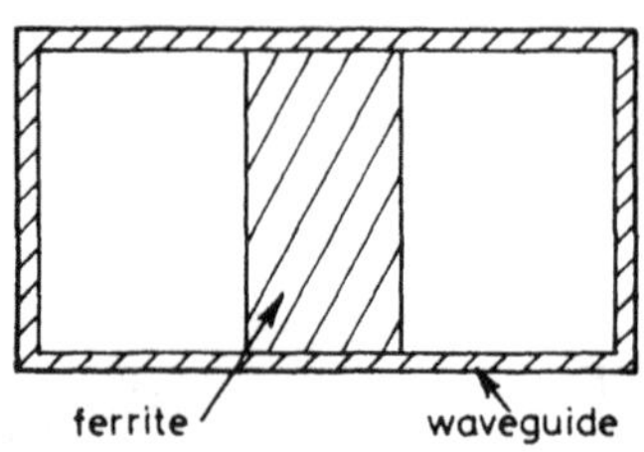

Fig. 7.21 *Longitudinally magnetised ferrite slab in rectangular waveguide*

the absorbing film is parallel to the broad face of the waveguide, the microwave power is absorbed in the film when the biasing magnetic field is above a certain minimum value. With the magnetic field off, the dominant mode is propagated through the device with little loss. With the magnetic field on, a perpendicular mode is generated in the ferrite which is absorbed in the film and the wave is attenuated. Reggia[33] quotes an 'on' state attenuation of 0·5 dB and an 'off' state isolation greater than 60 dB over a 30% bandwidth at about 9 GHz.

Hord, Rosenbaum and Boyd[34] have presented a theoretical analysis of the longitudinally magnetised ferrite phase changer of Reggia and Spencer.[28] They calculated the coupling between a slightly distorted dominant mode in the ferrite loaded waveguide and a perturbed cross-polarised mode in the ferrite. They analyse the ferrite loaded waveguide geometry of Fig. 7.21; a full-height ferrite slab being a reasonable approximation to the ferrite rod shown in Fig. 7.20 and being easier to analyse theoretically. They obtain reasonable agreement with experiment. Reggia and Mak[35] describe a latching version of the longitudinally magnetised phase changer but the dual-mode phase changer, described by Boyd[36] and in Section 5.3, is probably more compact. Ferrite phase control components are relatively bulky. In phased array aerial systems where the antenna feed is in waveguide and the power capability of the phase control devices needs to be quite large, ferrite devices are often the best. Where non-reciprocity is also required, there is no alternative to ferrite devices, but in an all solid-state aerial, phase control is probably best provided by diode switches.

7.8 References

1 INCE, W. J. and TEMME, D. H.: 'Phasers and time delay elements', *Advances in Microwaves*, 1969, **4**, pp. 2–189

2 INCE, W. J. and STERN, E.: 'Nonreciprocal remanence phase shifters in rectangular waveguide', *IEEE Trans.*, 1967, **MTT-15**, pp. 87–95

3 TREUHAFT, M. A. and SILBER, L. M.: 'Use of microwave ferrite toroids to eliminate external magnets and reduce switching power', *Proc. IRE*, 1958, **46**, p. 1538

4 WHICKER, L. R.: 'Ferrite control components. Vol. 2' (Artech House, 1974) (includes reprints of References 2, 3, 5, 8, 10, 11, 28, 29, 33–36)

5 SCHLOMANN, E.: 'Theoretical analysis of twin-slab phase shifters in rectangular waveguide', *IEEE Trans.*, 1966, **MTT-14**, pp. 15–23

6 GARDIOL, F. E.: 'Computer analysis of latching phase shifters in rectangular waveguide', *IEEE Trans.*, 1973, **MTT-21**, pp. 57–61

7 CLARK, W. P.: 'A technique for improving the figure-of-merit of a twin-slab nonreciprocal ferrite phase shifter', *IEEE Trans.*, 1968, **MTT-16**, pp. 974–975

8 INCE, W. J., TEMME, D. H. and WILLWERTH, F. G.: 'Toroid corner chamfering as a method of improving the figure of merit of latching ferrite phasers', *IEEE Trans.*, 1971, **MTT-19**, pp. 563–564

9 WEISS, J. A. and BRYANT, T. G.: 'Numerical determination of potential in inhomogeneous dielectrics by Earnshaw's theorem', *IEEE Trans.*, 1970, **MTT-18**, pp. 595–601

10 MIZOBUCHI, A. and KUREBAYASHI, H.: 'Nonreciprocal remanence ferrite phase shifters using the grooved waveguide', *IEEE Trans.*, 1978, **MTT-26**, pp. 1012–1016

11 DIBARTOLO, J., INCE, W. J. and TEMME, D. H.: 'A solid-state "flux drive" control circuit for latching-ferrite-phaser applications', *Microwave J.*, 1972, **15**, 9, pp. 59–64

12 SCHLOMANN, E.: 'Microwave behaviour of partially magnetised ferrites', *J. Appl. Phys.*, 1970, **41**, pp. 204–214

13 HOGAN, C. L.: 'The ferromagnetic Faraday effect at microwave frequencies and its applications', *Bell Syst. Tech. J.*, 1952, **31**, pp. 1–31

14 FOX, A. G., MILLER, S. E. and WEISS, M. T.: 'Behaviour and applications of ferrites in the microwave region', *Bell Syst. Tech. J.*, 1955, **34**, pp. 5–103
15 LAX, B. and BUTTON, K. J.: 'Microwave ferrites and ferrimagnetics' (McGraw-Hill, 1962), p. 590
16 GARDIOL, F. E.: 'Nonreciprocal phaseshift in ferrite loaded rectangular waveguides', *Arch. Elecktron. and Uebertragungstech.*, 1983, **37**, pp. 278–280
17 GARDIOL, F. E. and VANDER VORST, A. S.: 'Wave propagation in a rectangular waveguide loaded with an H-plane dielectric slab', *IEEE Trans.*, 1969, **MTT-17**, pp. 56–57
18 LANDRY, D. H. and PASSARO, W. C.: 'A four-bit latching ferrite switch', *IEEE Trans.*, 1965, **MTT-13**, pp. 866–867
19 SCHLOMANN, E.: 'On the theory of the ferrite resonance isolator', *IRE Trans.*, 1960, **MTT-8**, pp. 199–206
20 CLARRICOATS, P. J.B.: 'Microwave ferrites' (Chapman & Hall, 1961), p. 214
21 LAX, B. and BUTTON, K. J.: 'Microwave ferrites and ferrimagnetics' (McGraw-Hill, 1962), p. 578
22 GRIMES, E. S., BARTHOLOMEW, D. D., SCOTT, D. C. and SLOAN, S. C.: 'Broad-band ridge waveguide ferrite devices', *IRE Trans.*, 1960, **MTT-8**, pp. 489–492
23 BORONSKI, S.: 'Some properties and applications of ferrites at 3 cm wavelength', *Proc. IEE*, 1957, **104B**, Suppl. No. 6, pp. 331–337
24 WEISBAUM, S. and SEIDEL, H.: 'The field displacement isolator', *Bell Syst. Tech. J.*, 1956, **35**, pp. 877–898
25 WEISBAUM, S. and BOYET, H.: 'Field displacement isolators at 4, 6, 11 and 24 KMC', *IRE Trans.*, 1957, **MTT-5**, pp. 194–198
26 COMSTOCK, R. L. and FAY, C. E.: 'Operation of the field displacement isolator in rectangular waveguide', *IRE Trans.*, 1960, **MTT-8**, pp. 605–611
27 BUSH, D.: 'Contribution to discussion on microwave apparatus', *Proc. IEE*, 1957, **104B**, Suppl. No. 6, pp.368–369
28 REGGIA, F. and SPENCER, E. G.: 'A new technique in ferrite phase shifting for beam scanning of microwave antennas', *Proc. IRE*, 1957, **45**, pp. 1510–1517
29 WEISS, J. A.: 'A phenomenological theory of the Reggia-Spencer phase shifter', *Proc. IRE*, 1959, **47**, pp. 1130–1137
30 LAX, B. and BUTTON, K. J.: 'Microwave ferrites and ferrimagnetics' (McGraw-Hill, 1962), p. 349
31 RIZZI, P. A. and GATLIN, B.: 'Rectangular guide ferrite phase shifters employing longitudinal magnetic fields', *Proc. IRE*, 1959, **47**, pp. 446–447
32 WEISS, J. A.: 'The tetrahedral junction as a waveguide switch', *IRE Trans.*, 1960, **MTT-8**, pp. 120–121
33 REGGIA, F.: 'A new broad-band absorption modulator for rapid switching of microwave power', *IRE Trans.*, 1961, **MTT-9**, pp. 343–349
34 HORD, W. E., ROSENBAUM, F. J. and BOYD, C. R.: 'Theory of the suppressed-rotation reciprocal ferrite phase shifter', *IEEE Trans.*, 1968, **MTT-16**, pp. 902–910
35 REGGIA, F. and MAK, T.: 'Reciprocal latching phase modulator for microwave frequencies', *IEEE Trans.*, 1966, **MAG-2**, pp. 269–273
36 BOYD, C. R.: 'A dual-mode latching reciprocal ferrite phase shifter', *IEEE Trans.*, 1970, **MTT-18**, pp. 1119–1124

Chapter 8

Y-junction circulator

8.1 Introduction

The most usual form of the three-port junction circulator consists of a symmetrical Y-junction, as shown in Fig. 8.1, in either rectangular waveguide or stripline. There is a symmetrical distribution of ferrite at the centre of the junction which is magnetised perpendicular to the plane of the Y. Fig. 8.1 shows an *H*-plane Y-junction with a post of ferrite at the centre. The stripline Y-junction is shown in Fig. 8.11. The circuit symbol for a three-port circulator is

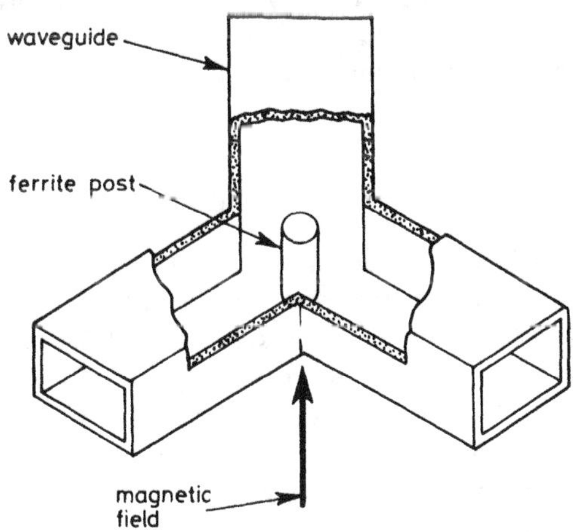

Fig. 8.1 *Waveguide Y-junction circulator*

shown in Fig. 1.1, but is repeated as Fig. 8.2 for convenience. For a circulator operating perfectly, a signal entering port 1 will be transmitted to port 2 without loss and nothing will come out of port 3. Similarly, a signal entering port 2 will be transmitted to port 3 etc.

An empty Y-junction cannot be matched without altering the impedance of the exit ports and such a device is no longer symmetrical. Provided that there is some gyratory material symmetrically placed within the junction, a symmetrical junction can be matched and will act as a circulator. Usually the gyratory effect is provided by the gyromagnetic effect of a ferrite material magnetised

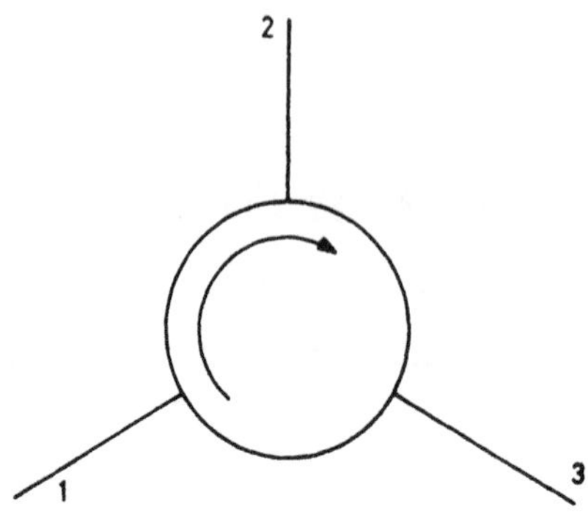

Fig. 8.2 *Circulator circuit symbol*

perpendicular to the plane of the Y. A very simple minded (and slightly inaccurate) description of the mode of operation of the Y-junction circulator in rectangular waveguide may be given by reference to the reciprocal phase changer described in Section 7.4 and Fig. 7.15a. Assume that the system shown in Fig. 7.15a is not constrained by any waveguide walls. If the effect of the ferrite on the left-hand side is to increase the wavelength of the wave and that of the ferrite on the right-hand side is to reduce the wavelength, the wavefront will turn to the right. In the Y-junction, the wave will tend to travel up one arm of the junction and not up the other.

An alternative description of the mode of operation of the Y-junction circulator (and probably more accurate) which may be applied to both waveguide and stripline circulators is in terms of its circuit properties. This approach has been particularly advocated by Helszajn.[1,2] An important property of the device is that perfect circulation occurs when the junction is matched. The ferrite loaded junction then behaves like a resonant cavity supporting two modes whose field patterns are contrarotating in the plane of the junction. With the ferrite unmagnetised, the resonant frequencies of the two modes are the same. When the ferrite is magnetised, however, the two resonant frequencies are different and a standing wave pattern is set up which presents a null in the field at the isolated port of the junction. In general, the junction circulator may have any number of ports, and the circulators described in Chapters 5 and 7 had four ports, but the rest of this chapter will be devoted to a description of the three-port circulator.

8.2 Scattering matrix theory

Scattering parameters have been used in the design of symmetrical n-port junction circulators by Auld,[3] Humphreys and Davies[4] and Davies.[5] Penfield[6] also gives a similar general analysis of three-port junctions and includes the circulator as a particular case. The use of scattering matrix theory in the design of circulators has been particularly advocated by Helszajn.[1,2,7] Aitken and McLean[8] give the same theory confined to three-port circulators and the analysis given here follows their presentation. If $\boldsymbol{a}$ and $\boldsymbol{b}$ are column vectors representing the input and output quantities, respectively, they are related by the scattering matrix

$$\boldsymbol{b} = \boldsymbol{S}\boldsymbol{a} \tag{8.1}$$

where, for a symmetrical nonreciprocal Y-junction, the scattering matrix is given by

$$\boldsymbol{S} = \begin{vmatrix} S_1 & S_2 & S_3 \\ S_3 & S_1 & S_2 \\ S_2 & S_3 & S_1 \end{vmatrix} \tag{8.2}$$

From energy considerations for a lossless junction, it can be shown that the scattering matrix is symmetrical and unitary[9]

$$\boldsymbol{S}^{\mathrm{T}}\boldsymbol{S}^* = \boldsymbol{I} \tag{8.3}$$

where $\boldsymbol{I}$ is the unit matrix and $\boldsymbol{S}^{\mathrm{T}}$ is the transpose of $\boldsymbol{S}$.

Substitution of eqn. 8.2 into eqn. 8.3 gives

$$|S_1|^2 + |S_2|^2 + |S_3|^2 = 1 \tag{8.4}$$

and

$$S_1 S_2^* + S_3 S_1^* + S_2 S_3^* = 0 \tag{8.5}$$

A junction is matched when the input reflection coefficient is zero, but the diagonal components of the scattering matrix are the reflection coefficients, so that this condition makes $S_1 = 0$. Then, from eqns. 8.4 and 8.5, either $S_2 = 0$ and $S_3 = 1$, or $S_3 = 0$ and $S_2 = 1$. Therefore, if the junction is matched, the power entering port 1 must go to port 2 and power entering port 2 must go to port 3 etc. In other words, if the junction is matched, it must be acting as a circulator. Then the scattering matrix of an ideal three-port junction circulator is

$$\boldsymbol{S} = \begin{vmatrix} 0 & 1 & 0 \\ 0 & 0 & 1 \\ 1 & 0 & 0 \end{vmatrix} \tag{8.6}$$

or

$$S = \begin{vmatrix} 0 & 0 & 1 \\ 1 & 0 & 0 \\ 0 & 1 & 0 \end{vmatrix} \tag{8.7}$$

depending on the direction of circulation through the junction.

Take the condition for an imperfect circulator, where S_1 and S_3 are small quantities, and let

$$S_2 = |S_2| \exp(j\theta_2)$$

then from eqn. 8.5

$$|S_2| [S_1 \exp(-j\theta_2) + S_3^* \exp(j\theta_2)] + S_3^* S_1^* = 0 \tag{8.8}$$

Provided that S_1 and S_3 are small, $|S_2| \simeq 1$ and $S_3 S_1^*$ is a second-order small term, so that the expression in the square brackets of eqn. 8.8 is equal to zero and

$$|S_1| \simeq |S_3| \tag{8.9}$$

Similarly, substituting into eqn. 8.4 gives

$$|S_2|^2 + 2|S_1|^2 \simeq 1$$

Therefore

$$|S_2| = \sqrt{(1 - 2|S_1|^2)} \tag{8.10}$$

whence, for S_1 small

$$|S_2| = 1 - |S_1|^2 \tag{8.11}$$

Therefore minimum insertion loss coincides with maximum isolation and minimum reflection coefficient looking into any of the three ports of the circulator. Also the power coupled into the isolated arm is equal to the power reflected. Unfortunately, Bosma[10] produces the expression

$$|S_2| = 1 - 2|S_1|^2$$

which is probably a printing error for eqn. 8.10, the form in which the equation occurs in earlier publications, and the same error is perpetuated by Helszajn.[1,2]

The above analysis assumes that the junction can be matched, and for this to happen, circulator action is necessary. Therefore, matching may be achieved by providing a nonreciprocal element such as a magnetised ferrite body in the junction. However, although a nonreciprocal element is required to provide circulator action, the bandwidth of Y-junction circulators is dependent on obtaining a broadband match and any frequency dependent properties of the nonreciprocal element may be compensated by matching circuits; so much so, that at one time, most commercially available Y-junction circulators contained matching screws or other variable matching elements that could be adjusted on test.

The eigenvalue equation for the scattering matrix may be written

$$\boldsymbol{S}\boldsymbol{u} = s\boldsymbol{u} \tag{8.12}$$

where $\boldsymbol{u}$ is an eigenvector and s is an eigenvalue. The eigenvalues will be complex and will have unit magnitude if the junction is lossless. By comparison between eqns. 8.1 and 8.12, it can be seen that the eigenvector, $\boldsymbol{u}$, represents the fields at the terminal planes and the eigenvalue, s, represents a reflection coefficient measured at each terminal plane. The field quantities at each terminal plane are then proportional to the elements of the respective eigenvector. For eigenvalues

$$|\boldsymbol{S} - s\boldsymbol{I}| = 0 \tag{8.13}$$

Solving eqn. (8.13) gives

$$\left.\begin{aligned} s_0 &= S_1 + S_2 + S_3 \\ s_1 &= S_1 + S_2 \exp(-j2\pi/3) + S_3 \exp(j2\pi/3) \\ s_{-1} &= S_1 + S_2 \exp(j2\pi/3) + S_3 \exp(-j2\pi/3) \end{aligned}\right\} \tag{8.14}$$

Therefore the elements of the scattering matrix may be expressed in terms of the eigenvalues, as quoted by Castillo and Davis[11] and Helszajn[1,2]

$$\left.\begin{aligned} 3S_1 &= s_0 + s_1 + s_{-1} \\ 3S_2 &= s_0 + s_1 \exp(j2\pi/3) + s_{-1} \exp(-j2\pi/3) \\ 3S_3 &= s_0 + s_1 \exp(-j2\pi/3) + s_{-1} \exp(j2\pi/3) \end{aligned}\right\} \tag{8.15}$$

When the scattering matrix is reduced to the condition for an ideal circulator, eqn. 8.6 or 8.7, the solution to eqn. 8.13 is

$$\left.\begin{aligned} s_0 &= 1 \\ s_1 &= \exp(j2\pi/3) \\ s_{-1} &= \exp(-j2\pi/3) \end{aligned}\right\} \tag{8.16}$$

which are the cube roots of 1.

Following Helszajn,[1,2] the eigenvectors may be obtained by direct substitution into eqn. 8.12. They are determined by the symmetry properties of the junction only and are given by

$$\boldsymbol{u}_0 = \frac{1}{\sqrt{3}} \begin{vmatrix} 1 \\ 1 \\ 1 \end{vmatrix} \tag{8.17}$$

$$\boldsymbol{u}_1 = \frac{1}{\sqrt{3}} \begin{vmatrix} 1 \\ \exp(-j2\pi/3) \\ \exp(j2\pi/3) \end{vmatrix} \tag{8.18}$$

$$\boldsymbol{u}_{-1} = \frac{1}{\sqrt{3}} \begin{vmatrix} 1 \\ \exp(j2\pi/3) \\ \exp(-j2\pi/3) \end{vmatrix} \tag{8.19}$$

where each eigenvector has been normalised and corresponds to one eigenvalue. Helszajn[2] gives similar eigenvalues and eigenvectors for symmetrical four- and five-port junction circulators. The eigenvector values may then be related to the electric field strengths in the terminal waveguides at the ports of the junction. For the three-port junction, there are three eigenvalues and three eigenvectors, corresponding to three different field patterns in the junction. $\boldsymbol{u}_0$ corresponds to a stationary mode in the junction while $\boldsymbol{u}_1$ and $\boldsymbol{u}_{-1}$ correspond to two contrarotating circularly polarised modes in the junction. The field solutions corresponding to the eigenvectors have been described in the literature.[5,11] Then the simultaneous application of all three eigen excitations to the junction gives

$$\boldsymbol{u}_0 + \boldsymbol{u}_1 + \boldsymbol{u}_{-1} = \begin{vmatrix} 1 \\ 0 \\ 0 \end{vmatrix} \tag{8.20}$$

which is equivalent to excitation of one port. Owen[12] gives expressions for transmission and return loss arising from eqn. 8.20.

For a loss-free junction, the scattering matrix is unitary,[9] as shown by eqn. 8.3, and the eigenvalues lie on the unit circle of the complex plane. Helszajn[7] has shown how adjustment of the eigenvalues creates a junction which is a circulator. Owen[12] shows that, for circulator action to occur in a three-port circulator, it is only necessary to adjust the phase angles between the three eigenvalues. Owen[12] also describes how to measure the eigenvalues.

Then, only two adjustments are needed in the design of a three-port junction circulator. Helszajn[1,2] suggests that these two adjustments are the shape and size of the ferrite in the junction and the strength of the biasing magnetic field. In a waveguide junction circulator, there is a lot of scope for adjusting the shape and size of the ferrite in the junction and this is discussed further in the next section. For the stripline or microstrip circulator, however, the radius of the ferrite is usually the only adjustment possible in addition to the strength of the biasing magnetic field. Davies[5] and Castillo and Davis[11] give formulas relating the phase angle between the eigenvalues to the waveguide and ferrite properties at the junction for some simple shapes of full-height ferrite at the junction. Helszajn, in a good tutorial paper,[13] has given a theoretical description of the mode of operation of a circulator in terms of s-parameters, admittance parameters or ABCD matrices and equivalent circuits. That paper[13] is recommended for further information on the application of circuit techniques to the design of circulators.

8.3 Junction cavity

The form of the waveguide Y-junction circulator that is simplest to analyse is a circular cavity at the junction coupled through irises to the connecting waveguides, as shown in Fig. 8.3, where a centred ferrite rod is magnetised perpendicular to the plane of the Y. We assume that the ferrite loaded cavity is

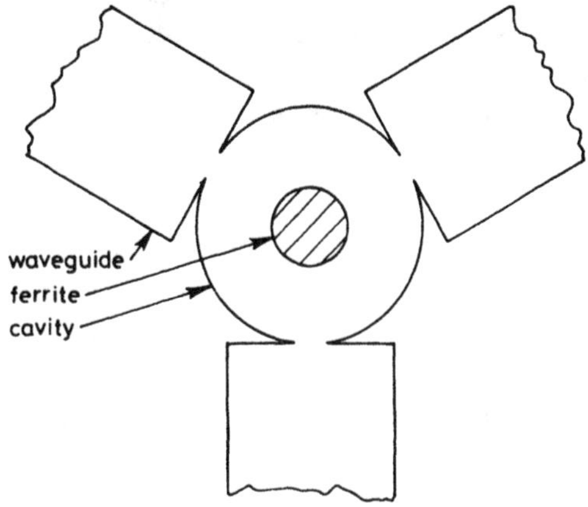

Fig. 8.3 *Waveguide Y-junction cavity, iris coupled to its input waveguides*

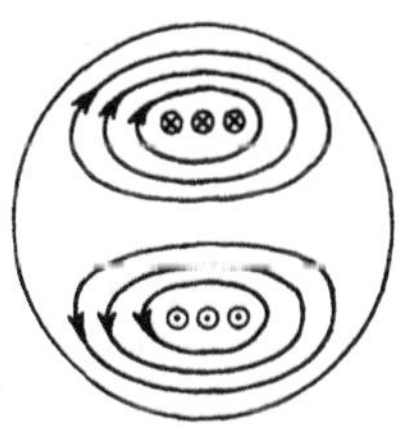

Fig. 8.4 *Field pattern for the TM_{110}-mode in a circular cavity*

resonant in the TM_{110}-mode, which is easily excited by the TE_{10}-mode in the rectangular waveguide. The field pattern in the cavity is shown in Fig. 8.4. Such a pattern can be generated by the sum of two similar contrarotating modes. The r.f. magnetic field in the ferrite at the centre of the cavity will be circularly polarised for each hand of rotation. With the ferrite unmagnetised, both contra-rotating modes will be resonant at the same frequency and the field pattern in the junction cavity will be as shown in Fig. 8.5*a*. The output ports will see two signals of about equal magnitude but the junction will be unmatched. If the standing wave pattern is rotated through 30°, as shown in Fig. 8.5*b*, port 3 is then situated at a voltage null of the cavity mode and the fields at the other two ports are equal in magnitude. The device behaves like a transmission cavity between ports 1 and 2 with port 3 isolated.

The standing wave patterns shown in Fig. 8.5 are stationary in the cavity with all fields varying as exp ($j\omega t$). They are generated by the two contrarotating modes in the cavity. If the ferrite is subjected to a d.c. magnetic bias field, the two contrarotating modes are no longer resonant at the same frequency, thus the ferrite loaded cavity is excited at a frequency intermediate between the resonant frequencies of the two modes. The impedance of the mode having the higher resonant frequency will have an inductive reactance component and that of the other mode will have a capacitive reactance component. If the frequency is chosen so that the two reactive components are equal, the total impedance will be real. If the degree of mode splitting is chosen so that the phase angle of

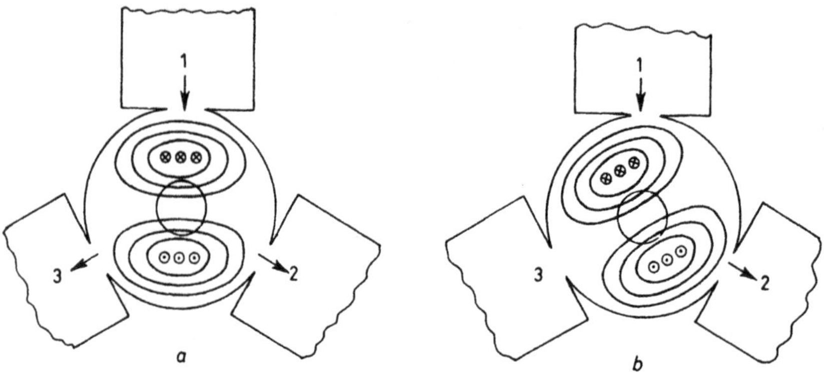

Fig. 8.5 *Field pattern of Fig. 8.4 in the cavity of the Y-junction shown in Fig. 8.3*
a Unmagnetised
b Supporting circulator action

impedance of each mode is 30° at the operating frequency, then the standing wave pattern will be rotated 30° from its position when no mode splitting occurs, as shown in Fig. 8.5*b*. The mode which has an inductive reactance component will have its voltage maximum (or E-field maximum) leading the current (or H-field maximum) by 30° at the input port. 30° in time phase is also 30° in space since this mode rotates one revolution in each cycle. Similarly, the mode having a capacitive reactance component will have its voltage maximum lagging the current by 30°, which means that the standing field pattern will be rotated 30° in the direction of rotation of the mode having an inductive component of input impedance, as shown in Fig. 8.5*b*. The standing field pattern now presents a null in the field at port 3 and no power is transmitted to that port.

Practical waveguide Y-junction circulators do not have coupling irises between the junction cavity and the input and output waveguides, but the waveguide is close coupled to the junction as shown in Fig. 8.1. It is difficult to define the boundary of the cavity of the junction, and calculation of the operating frequency becomes difficult even if it is possible to obtain expressions for the fields in the ferrite and the air at the junction.

Additional broadbanding may be obtained by impedance transformation.

Theoretically there is no limit to the bandwidth that may be obtained by stepped impedance transformers or filters external to the junction, but simple impedance transformers can also be incorporated into the junction itself. One often used method is to mount the ferrite element in reduced height, and therefore low impedance, waveguide. Often the step transformer to the reduced height is also within the junction region. Some possible geometries are shown in Fig. 8.6.

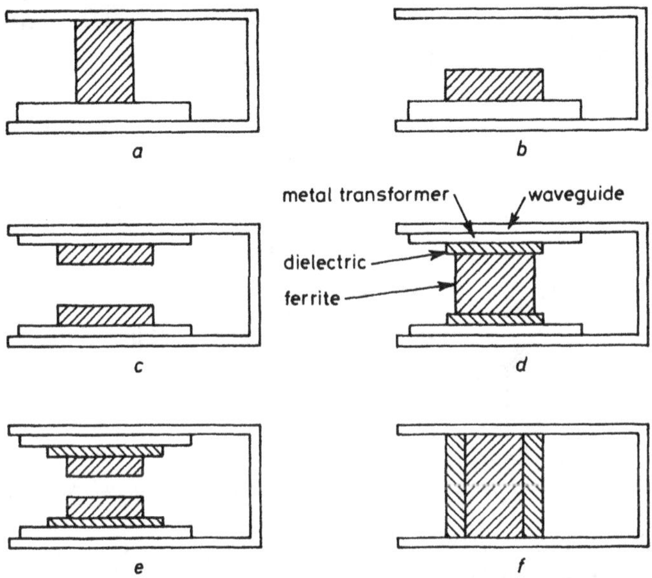

Fig. 8.6 *Various partial-height ferrite structures used in Y-junction circulators*
a Ferrite cylinder and metal matching transformer
b Ferrite disk and metal matching transformer
c Two ferrite disks and two metal disks
d Ferrite cylinder supported on two dielectric disks and two metal matching disks
e Two each of ferrite disks, dielectric disks and metal disks
f A ferrite cylinder surrounded by a matching dielectric tube

Helszajn and Tan[14–16] have investigated the design of partial height ferrite disks in a waveguide junction to give circulator action at X-band. They show how the circulator may be designed by two parameter adjustment as suggested in the discussion of eigenvalues in the last section. This is accomplished by measurements of the impedance presented to the input waveguide by the ferrite loaded junction. The unmagnetised ferrite is shown to operate in the TM_{11} open dielectric resonator mode. Reference 16, particularly, uses equivalent circuit theory in the design of a circulator.

Other shapes and sizes of ferrite in rectangular waveguide *H*-plane junctions have also been used in order to increase the bandwidth of junction circulators. If the ferrite post is of triangular cross-section, with the point of the triangle towards the input waveguide, as shown in Fig. 8.7, the ferrite behaves partly like a tapered matching transformer. The geometry shown in Fig. 8.7 was first suggested by Penney[17] and gave useful bandwidth circulators at the time. The idea has been extended by Khilla[18] using a tapered triangular ferrite post to provide near waveguide bandwidth circulators. The full-height triangular post has been analysed by Akaiwa[19] and Khilla,[20] who show theoretically that the full-height triangular post ought to have a bandwidth about twice that of the full-height cylindrical post of circular cross-section, when used in an *H*-plane waveguide Y-junction. An alternative approach was used by Aitken and McLean,[8] who determined experimentally a ferrite configuration which gave an

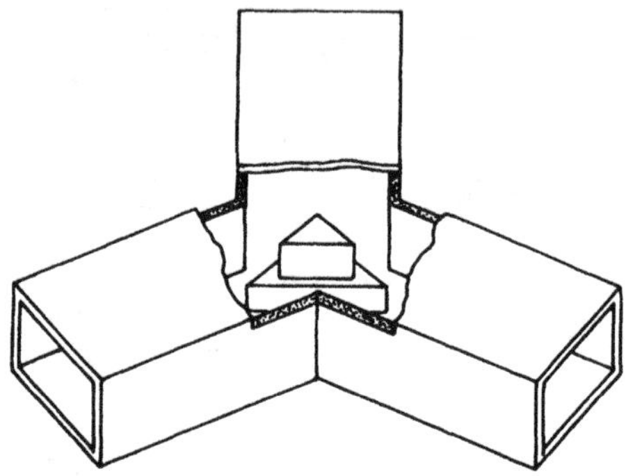

Fig. 8.7 *Circulator using a triangular ferrite element and triangular metal matching section*

approximately constant input impedance over the required bandwidth, and then reduced the waveguide height at the junction to match that constant input impedance. Helszajn and Sharp[21,22] supply further useful design information for disk and triangular ferrite loaded Y-junctions, and Helszajn[22] gives a theoretical design procedure and experimental results for a circulator with 10% bandwidth. He provides a step-by-step design approach which determines all the dimensions of a device capable of providing circulators with 20 dB isolation over 30% bandwidth. Y-junction circulators are available with near waveguide bandwidth in most standard sizes of waveguide having an isolation greater than 20 dB and insertion loss less than 0·3 dB; better performance can usually be obtained over a reduced bandwidth. Octave bandwidth Y-junction circulators are available in double-ridge waveguide.

8.4 *E*-plane junction

The *E*-plane Y-junction with a gyrator material at the junction is also a circulator. The geometry of the circulator is shown in Fig. 8.8 with the biasing magnetic field applied perpendicular to the plane of the ferrite disks, i.e. perpendicular to the plane of the Y-junction. The two ferrite disks are placed next to the narrow walls of the waveguide at the centre of the Y-junction. One advantage of the *E*-plane Y-junction circulator is its ability to handle higher peak powers than the *H*-plane Y-junction circulator because the ferrite disks are in a region of lower power flow in the waveguide. Circulation in an *E*-plane junction was first demonstrated by Yoshida,[23] and Davis and Longley[24] described the design of an *E*-plane Y-junction circulator having a 10% bandwidth at X-band. Helszajn and McDermott[25] show that the initial design of the *E*-plane circulator depends on the open dielectric resonances of the ferrite disks and that the mode chart for the *H*-plane circulator using ferrite disks[14] also applies to the *E*-plane circulator. They[25] also give a mode chart for the *E*-plane circulator where the spacing between the ferrite disks is allowed to vary. Therefore the waveguide width is reduced by metal inserts at the junction, as shown in Fig. 8.8. El-Shandwily, Kamal and Abdallah[26] give a full field theory analysis of the *E*-plane junction circulator with no metal inserts, and therefore no reduction in waveguide broad dimension at the junction. They give results for a full-height ferrite rod at the junction and for two ferrite disks adjacent to the walls with a dielectric cylinder between them.

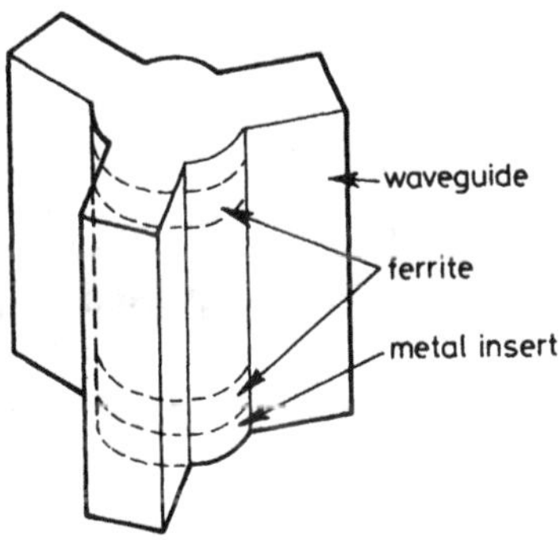

Fig. 8.8 *E-plane waveguide Y-junction circulator*

A very compact *E*-plane circulator can be constructed using suitable ferrite disks at the centre of an *E*-plane T-junction. Its length is then only slightly longer than the narrow dimension of the waveguide.[27] The use of an *E*-plane junction has also been suggested as a method of reducing the overall length of an isocirculator.[28] The isocirculator is described in the next section. For millimetre wave applications, the *E*-plane Y-junction circulator has been investigated by Solbach.[29] He uses a very simple design with only one ferrite disk and without the metal inserts shown in Fig. 8.8. It is discussed in Section 10.6.

8.5 Isocirculator

The name isocirculator is a contraction of isolator–circulator. It consists of a circulator with a matched termination on the third arm so that it acts as an isolator. It is shown schematically in Fig. 8.9. It is often used in stripline or microstrip systems where it is easier to use a standard Y-junction circulator with a load on one arm than to use a device specifically designed as an isolator which also may be more bulky. It is also used in high-power systems where the unwanted power can be dissipated in a specially designed load and the circulator has to dissipate only the power absorbed owing to its internal losses. The bandwidth of an isocirculator is a function of the performance of the matched termination as well as that of the circulator itself.

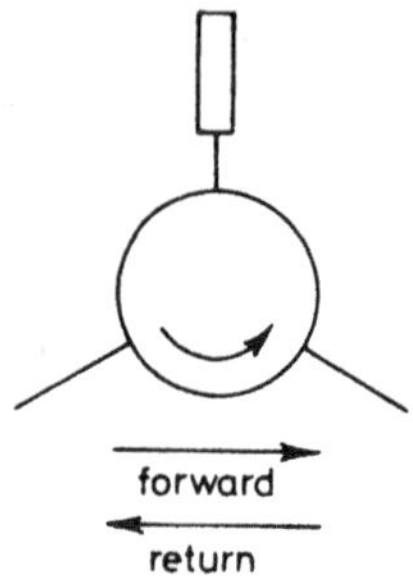

Fig. 8.9 *Circuit representation of the isocirculator*

One particular application of the isocirculator is in the manufacture of very short waveguide isolators. Consider the waveguide H-plane circulator shown in Fig. 8.1. For many applications, two of the output waveguides may have 30° bends so that the external ports of the device appear to be part of a T-junction.

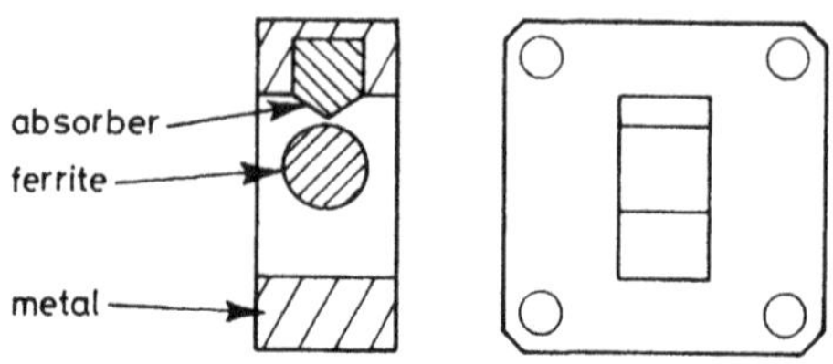

Fig. 8.10 *Compact waveguide isocirculator*

Then it is found that a T-junction as it stands will also act as a circulator. The matched termination is inserted into the side arm of the T-junction. If the absorbing material of the matched termination has a large value of effective permittivity and completely fills the side arm, the broad dimension of that side-arm waveguide may be reduced so that higher modes do not propagate. By this means, an isocirculator is produced which is not much longer than the diameter of the ferrite element which is producing the circulation. This is shown schematically in Fig. 8.10. These are sometimes called flange isolators,[30] as it is often possible to keep the cross-section dimensions to that of the waveguide coupling flange and the length to between 10 and 20 mm. Then, the performance of the isolator is more a function of the performance of the matched termination in the side arm than of the circulator itself. A compact tapered load or matched termination cannot usually be designed for satisfactory broadband performance, thus the compact isocirculator usually has less than 5% bandwidth.

An isocirculator may be based also on the ferrite loaded *E*-plane T-junction and will have an overall length about the same as the narrow dimension of the waveguide.

8.6 Matching networks

One important property of the three-port junction circulator is that ideal circulator action is obtained when the junction is matched. Therefore, one method of obtaining broadband performance is to use external matching networks on the junction. The networks often consist of a series of quarter-wave transformers. Both waveguide and stripline stepped impedance filters are normally used. The general theory of impedance matching networks is given by Matthaei, Young and Jones.[31] Helszajn[2,32] gives a particular application of the use of Chebyshev filter networks to give equal-ripple performance in the required band of operation. It is necessary to obtain first the input admittance of the unmatched junction and then to apply matching network theory to match that admittance. Levy and Helszajn[33] give tables of data for matching circulators. In stripline, once the diameter of the ferrite disk and the biasing magnetic field strength are decided, any matching has to be added external to the junction region. However, in waveguide, the matching circuits may be brought into the junction by using a region of reduced-height waveguide at the centre of the junction,[8] as discussed in Section 8.3 and shown in Figs. 8.6 and 8.7. By the use of matching techniques, stripline circulators are produced with octave bandwidths.

Another interesting application of matching techniques is the use of evanescent-mode below cut-off resonators in conjunction with a below cut-off ferrite loaded junction to produce a circulator with an extremely high attenuation outside the passband. The general principles of evanescent-mode components in below cut-off waveguide has been well described by Craven[34] and the design of filters by Craven and Mok.[35] Schieblich and Schunemann[36] have shown that evanescent-mode filters can be used as matching networks with a below cut-off ferrite loaded Y-junction to construct an evanescent-mode circulator. As Craven has pointed out,[30] below cut-off waveguide does not support progressive waves but only stored fields. Therefore, below cut-off waveguide approximates to a lumped element circuit and will be particularly amenable to the lumped element circuit analysis suggested by Helszajn.[2]

Schieblich and Schunemann[36] produced an evanescent-mode circulator in

X-band with an insertion loss of 0·4 dB and isolation better than 20 dB across a 20% band using a broadband transformer to feed into the evanescent-mode section. When the circulator was combined with evanescent-mode bandpass filter matching networks, the bandwidth was limited by the possible bandwidth of the evanescent mode filters to about 13%. However, the out-of-band attenuation exceeded 40 dB on all ports, which makes such a circulator well suited for applications in communication systems.

8.7 Stripline

Although most stripline and microstrip devices will be discussed in the next chapter, so much of the theory of the stripline Y-junction circulator is similar to the theory discussed earlier in this chapter that it will be analysed here. The stripline Y-junction circulator is shown in Fig. 8.11. It is a symmetrical junction

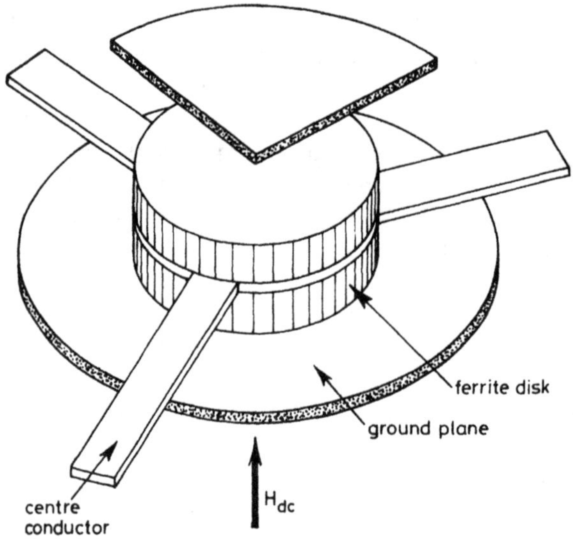

Fig. 8.11 *Stripline Y-junction circulator*

of three transmission lines having two ferrite disks at the junction, one each side of the centre conductor. The centre conductor is extended at the junction to form a circle of the same diameter as the ferrite disks. There is a biasing magnetic field perpendicular to the disks, i.e. perpendicular to the plane of the junction. As discussed in Section 8.2, only the adjustment of two variables is needed to match the junction and to provide circulation. Those two variables are the diameter of the ferrite disks and the strength of the biasing magnetic field.

As with the waveguide junction, the mode of operation may be described in terms of two contrarotating modes in the ferrite disks. As the fields in the stripline, as shown in Fig. 9.2, consist of an electric field perpendicular to the plane of the junction and a magnetic field paralel to the plane of the junction, the gyratory mode in the ferrite will approximate to that of the TM_{110}-mode in an unbounded dielectric cavity. The mode is shown in Fig. 8.12*a* when it is excited from port 1. When the ferrite is unmagnetised, the two contrarotating modes are rotating at the same rate and the field pattern is as shown in Fig. 8.12*a*. However, when the ferrite is magnetised, the two modes rotate at different speeds and the standing mode pattern is rotated through 30°, as described already for the waveguide circulator in Section 8.3. The new mode pattern is shown in Fig. 8.12*b*. It is seen from Fig. 8.12*a* that, in the unmagnetised condition, power will be coupled to each of ports 2 and 3 equally. In the magnetised condition, however, Fig. 8.12*b* shows that power is coupled from port 1 to port 2 and port 3 is at a node of the standing field pattern and is isolated.

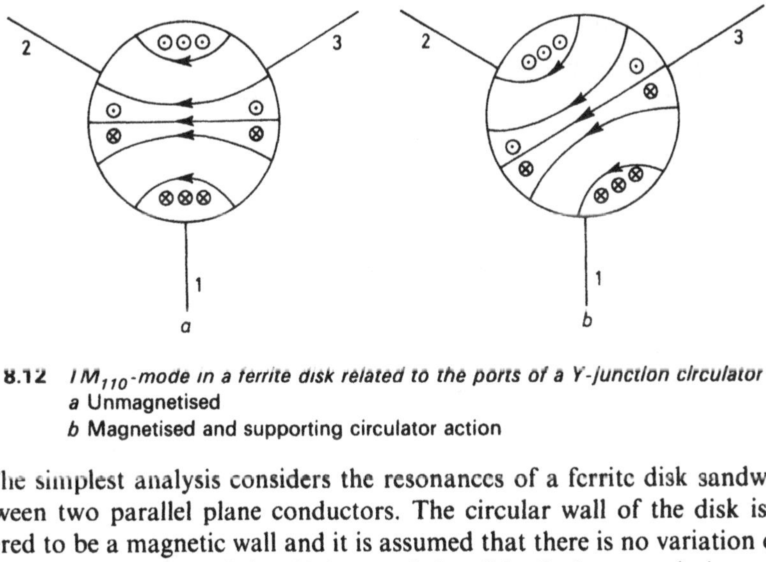

Fig. 8.12 *TM_{110}-mode in a ferrite disk related to the ports of a Y-junction circulator*
a Unmagnetised
b Magnetised and supporting circulator action

The simplest analysis considers the resonances of a ferrite disk sandwiched between two parallel plane conductors. The circular wall of the disk is considered to be a magnetic wall and it is assumed that there is no variation of the fields in the direction of the thickness of the disk. Such an analysis was first considered by Bosma[37] but some of his results were not in agreement with experiment. His results were refined[10] following the results of experiment, and a practical solution is proposed, which satisfies the theoretical considerations and agrees well with experimental results. Fay and Comstock[38] present a further extension of Bosma's approach to the design of stripline circulators. Their approach is consolidated in von Aulock and Fay.[39] In a review paper, Bosma[40] summarises the effects of material properties and bias magnetic field on the performance of junction circulators.

Following von Aulock and Fay,[39] a simple analysis is given of the electromagnetic fields in the ferrite disk. The disk may be considered as a very short length of longitudinally magnetised ferrite rod with a magnetic wall at the outside radius of the rod and no variation of the fields across the thickness of the disk. The geometry for one disk is shown in Fig. 8.13. The wave equation in the ferrite may easily be obtained from eqns. 3.1 to 3.6 when $\beta = 0$ or by putting $\beta = 0$ into eqn. 3.16. Therefore, in cylindrical polar co-ordinates

$$\frac{\partial^2 E_z}{\partial r^2} + \frac{1}{r}\frac{\partial E_z}{\partial r} + \omega^2 \varepsilon \mu_e E_z = 0 \qquad (8.21)$$

where μ_e is given by eqn. 6.15. The other field components are given by elimination from eqns. 3.26 and 3.27 or by suitable simplification of eqns. 3.30 and 3.31. Then

$$H_r = \frac{1}{\omega \mu_e}\left(\frac{\kappa}{\mu}\frac{\partial E_z}{\partial r} + \frac{j}{r}\frac{\partial E_z}{\partial \theta}\right) \qquad (8.22)$$

$$H_\theta = \frac{1}{\omega \mu_e}\left(\frac{\kappa}{\mu}\frac{1}{r}\frac{\partial E_z}{\partial \theta} - j\frac{\partial E_z}{\partial r}\right) \qquad (8.23)$$

Since the fields must be finite at the origin, only Bessel functions of the first kind occur in the solution to eqn. 8.21. A full solution is then

$$E_z = \sum_{n=-\infty}^{\infty} A_n J_n(\gamma r) \exp(jn\theta) \qquad (8.24)$$

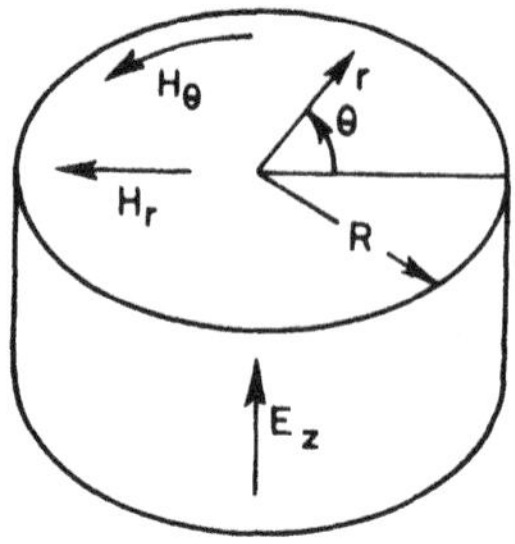

Fig. 8.13 *Ferrite disk showing the notation and dimensions used in the analysis*

where $\gamma^2 = \omega^2 \varepsilon \mu_e$. Bosma[37] used this expression and determined the coefficients A_n from the boundary conditions at the periphery of the disk. He considered a constant magnetic field component H_θ over the width of the input and output ports 1 and 2 and $H_\theta = 0$ everywhere else on the boundary. As mentioned above, his results were not in agreement with experiment. Fay and Comstock[38] measured the electric field along the edge of a loosely coupled ferrite disk in a circulator and found negligible contribution from higher-order modes. Higher-order modes did appear in a tightly coupled broadband circulator, however. Since only the lowest-order modes appear to be necessary to justify circulator action in terms of the fields given in Fig. 8.12, Fay and Comstock[38,39] simplify the analysis by concentrating on just the lowest-order modes, $n = \pm 1$. Helszajn,[41] however, does analyse the fields in the ferrite starting with eqn. 8.24 and the boundary conditions suggested by Bosma.[37]

Then, for the dominant mode, from eqns. 8.22 to 8.24

$$E_z = A_\pm J_1(\gamma r) \tag{8.25}$$

$$H_r = \frac{A_\pm}{\eta_e}\left(\frac{\kappa}{\mu} J_1'(\gamma r) \mp \frac{J_1(\gamma r)}{\gamma r}\right) \tag{8.26}$$

$$H_\theta = \frac{jA_\pm}{\eta_e}\left(-J_1'(\gamma r) \pm \frac{\kappa}{\mu}\frac{J_1(\gamma r)}{\gamma r}\right) \tag{8.27}$$

where the expression $\exp(\pm j\theta)$ is assumed in each of the field expressions, eqns. 8.25 to 8.27, and $\eta_e = \sqrt{(\mu_e/\varepsilon)}$. For the simplest condition, the resonance can be defined by the condition $H_\theta = 0$ at the edge of the disk, $r = R$. Then, when $\gamma R = x$

$$J_1'(x_\pm) = \mp \frac{\kappa}{\mu}\frac{J_1(x_\pm)}{x_\pm}$$

therefore

$$\frac{\kappa}{\mu} = \mp \frac{x_\pm J_1'(x_\pm)}{J_1(x_\pm)} \tag{8.28}$$

where $x_\pm = \omega_\pm R\sqrt{(\varepsilon\mu_e)}$ defines the two resonances for the two rotating modes. The resonance frequency for the unmagnetised disk, when $\kappa/\mu = 0$, is given by

$$J_1'(x_0) = 0$$

therefore $x_0 = 1{\cdot}84$ and

$$\omega_0 = \frac{1{\cdot}84}{R\sqrt{(\varepsilon\mu_e)}}$$

If, in the magnetised condition, the frequency splitting is small, the right-hand side of eqn. 8.28 can be shown to be approximately equal to[39]

$$\pm\left(\frac{\omega_\pm - \omega_0}{\omega_0}\right)(x_0^2 - 1)$$

therefore

$$\omega_\pm = \omega_0\left(1 \pm 0{\cdot}42\,\frac{\kappa}{\mu}\right) \tag{8.29}$$

From the point of view of the ferrite itself, the bandwidth of the circulator is going to lie between the two split resonances. However, eqn. 8.29 is an approximate expression assuming that the difference in the resonance frequencies is small so it is an equation that can give no more than a vague indication of the performance of a broadband circulator. The further stages in the design of a circulator are given in the literature.[38,39,41]

Wu and Rosenbaum[42] have applied the theory of Fay and Comstock[38] to the design of broadband microstrip circulators. Further details of microstrip circulators are given in the next chapter in Section 9·4; however, Wu and Rosenbaum do show that circulation is possible over a broadband when $|\kappa/\mu|$ lies between a half and one. They also give expressions for S_1, S_2, S_3 and the input impedance in terms of the physical parameters of the microstrip junction. Helszajn[41] also shows that circulator action is possible using the $n = 2$ nodes in the ferrite disk. He shows that circulation is in the opposite direction from that of the normal mode which uses the $n = 1$ field patterns. However, he[43] subsequently shows that the higher-order modes do not always produce a practical circulation condition.

At the lower microwave frequencies, the construction of a stripline circulator may be as shown in Fig. 8.11, with the centre conductor suspended in air, except at the junction where it is supported by the ferrite disks. All the theory given in the section has assumed that the geometry shown in Fig. 8.11 is being analysed and the effect of any dielectric substrate outside the ferrite disks has been ignored. Helszajn[44] has suggested that for UHF, when the ferrite disks become large, the outside region of the ferrite disk may be replaced with a dielectric ring. He gives the general theory of stripline circulators consisting of ferrite disks and dielectric rings in terms of the impedance matrix of the junction. Some experimental results show good agreement with the theory. Helszajn, James and Nisbet[45] have similarly analysed the effect of replacing the ferrite disks in Fig. 8.11 with planar ferrite sections of triangular shape. They verified their calculated field pattern in the triangular ferrite element with experimental measurements of the electric field at the edge of the element. They show that the bandwidth of a circulator using triangular ferrite elements with the apex of the triangle aligned with the input ports is three times that of a circulator using circular ferrite disks. Lyon and Helszajn[46] have extended the investigation to the analysis of any planar shape of ferrite at the junction. They describe a general purpose finite element analysis for finding the Z-matrix of a planar circulator using arbitrary shaped resonators. A general purpose computer program has been written to find the circulation properties of a Y-junction loaded with any planar shape ferrite element. Results for disk, triangle and irregular hexagon shaped ferrites are given. The program may also be used to plot the frequency response of the junction.

One problem with the analysis described in the last paragraph is that, at most normal and higher microwave frequencies, the centre conductor in the stripline has to be supported on a dielectric substrate and then it becomes difficult to insert arbitrary shapes of ferrite into the substrate to form a circulator. The circulator usually consists of ferrite disks fitting into circular holes in the substrate, or a whole section of stripline circuit may be accommodated on a planar ferrite substrate which is then magnetised in the region of the Y-junction to provide circulator action. With this in mind, Nisbet and Helszajn[47] and Helszajn and Lyon[48] have analysed the performance of different shapes of metallisation, such as disks and triangles, on dielectric and on unmagnetised ferrite substrates.

8.8 References

1 HELSZAJN, J.: 'Principles of microwave ferrite engineering', (Wiley, 1969), pp. 153–192

2 HELSZAJN, J.:'Nonreciprocal microwave junctions and circulators', (Wiley, 1975)

3 AULD, B. A.: 'The synthesis of symmetrical waveguide circulators', *IRE Trans.*, 1959, **MTT-7**, pp. 238–246

4 HUMPHREYS, B. L. and DAVIES, J. B.: 'The synthesis of N-port circulators', *IRE Trans.*, 1962, **MTT-10**, pp. 551–554

5 DAVIES, J. B.: 'An analysis of the m-port symmetrical H-plane waveguide junction with central ferrite post', *IRE Trans.*, 1962, **MTT-10**, pp. 596–604

6 PENFIELD, P.: 'A classification of lossless three-ports', *IRE Trans.*, 1962, **CT-9**, pp. 215–223

7 HELSZAJN, J.: 'The adjustment of the m-port single-junction circulator', *IEEE Trans.*, 1970, **MTT-18**, pp. 705–711

8 AITKEN, F. M. and McLEAN, R.: 'Some properties of the waveguide Y-circulator', *Proc. IEE*, 1963, **110**, pp. 256–260

9 MONTGOMERY, C. G., DICKE, R. H. and PURCELL, E. M.: 'Principles of microwave circuits' (McGraw-Hill, 1948), p. 148

10 BOSMA, H.: 'On stiripline Y-circulation at UHF, *IEEE Trans.*, 1964, **MTT-12**, pp. 61–72

11 CASTILLO J. B. and DAVIS, L. E.: 'Computer-aided design of three-port waveguide junction circulators', *IEEE Trans.*, 1970, **MTT-18**, pp. 25–34

12 OWEN, B.: 'The identification of modal resonances in ferrite loaded waveguide Y-junctions and their adjustment for circulation', *Bell Syst. Tech. J.*, 1972, **51**, pp. 595–627

13 HELSZAJN, J.: Unified approach to lumped-element stripline and waveguide junction circulators', *IEE J. MOA.*, 1976, **1**, pp. 18–26

14 HELSZAJN J. and TAN, F. C. F.: 'Mode charts for partial-height ferrite waveguide circulators', *Proc. IEE*, 1975, **122**, pp. 34–36

15 HELSZAJN, J. and TAN, F. C. F.: 'Susceptance-slope parameter of waveguide partial-height ferrite circulators', *Proc. IEE*, 1975, **122**, pp. 1329–1332

16 HELSZAJN, J. and TAN, F. C. F.: 'Design data for radial-waveguide circulators using parital height ferrite resonators', *IEEE Trans.*, 1975, **MTT-23**, pp. 288–298

17 PENNEY, J. A.: 'Experiments in the design of some three-port circulators', Digest of Conf. on Components for Microwave Circuits, IEE, London, 1962, pp. 17–20

18 KHILLA, A. M.: 'Ausbau von breitbandigen H-Ebene-Hohlleiter-Y-Zirkulatoren. (Design of wideband H-plane waveguide Y-circulators)' *Arch. Elektron. & Uebertragungstech.*, 1982, **36**, pp. 258–260

19 AKAIWA, Y.: 'Mode classification of a triangular ferrite post for Y-circulator operation', *IEEE Trans.*, 1977, **MTT-25**, pp. 59–61

20 KHILLA, A. M.: 'Y-circulator with full-height prism ferrite post', *Arch. Elektron. & Uebertragungstech.*, 1981, **35**, pp. 93–96

21 HELSZAJN, J. and SHARP, J.: 'Resonant frequencies, Q-factor, and susceptance slope parameter of waveguide circulators using weakly magnetised open resonators', *IEEE Trans.*, 1983, **MTT-31**, pp. 434–441

22 HELSZAJN, J.: 'Design of waveguide circulators with Chebyshev characteristics using partial-height ferrite resonators', *IEE Trans.*, 1984, **MTT-32**, pp. 908–917

23 YOSHIDA, S.: 'E type X circulator; E type T circulator', *Proc. IRE*, 1959, **47**, pp. 2017–2018

24 DAVIS, L. E. and LONGLEY, S. R.: 'E-plane 3-port X-band waveguide circulators', *IEEE Trans.*, 1963, **MTT-11**, pp. 443–445
25 HELSZAJN, J. and McDERMOTT, M.: 'Mode chart for E-plane circulators', *IEEE Trans.*, 1972, **MTT-20**, pp. 187–188
26 EL-SHANDWILY, M. E., KAMAL, A. A. and ABDALLAH, E. A. F.: 'General field theory treatment of E-plane waveguide junction circulators. Pt. I. Full height ferrite configuration. Pt. II. Two-disk ferrite configuration', *IEEE Trans.*, 1977, **MTT-25**, pp. 784–803
27 SIVERS LAB: 'Short waveguide circulator', *Microwave J. (European Ed.)*, 1966, **9**, 12, p. 38E
28 KURUSHIN, E. P. and SHEKHMAMET'EV, K. Kh.: 'Miniature ferrite isolator for the millimeter band using an E-plane T circulator', *Radiotek.*, 1978, **33**, 10, p. 83 (English translation in *Telecom. & Rad. Eng.*, **32/33**, 10, p. 132)
29 SOLBACH, K.: 'E-plane circulators aid mm-wave design', *Microwaves & RF*, 1983, **22**, 13, pp. 73–78
30 BOCHKAREV, A. I. and KURUSHIN, E. P.: '4-millimeter band ferrite isolator-flange', *Radiotek.*, 1982, **37**, 3, pp. 80–81 (English translation in *Telecom. & Rad. Eng.*, **36/37**, 3, pp. 131–132)
31 MATTHAEI, G. L., YOUNG, L. and JONES, E. M. T.: 'Microwave filters, impedance matching networks and coupling structures' (McGraw-Hill, 1964)
32 HELSZAJN, J.: 'The synthesis of quarter-wave coupled circulators with Chebyshev characteristics', *IEEE Trans.*, 1972, **MTT-20**, pp. 764–769
33 LEVY, R. and HELSZAJN, J.: 'Short-line matching networks for circulators and resonant loads', *Proc. IEE*, 1983, **130H**, pp. 385–390
34 CRAVEN, G.: 'Waveguide below cutoff: A new type of microwave integrated circuit', *Microwave J.*, 1970, **13**, 8, pp. 51–58
35 CRAVEN, G. and MOK, C. K.: 'The design of evanescent mode waveguide bandpass filters for a prescribed insertion loss characteristic', *IEEE Trans.* 1971, **MTT-19**, pp. 295–308
36 SCHIEBLICH, C. and SCHUNEMANN, K.: 'Circulators in below-cutoff waveguides', *Arch. Elektron. & Uebertragungstech.*, 1983, **37**, pp. 50–55
37 BOSMA, H.: 'On the principle of stripline circulation', *Proc. IEE*, 1962, **109B**, Supplement No. 21, pp. 137–146
38 FAY, C. E. and COMSTOCK, R. L.: 'Operation of the ferrite junction circulator', *IEEE Trans.*, 1965, **MTT-13**, pp. 15–27
39 VON AULOCK, W. H. and FAY, C. E.: 'Linear ferrite devices for microwave applications' (Academic Press, 1968), pp. 116–129
40 BOSMA, H.: 'A general model for junction circulators; choice of magnetisation and bias field', *IEEE Trans.*, 1968, **MAG-4**, pp. 587–596
41 HELSZAJN, J.: 'Nonreciprocal microwave junctions and circulators' (Wiley, 1975), p. 172
42 WU, Y. S. and ROSENBAUM, F. J.: 'Wide-band operation of microstrip circulators', *IEEE Trans.*, 1974, **MTT-22**, pp. 849–856
43 HELSZAJN, J.: 'Complex gyrator circuits of planar circulators using higher order modes in a disk resonator', *IEEE Trans.*, 1983, **MTT-31**, pp. 931–938
44 HELSZAJN, J.: 'Composite-junction circulators using ferrite disks and dielectric rings', *IEEE Trans.*, 1974, **MTT-22**, pp. 400–410
45 HELSZAJN, J., JAMES, D. S. and NISBET, W. T.: 'Circulators using planar triangular resonators', *IEEE Trans.*, 1979, **MTT-27**, pp. 188–193
46 LYON, R. W. and HELSZAJN, J.: 'A finite element analysis of planar circulators using arbitrarily shaped resonators', *IEEE Trans.*, 1982, **MTT-30**, pp. 1964–1974
47 NISBET, W. T. and HELSZAJN, J.: 'Mode charts for microstrip resonators on dielectric and magnetic substrates using a transverse-resonance method', *IEE J. MOA*, 1979, **3**, pp. 69–77
48 HELSZAJN, J. and LYON, R. W.: 'Mode charts of magnetised and demagnetised planar hexagonal resonators on ferrite substrates using finite elements', *Proc. IEE.*, 1984, **131H**, pp. 420–422

Chapter 9

Stripline and microstrip devices

9.1 Introduction

Stripline is often used as a generic term for any transmission line consisting of a thin planar conductor supported on a dielectric sheet parallel to an earthed metallic plane. Symmetrical stripline has the conductor in the region between two earthed metallic conductors as shown in Fig. 9.1, and is sometimes called

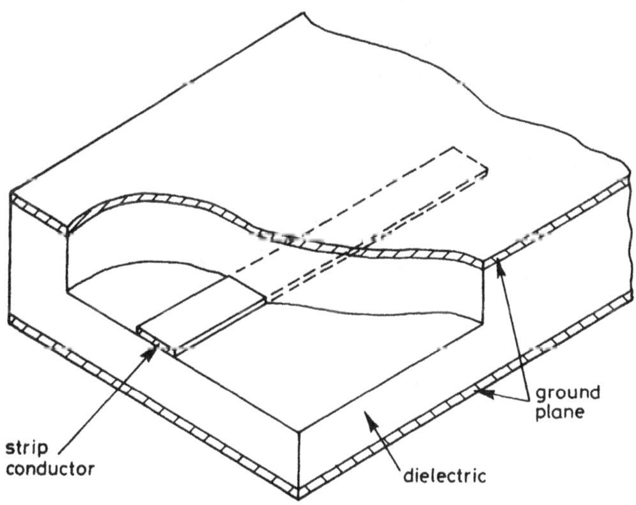

Fig. 9.1 *Triplate stripline*

triplate line. It has the advantage that the microwave field is enclosed and is symmetrical and can be analysed from electrostatic principles assuming a pure TEM-mode. It can be constructed with a printed conductor sandwiched between two dielectric plates and their ground planes. The line with its electric and magnetic fields is shown in Fig. 9.2. A good summary of stripline design theory is given by Gupta, Garg and Chadha,[1] who quote all the formulas needed to determine the characteristic impedance and loss of the line and the corresponding inverse design formulas. The most accurate empirical formulas have been given by Wheeler,[2] who also publishes some curves which may be used for design. Some curves of characteristic impedance for some particular standard sizes of stripline substrate are given in the 'Microwave Engineers' Handbook'.[3]

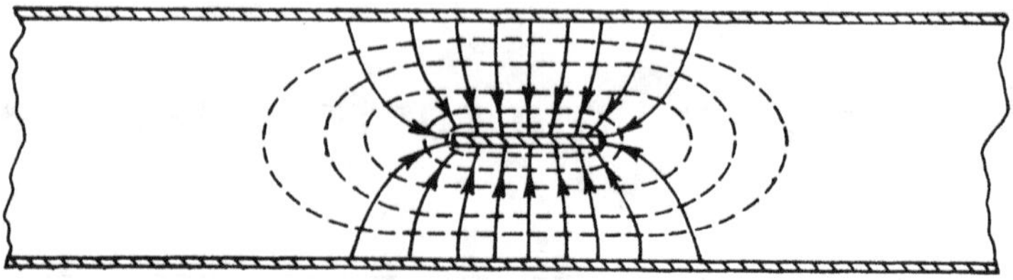

Fig. 9.2 *Triplate stripline and its fields*
—— electric
---- magnetic

An earlier presentation of most of the formulas given by Gupta *et al.*[1] was also given by Bahl and Garg.[4] For the lower microwave frequencies in the UHF band, stripline may be constructed with a rigid inner conductor and air dielectric, which removes any losses due to the dielectric substrate material. For higher frequencies, a similar effect is obtained using suspended stripline, as shown in Fig. 9.3, where the conductor is a thin metal strip on each side of a dielectric substrate and very little microwave power is carried by the dielectric material.

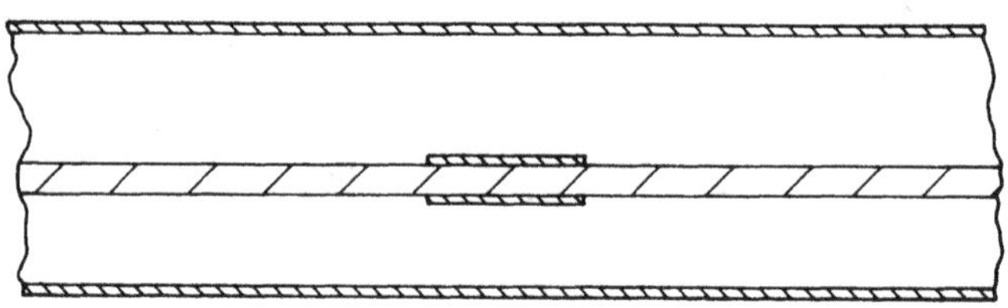

Fig. 9.3 *Suspended stripline*

9.2 Microstrip

An alternative form of stripline is the unsymmetrical line shown in Fig. 9.4, called *microstrip*. It is used much more than symmetrical stripline for microwave components and systems. It is easier to manufacture and it simplifies the mounting of semiconductor devices and is compatible with integrated circuits. In fact, in microwave integrated circuits, the interconnections formed by strip conductors on the insulating GaAs substrate are sections of microstrip line. The electromagnetic fields in microstrip are difficult to analyse theoretically, because the fields exist in the composite dielectric and air space, as shown in Fig. 9.5. It is seen that the fields experience a discontinuity at the air–dielectric interface. If an attempt is made to analyse the performance of microstrip using static field calculations, the electric field is affected by the air–dielectric interface, whereas the magnetic field is not so affected. Then an a.c. field analysis is necessary. For the lower microwave frequencies, it is satisfactory to assume that the field in microstrip arises from a TEM-mode and that there are no longitudinal components of the microwave fields. However, at higher microwave frequencies this assumption is not valid and a full-field analysis which takes account of dispersion due to the microstrip is necessary.

Again a good summary of microstrip design theory is given by Gupta, Garg and Chadha[1] and also by Edwards.[5] They also give formulas which take account

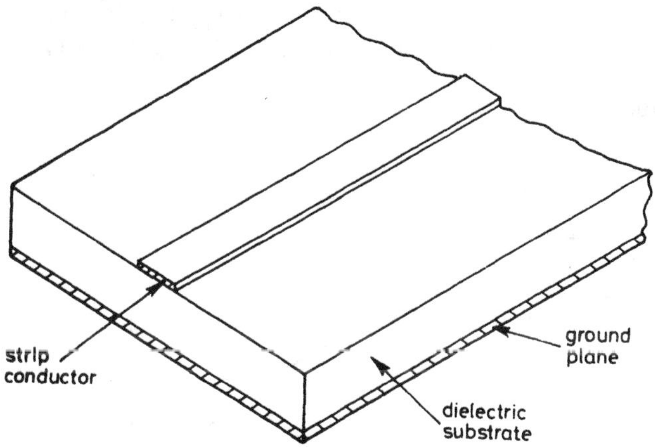

Fig. 9.4 *Microstrip line*

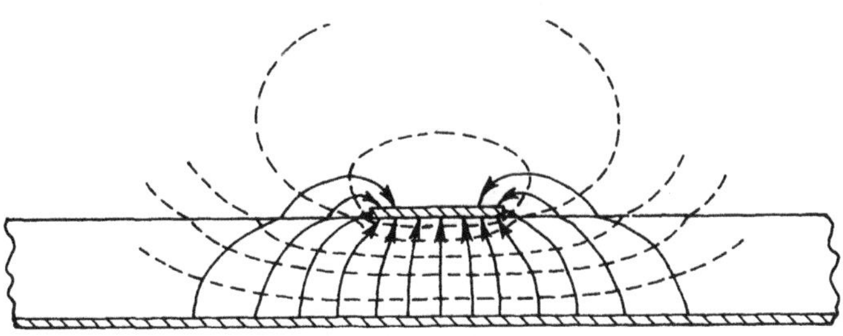

Fig. 9.5 *Microstrip and its fields*
—— electric
---- magnetic

of the effect of dispersion. A closed form expression for the TEM-mode of propagation has been given by Wheeler[6] and some curves of characteristic impedance calculated from Wheeler's expressions are given in Fig. 9.6. If Z_1 is the characteristic impedance in a microstrip line for which $\varepsilon_r = 1$, then the waveguide wavelength in a microstrip line of the same dimensions and having a dielectric substrate is given by

$$\lambda = \lambda_0 \frac{Z_0}{Z_1} \tag{9.1}$$

Closed form equations suitable for computer analysis and design have been given in each of the previously mentioned references and also by Bahl and Trivedi[7] and Hammerstad and Jenson.[8] A particular set of expressions to describe dispersion in microstrip has been given by Pramanick and Bhartia.[9,10] All these results produce relatively simple algebraic expressions that can be easily programmed on to a computer and used with computer-aided analysis or design. Otherwise it is necessary to use a field analysis to accurately describe the microstrip line, such as that given by Mirshekar-Syahkal and Davies.[11]

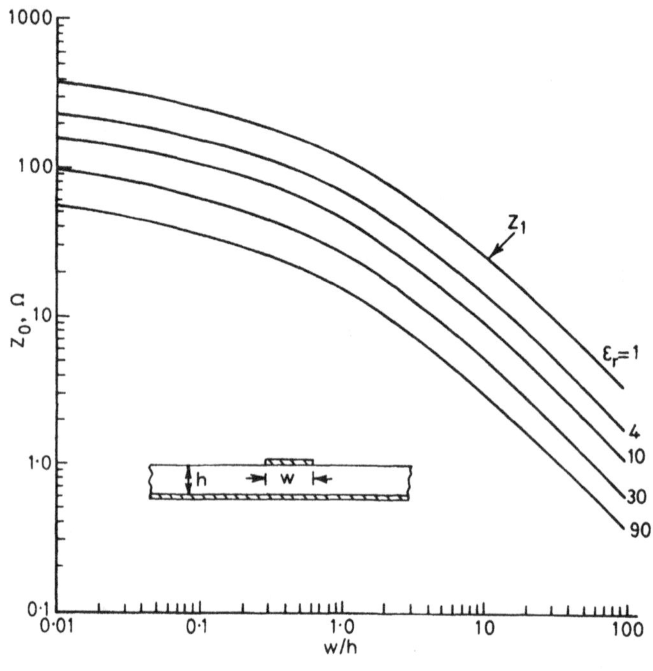

Fig. 9.6 *Microstrip characteristic impedance for various values of ε_r, calculated from a formula given by Wheeler*[6]

There are other shapes of stripline which are sometimes used because of particular properties, four of which are shown in Fig. 9.7. Inverted and suspended microstrip are used in situations where the dielectric loss needs to be especially low. Slotline uses the same basic substrate configuration as microstrip except that propagation is guided by a slot in a continuous sheet conductor. Finline is used in systems where propagation is mostly in waveguide.

9.3 Ferrite substrate

Except for the stripline circulator shown in Fig. 8.11, all the applications of ferrite materials to stripline circuits seem to be realised in microstrip, slotline or finline. The first applications were reported by Lewin,[12] who used side stubs to create a circularly polarised magnetic field with a small ferrite insert to produce a resonance isolator, and Brodwin,[13] who analysed propagation in longitudinally-magnetised ferrite-filled parallel-plate waveguide and applied the results to ferrite-filled microstrip line. Pucel and Masse[14] apply the analytical expressions of Wheeler[15] to give formulas for propagation coefficient and characteristic impedance of microstrip lines on unmagnetised ferrite substrates.

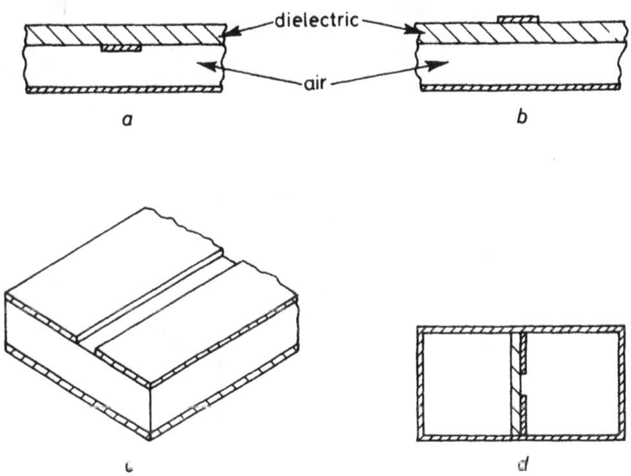

Fig. 9.7 *a* Inverted microstrip
b Suspended microstrip
c Slotline
d Finline

Edwards[5] applies the formulas of Pucel and Masse[14] and incorporates the effect of longitudinally magnetised ferrite by using μ_e from eqn. 6.15. Nisbet and Helszajn[16] give all the expressions needed for the design of microstrip lines on demagnetised ferrite substrates and give a comprehensive list of references for both dielectric-filled microstrip and ferrite-filled microstrip.

As in the dielectric supported microstrip, full-field analyses have been made for the microstrip line on ferrite substrate. Minor and Bolle[17] calculate the modes in a microstrip on a ferrite substrate enclosed in a conducting box, i.e. shielded microstrip. The ferrite is magnetised perpendicular to the direction of propagation and parallel to the ferrite–air interface and their analysis shows that there is no differential phase change for this direction of transverse magnetisation. Hofmann[18] analyses the fields for all three possible magnetisations of the ferrite substrate, as shown in Fig. 9.8, and gives curves of the variation of phase constant with frequency. He shows that magnetising the ferrite causes reciprocal phase change for longitudinal magnetisation and for magnetisation perpendicular to the air–ferrite interface but very little phase change for transverse magnetisation parallel to the air–ferrite interface. He does not include any nonreciprocal effects in his calculations. Lerer, Mikhalevskiy and Shchuchinskiy[19] give the complete formulas for the numerical analysis of a microstrip line on a ferrite substrate which is magnetised in any of the three directions shown in Fig. 9.8. Yeh and Bolle[20] give some results of characteristic impedance and field patterns of the shielded microstrip on a ferrite substrate at 1 GHz employing the same program used by Minor and Bolle.[17] Their ferrite is transversely magnetised in the plane of the microstrip in its remanent condition; because of the low magnetisation in the ferrite and the low frequency, they obtain nonreciprocal phase change owing to the magnetisation on the line.

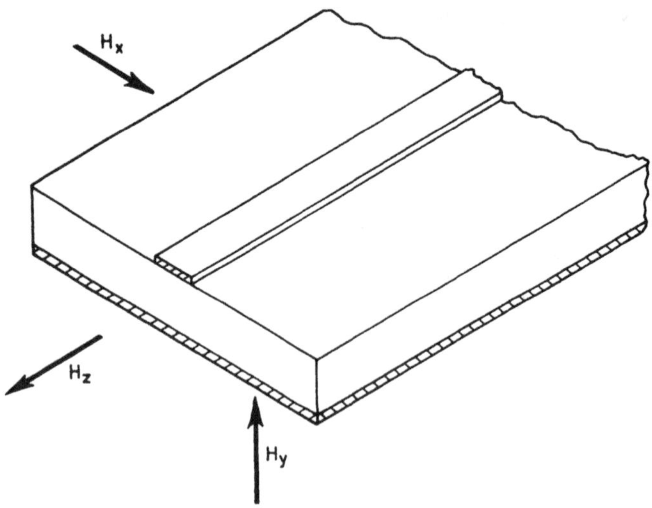

Fig. 9.8 *Directions of magnetisation in a ferrite loaded microstrip*
H_z longitudinal magnetisation
H_x transverse magnetisation in the plane of the substrate
H_y magnetisation perpendicular to the plane of the substrate

Hofmann,[18] in his theoretical analysis shows how μ_e occurs in the expressions for the wave equation due to the magnetised ferrite material and gives a slight justification for the use of μ_e in the expressions of Pucel and Masse[14] as given by Edwards.[5] Janiczak and Kitlinski[21] calculate the propagation conditions for two parallel coupled microstrip lines on a ferrite substrate which is magnetised perpendicular to the plane of the substrate. They analyse the effect of having two lines of different widths and give calculated values of propagation constant for one direction of propagation only, because the nonreciprocal phase difference is negligible. Lange[22] gives a further analysis of the shielded suspended microstrip with the ferrite substrate magnetised transversely in the plane of the substrate.

In very many situations, the microstrip line on a ferrite substrate is only being used for interconnection to a nonreciprocal or control ferrite device, and the ferrite is probably unmagnetised except at the device itself. Then the equations given by Pucel and Masse[14] or Nisbet and Helszajn[16] describe the electrical characteristics of the microstrip.

For microstrip circuits incorporating ferrite devices such as circulators, it is possible to fabricate the complete circuit on a ferrite substrate and to provide the external magnetisation only where needed to magnetise the ferrite. However, microstrip on a ferrite substrate is more lossy than one on a dielectric substrate such as alumina, and it is possible to incorporate a ferrite disk as required into a hole in the alumina substrate. One problem, however, is the difficulty of giving adequate mechanical strength to the conductors where they bridge the transition between ferrite and dielectric substrates. A suggested solution is to create a composite substrate material consisting of a non-magnetic (as opposed to an unmagnetised) ferrite dielectric material which is structurally compatible with the usual microwave ferrite material. The magnetic ferrite disks are inserted into the non-magnetic ferrite substrate before firing so that a mechanically continuous substrate material is produced having regions with different microwave properties. The system does work but appears to be complicated and expensive and unlikely to be economic. One disadvantage of the all-ferrite substrate for a microstrip circuit is that the dielectric loss of unmagnetised ferrite is generally larger than that of the dielectric substrates usually used for microstrip. For low-loss circuits then, it is advantageous to have the minimum length of microstrip interconnection on ferrite substrates.

A large area of ferrite or YIG substrate is considerably more expensive than a similar substrate made of dielectric, such as alumina. A cost-effective alternative is a thin ferrite layer on the surface of a dielectric substrate. Kal, Bhattacharya and Chakraborti[23] have shown that nonreciprocal components can be made using such a composite substrate. They use a ferrite paste which is fired to give a thick film layer on an alumina substrate. They report experimental results for a field displacement isolator (see Section 9.7) on a YIG film. The forward loss is high, generally greater than 5 dB for 20 dB isolation, which is possibly due to the high porosity of the fired YIG film compared with bulk material. The losses of the YIG film were lower than those achieved using fired ferrite films. Such a film could be screen printed to give the ferrimagnetic film only at the region of the ferrite device. Lange[22] has analysed a suspended microstrip geometry which could easily be modified to give results for the thin ferrite film on the surface of a dielectric substrate. Mirshekar-Syahkal and Davies,[11] in their analysis of microstrip, allow for a layered dielectric substrate so that their analysis can be used to give results for the effect of an unmagnetised ferrite film on a dielectric substrate.

Here are some general comments on the use of microstrip. They are not peculiar to ferrite substrates. Generally, as with any microwave system, low loss is needed. In microstrip, low loss is achieved by using low-loss substrate materials and low-resistivity conductor materials. Low loss is also achieved by reducing the current density in the microstrip conductor, which is the result of using wide conductors, i.e. low impedance lines, and thick substrates because the lines are then even wider. The conductor thickness needs to be at least four skin depths at the lowest operating frequency. The surface of the substrate must be very smooth with a surface roughness ~ 1 μm, and a flatness better than 1 in 1000.

9.4 Slotline and finline

Particularly for millimetre wave devices (see Chapter 10) slotline and finline have been investigated as a transmission medium. They are shown in Fig. 9.7. Lange[22] investigates the properties of shielded suspended slotline with a ferrite substrate magnetised transversely in the plane of the substrate. The mathematical expressions for slotline are obtained using a similar model to that used for microstrip. He shows that there is nonreciprocal field displacement for the

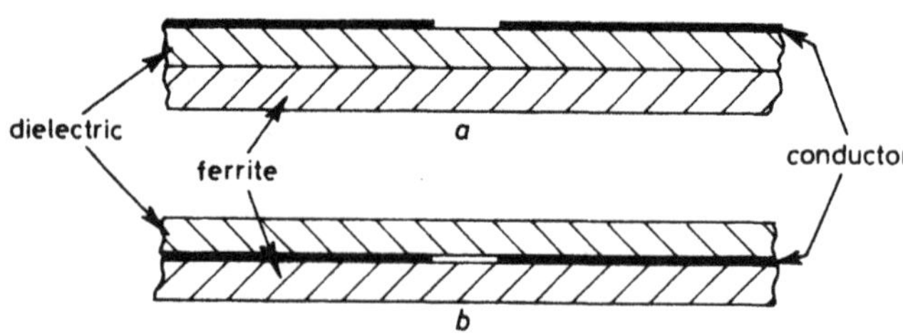

Fig. 9.9 *Slotline on a composite ferrite dielectric substrate*
a Conductor on dielectric on ferrite
b Conductor sandwiched between the dielectric and the ferrite

fundamental mode fields in slotline on a ferrite substrate. Zieniutycz[24] and Bock[25] have investigated slotline with the ferrite dielectric layered substrate shown in Fig. 9.9*a*, which gives a nonreciprocal phase difference almost constant with change of frequency. However, Bock[25] also shows that the slotline geometry shown in Fig. 9.9*b* gives a much larger nonreciprocal phase difference more variable with frequency. Zieniutycz[26] also gives results for the ferrite dielectric layered substrate shown in Fig. 9.9*a* used in finline. Hayashi and Mittra[27] have already shown that the simpler geometry of finline on a ferrite substrate, as shown in Fig. 9.7*d*, does not exhibit adequate nonreciprocal phase difference.

9.5 Y-junction circulator

The Y-junction circulator in symmetrical stripline has been described in Section 8.7 and is shown in Fig. 8.11. All the design theory is given in that section or in von Aulock and Fay.[28] The stripline circulator similar to that shown in Fig. 8.11 is used for discrete devices for connection to coaxial lines and often the stripline connection to the coaxial connector is air spaced and unsupported except by the connectors and the ferrite disks. In most other applications, the circulator is part of a microstrip circuit.

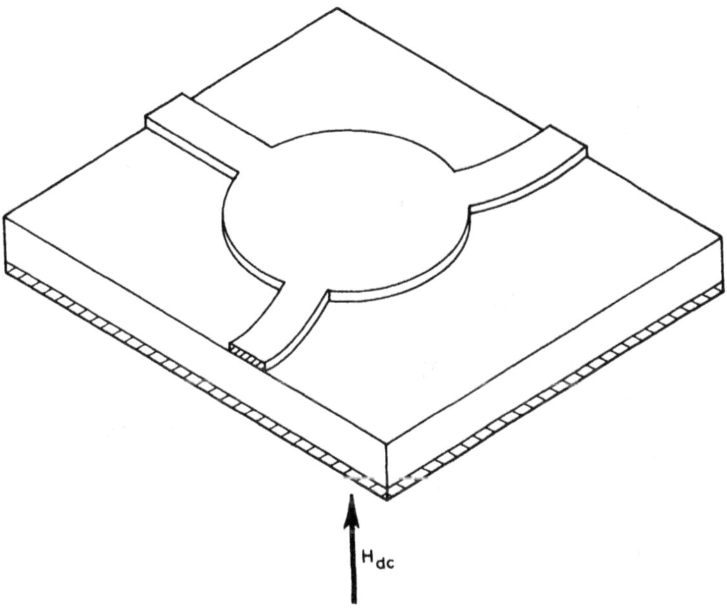

Fig. 9.10 *Y-junction circulator in microstrip*

The Y-junction circulator in microstrip is shown in Fig. 9.10. It may consist of a ferrite disk inserted into a circular hole in a dielectric substrate but more often it consists of a circular conductor at the junction on a uniform ferrite substrate. The biasing magnetic field is usually applied at the junction region alone and the rest of the ferrite substrate is unmagnetised. Then the theory of microstrip conductors, as discussed in the last section, is relevant. The theory of the design of symmetrical stripline circulators, as given in Section 8.7 and von Aulock and Fay,[28] appears to be equally applicable to the microstrip circulator. In the case of the uniform ferrite substrate, the diameter of the equivalent ferrite disk is determined by the shape and size of the microstrip conductor at the junction. To further this design, Nisbet and Helszajn[29] have analysed the performance of different shapes of metallisation, such as disks and triangles, on unmagnetised ferrite substrates.

The theory of von Aulock and Fay[28] has been applied by Wu and Rosenbaum[30] to the design of wideband microstrip circulators. They give expressions for the elements of the scattering matrix, see eqn. 8.2, and the input impedance of the circulator in terms of the physical parameters of the device. They show that the radius of the ferrite disk and the width of the microstrip conductors connecting to the junction are the design parameters which may be easily varied. They also show that satisfactory circulation is obtainable when $|\kappa/\mu|$ lies anywhere between a half and one. Simplifying eqn. 2.15 for operation well below resonance where $\omega \gg \omega_0$, gives

$$\left|\frac{\kappa}{\mu}\right| = \frac{\omega_m}{\omega} \tag{9.2}$$

where the terms are defined in eqn. 2.14. Then the operating frequency of the circulator may lie anywhere between ω_m and $2\omega_m$ and it will have a bandwidth of an octave. They show that at around 10 GHz their optimum design had wide interconnecting conductor strips and a relatively small ferrite disk at the junction. Khilla[31] analyses a similar geometry by the point matching technique to give an improved design. Both these designs provide comparatively broadband devices without any external matching except that necessary to accommodate the wide microstrip conductors into the rest of the circuit. If the condition $\frac{1}{2} < |\kappa/\mu| < 1$ cannot be attained, then any broadband performance has to be obtained by broadband impedance matching a narrowband circulator, as discussed in Section 8.6.

As mentioned in the last section, there is a problem of providing an adequate mechanical continuity for the microstrip conductor across the discontinuity between a ferrite disk and a dielectric substrate. One solution is to construct a circulator or circulators together with other components as necessary on a uniform ferrite substrate which is only magnetised at the circulator junctions. It is then possible to provide substantial connections at the edge of the substrate. Another alternative is to use *drop-in* circulators. These consist of a circular disk package of the junction alone, together with its magnet, ground plane, ferrite disk and conductor pattern having three conductor tabs for connection to the rest of the circuit. A further alternative is to add a ferrite disk to the back surface of a continuous dielectric substrate having the required conductor pattern of the Y-junction already on it. Such a system has been investigated by Hall, Prior and Ramsdale[32] and is shown in Fig. 9.11. Such circulators show a narrower bandwidth than inserted puck or all ferrite substrate devices. Hall *et al.*[32] used two-stage impedance matching to produce a device having a 9% bandwidth at 9 GHz.

Y-junction circulators have also been constructed in finline. Braas and Schieblich[33] constructed a Y-junction circulator using a single ferrite disk with unsymmetrical finline, as shown in Fig. 9.12*a*. The device was rather narrowband. The slots open at the edge of the ferrite disk. An alternative geometry is suggested by Goebel and Schieblich,[34] where the slots join the opening at the ferrite disk tangentially, as shown in Fig. 9.12*b*. Systematic experimental investigation showed that the optimum coupling angle was the same as the calculated optimum given by Wu and Rosenbaum.[30] It seems that the tangential connection to the ferrite disk enables wide coupling angles to be obtained. An operating bandwidth of 35% was obtained at about 10 GHz.

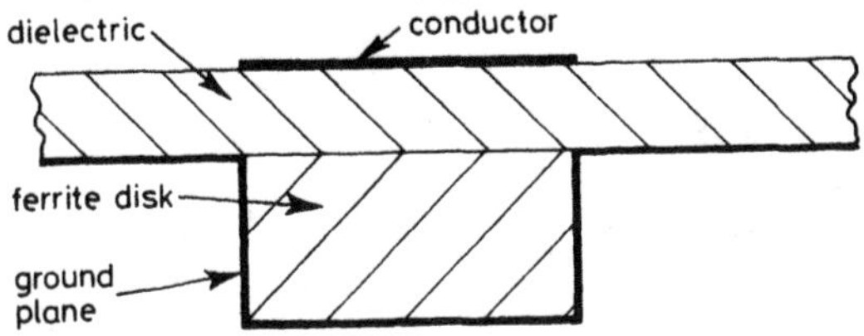

Fig. 9.11 *Microstrip Y-junction circulator with an external ferrite disk*

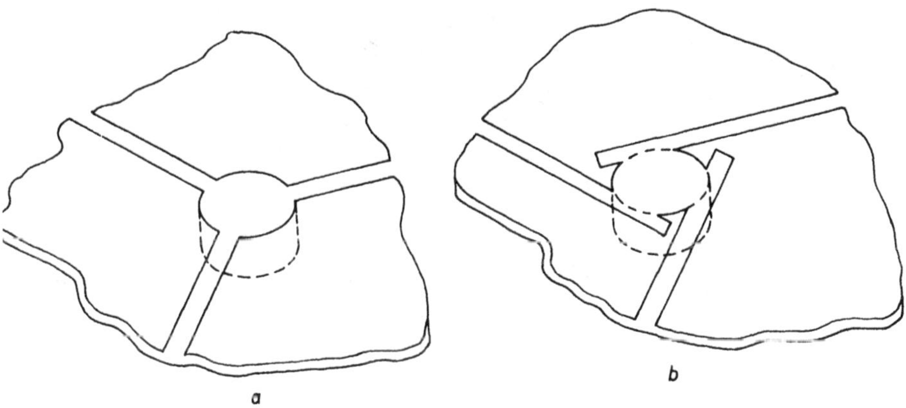

Fig. 9.12 *Slot conductor pattern for a finline Y-junction circulator*
a Conventional end coupling to the junction
b Tangential coupling to the junction. The junction region on the ferrite disk is a region without conductor

A multiport circulator may be constructed by cascading a number of three-port Y-junction circulators. However, in a communication system, the different path lengths for different channels may be undesirable. Denisov[35,36] has produced a design procedure for symmetrical five-port and multiport junction circulators consisting of a symmetrical multiport junction in stripline with ferrite disks at the junction, similar to that shown in Fig. 8.11.

9.6 Variable phase changer

The phase change effect due to the various possible directions of magnetisation of a ferrite substrate supporting a microstrip line has been discussed in Section

9.3. Hofmann[18] shows that longitudinal magnetisation, H_z of Fig. 9.8, causes reciprocal phase change, and Yeh and Bolle[20] show that transverse magnetisation in the plane of the substrate, H_x of Fig. 9.8, causes nonreciprocal phase change. However, the magnetically variable phase change is small. The phase change of the longitudinally magnetised microstrip is increased if coupled striplines are used in the form of a meanderline, as shown by Fig. 9.13. The

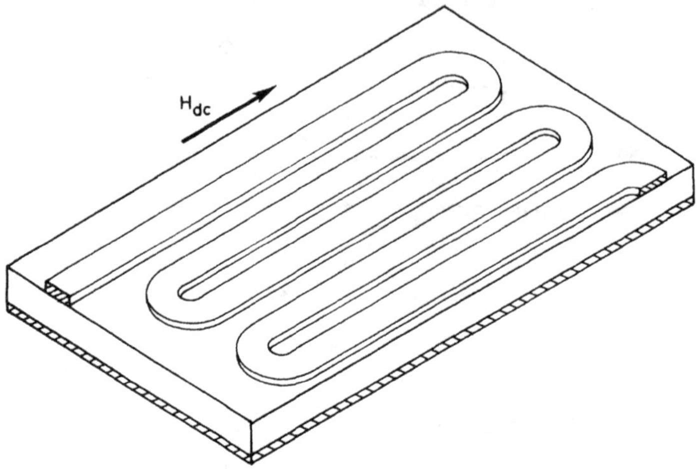

Fig. 9.13 *Microstrip meanderline phase changer having longitudinal magnetisation of the ferrite substrate*

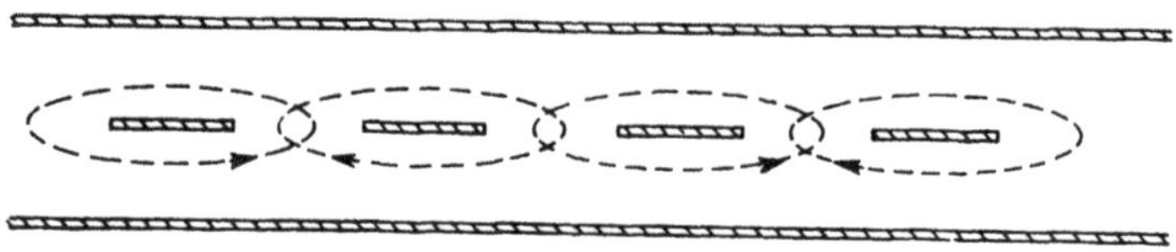

Fig. 9.14 *Section through a stripline meanderline phase changer showing the microwave magnetic field*

phase change is also nonreciprocal. There is close coupling between adjacent conductors and each section of line is a quarter wavelength long. At a point midway along one of the strips, the microwave magnetic field is shown by the dashed line loops in Fig. 9.14. These magnetic fields are perpendicular to one another half way between the two lines, and, as there is a quarter wavelength between them, the two fields are 90° out of phase. Therefore, this resultant magnetic field is circularly polarised in the plane of the magnetic field loops shown in Fig. 9.14, and the hand of circular polarisation is different, thus the effective permeability of the ferrite substrate is different, for waves travelling through the meander in different directions. Such a device is a variable nonreciprocal phase changer, dependent on the strength of the biasing magnetic field. If the spacing between the conductors in the meander is increased so that they are no longer coupled, the device becomes a variable reciprocal phase changer.

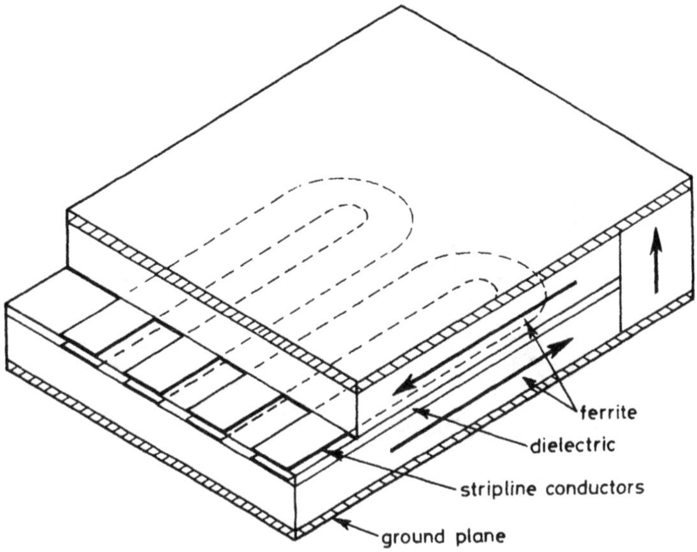

Fig. 9.15 *Part section through a stripline meanderline phase changer showing the ferrite magnetisation suitable for latching operation*

A form of latching nonreciprocal phase changer is shown in Fig. 9.15, where the meanderline is in the middle of a stripline ferrite sandwich and the two sides of ferrite are magnetised in opposite directions. The magnetic circuit may be completed outside the region of the meanderline to make a latching ferrite nonreciprocal phase changer. Such a device was first described by Jones,[37] who used a suspended stripline construction to form the stripline meander and ferrite slabs to fill what is shown as air space in Fig. 9.3. Subsequent experimental results show that a better performance is obtained if the meanderline conductors are spaced away from the ferrite, as shown in Fig. 9.16. A device of this geometry has been analysed by Hansson, Aditya and Larsson,[38] who give a design procedure for a latching ferrite-dielectric meanderline stripline phase changer.

9.7 Field-displacement isolator

For propagation along a very wide stripline or microstrip conductor on a ferrite substrate magnetised perpendicular to the plane of the substrate, the electromagnetic field is displaced and is strongest along one edge of the conductor. The effect is nonreciprocal because the reverse wave will cling to the opposite edge of the conductor. By loading one edge of the conductor with an absorbing material, an isolator can be constructed which is extremely wideband. Because the wave appears to cling to one edge of the stripline conductor, the isolator is often called an *edge-mode isolator*.

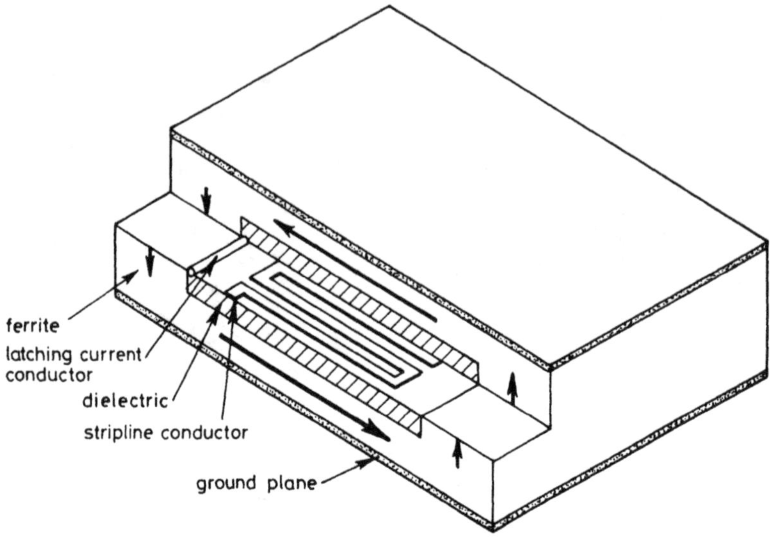

Fig. 9.16 *Part section through a latching stripline meanderline phase changer using a composite dielectric ferrite substrate*

Hines[39] was the first to investigate the field-displacement effect in stripline. The simple analysis given here largely follows his exposition. Assume a very wide conductor strip so that only one edge of the conductor needs to be considered. Neglect the effect of the fringing fields at the edge of the conductor. The wave equation in a transversely magnetised ferrite material is given by eqns. 6.14 to 6.17, which are equally applicable to the wide microstrip geometry shown in Fig. 9.17. It is assumed that a TE_{10}-mode type is able to propagate. The fields decay to zero at $x = \infty$, so that, in eqn. 6.17, $A = 0$ and the electric field is given by

$$E_y = B \exp(-jk_f x) \exp j(\omega t - \beta z) \tag{9.3}$$

As in earlier expressions, the factor $\exp j(\omega t - \beta z)$ will be assumed in all subsequent expressions for the field quantities. From eqns. 6.22 and 6.25 the other field quantities are given by

$$H_x = -\frac{B}{\omega\mu_e}\left[j\frac{\kappa}{\mu}k_f + \beta\right]\exp(-jk_f x) \tag{9.4}$$

$$H_z = -\frac{jB}{\omega\mu_e}\left[jk_f + \frac{\kappa}{\mu}\beta\right]\exp(-jk_f x) \tag{9.5}$$

At the boundary of the conductor strip, there cannot be any longitudinal component of the magnetic field, therefore $H_z = 0$ at $x = 0$. Then

$$jk_f + \frac{\kappa}{\mu}\beta = 0 \tag{9.6}$$

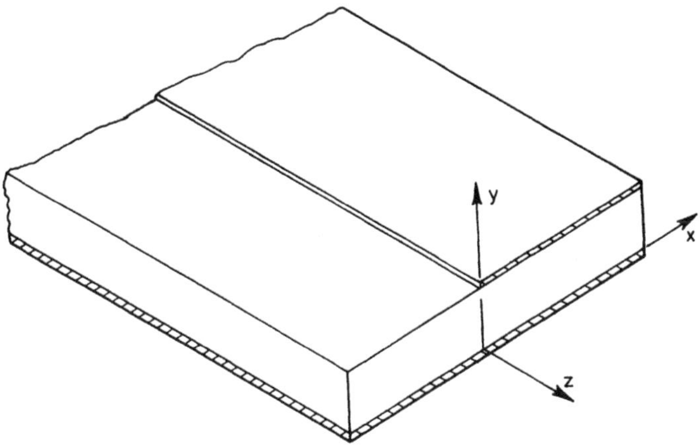

Fig. 9.17 *One edge of a wide microstrip conductor showing its relationship to a system of rectangular co-ordinate axes*

Solving for β and k_f between eqns. 6.16 and 9.6 gives

$$\beta^2 - \omega^2\varepsilon\mu \tag{9.7}$$

and

$$-k_f^2 = \left(\frac{\kappa}{\mu}\right)^2\omega^2\varepsilon\mu \tag{9.8}$$

jk_f is a real number and there is an exponential decay of the fields away from the edge of the conductor strip. Note that $H_z = 0$ everywhere and $\nabla \cdot \boldsymbol{H} \neq 0$ but $\nabla \cdot \boldsymbol{B} = 0$, because of the gyromagnetic effect of the ferrite, and Maxwell's equations are satisfied. Then B_z is obtained by substituting $H_z = 0$ into eqn. 6.4

$$B_z = -j\kappa H_x \tag{9.9}$$

The magnetic flux lines are then as shown in Fig. 9.18. As far as the E- and H-fields are concerned, the wave is a TEM-mode and it has the propagation constant of a non-gyromagnetic ferrite material for the ferrite-filled stripline. For a microstrip line, a filling factor correction will need to be applied to make allowance for the rather more dominant fringing field effects in microstrip. De Santis[40] has investigated the effect of the fringing-field on field displacement devices both theoretically and experimentally. He provides a numerical correction term which varies for different shapes of mode guiding edge.

For operating conditions well away from resonance, $\omega_0 \ll \omega$, and eqn. 2.15 shows that

$$\frac{\kappa}{\mu} \simeq \frac{\omega_m}{\omega} \tag{9.10}$$

and then

$$-k_f^2 = \omega_m^2 \varepsilon \mu \tag{9.11}$$

so that there is a constant rate of decay of the fields in a transverse direction at all frequencies. Hines[39] also shows that this edge-guided wave has a constant

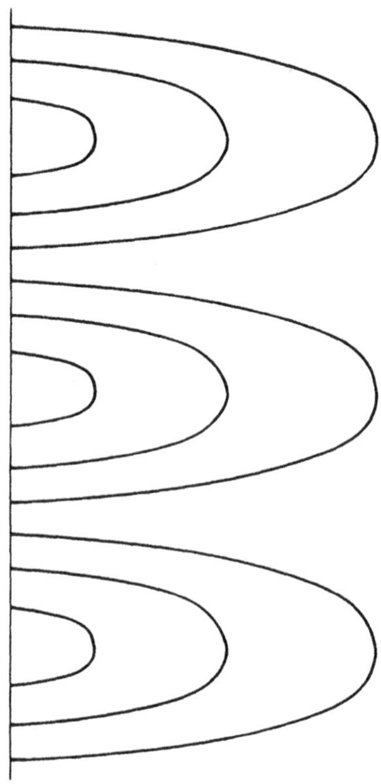

Fig. 9.18 *Magnetic flux lines of the edge-guided mode in wide microstrip*

characteristic impedance in the line independent of the frequency. He proposed the construction of a very-wide-band isolator similar to that shown in Fig. 9.19. Courtois, Forterre and Chiron[41] developed a similar wideband isolator using a bulk load material for part of the microstrip substrate, as shown in Fig. 9.20.

They obtained good isolator performance over two octaves or more. Forterre, Chiron and Courtois[42] show that similar performance is obtainable over a wide range of frequencies. Courtois *et al.*,[43] also show that this field-displacement

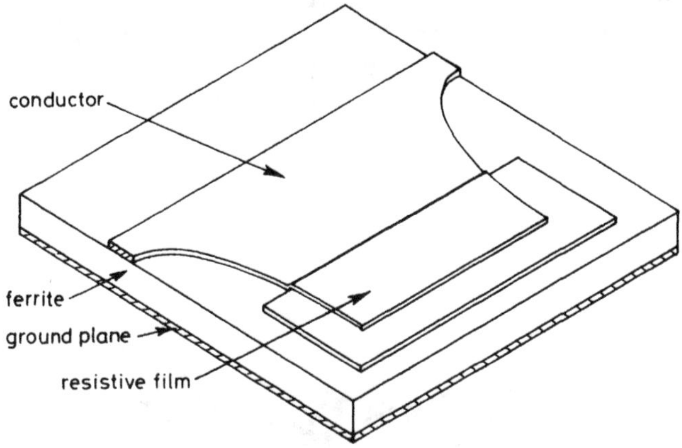

Fig. 9.19 *Microstrip field-displacement isolator using a resistive film*

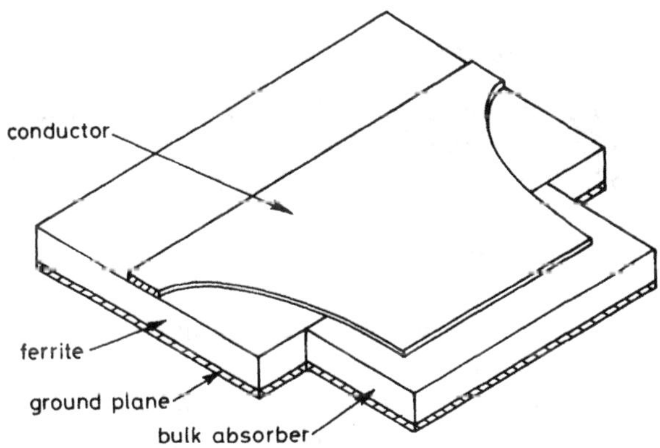

Fig. 9.20 *Microstrip field-displacement isolator using a bulk absorber*

isolator has a bandwidth of at least an octave when operating in the high bias field region, below its resonance frequency, at about 500 MHz. Dydyk[44] gives a summary of the design procedures for the design of field-displacement isolators. He reports the design of mode supressors enabling an isolator to operate above the frequency when higher-order modes would be able to propagate in the ferrite. For the normal edge-guided mode, the currents in the conductors are in the direction of propagation whereas the higher-order modes will generate some transverse current in the strip conductor and the ground plane. If the ground plane conductors are slotted in the direction of propagation, generation of the higher-order modes will be inhibited, and, if generated, they will be absorbed.

Araki, Koyama and Naito[45] have proposed an alternative form of the field-displacement isolator where the isolated signal is absorbed in the ferrite. They remove the resistive film shown in Fig. 9.19 and short the conductor to the ground plane, as shown in Fig. 9.21. The isolator presents a match to the isolated wave so it is assumed that higher-order modes are generated which are absorbed in the ferrite. As described by Araki *et al.*,[45] the isolator had a symmetrical wide microstrip conductor pattern with one edge shorted to the

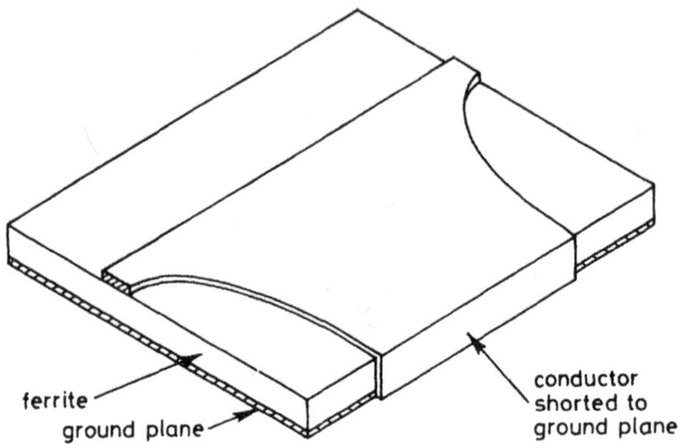

Fig. 9.21 *Microstrip field-displacement isolator with no absorber*

ground plane so that without the shorted connection there would have been no isolation. In its more conventional form, shown in Fig. 9.21, the curved edges aid conversion to bulk modes in the ferrite. De Santis[46] has shown that energy is radiated from the edge mode on a curved edge into the volume of the ferrite substrate, thus aiding the generation of higher-order modes in the ferrite due to the reverse wave.

Bolle[47,48] and Talisa and Bolle[49,50] have investigated the modes that can propagate in ferrite loaded stripline in much more detail than Hines,[39] and have identified a number of other possible modes of propagation. They have analysed various combinations of transversely magnetised ferrite with dielectric and bulk absorbing material and show that, for these geometries, the mathematical description of Hines[39] and that given above is a gross simplification. However, successful design of isolators appears to be proceeding using these simpler formulas. The work of Bolle and Talisa does facilitate the identification of spurious modes when these occur and are the cause of poor isolator performance.

9.8 Field-displacement phase changer and circulator

Hines[39] proposed that if one edge of a wide stripline using transversely magnetised ferrite was loaded with a dielectric material of high permittivity as shown in Fig. 9.22, such a device would show nonreciprocal phase change. Edge-guided mode waves, propagating in opposite directions, would cling to opposite edges of the stripline conductor and have appreciable quantities of their fields in materials of different permittivity. Bolle and Talisa[51] have analysed a similar geometry consisting of air, ferrite, dielectric and air regions and shown that

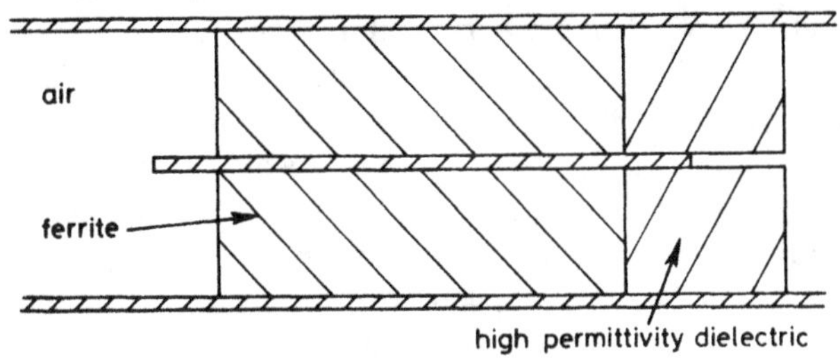

Fig. 9.22 *Field-displacement phase changer*

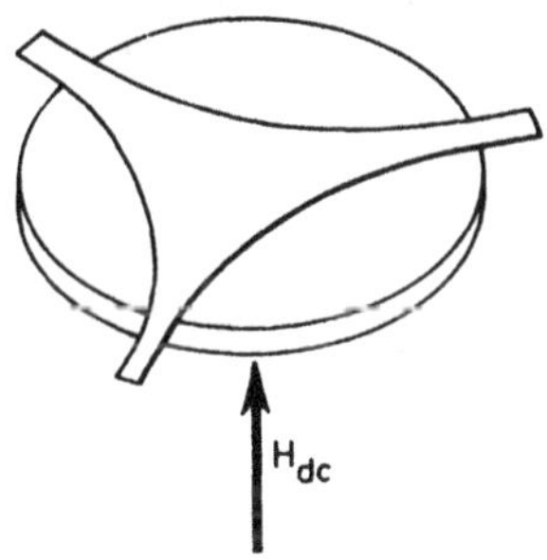

Fig. 9.23 *Field-displacement circulator*

useful phase change ought to occur with a bandwidth of one and a half octaves. Lavrova and Nikol'skii[52] have used a numerical method to investigate the electromagnetic field distribution due to a wide microstrip conductor on a larger ferrite substrate. They give a number of tables of calculated field distributions and one diagram of the transverse electric field strength which shows a small amount of field displacement for the particular conductor width used, $w/h = 2$, see Fig. 9.6. Helszajn and Downs[53] have used the transverse mode matching technique described by Gardiol[54] and in Section 6.5, to analyse the fields in a parallel plate waveguide loaded with ferrite and dielectric slabs. They demonstrate field displacement and show that the field exhibits planes of circular polarisation at the edges of the ferrite region.

De Santis and Pucci[55] have proposed that the field-displacement effect in stripline or microstrip can be used to construct a very wideband circulator. The geometry is shown in Fig. 9.23. However, there are problems. De Santis[46] has shown that, whenever the guiding edge is curved concavely, as in Fig. 9.23, energy is radiated from the edge mode into the volume of the ferrite substrate.

The amount of radiation increases as the radius of curvature of the edge decreases. Therefore, if it is possible to support a bulk mode in the ferrite under the conductor at the centre of the circulator, circulator action will be impared. De Santis and Pucci[55] suppress these unwanted volume modes. They use a circular hole in the ferrite substrate at the centre of the circulator to produce a circulator which provides 1 dB insertion loss and 20 dB isolation over the band 10·5 to 12·5 GHz. They[56] then use a particular inhomogeneous magnetic bias on a continuous ferrite substrate to produce a circulator with similar performance over the band 8 to 12 GHz. However, it does not seem that the very large bandwidths obtainable with field displacement isolators are realisable in field-displacement circulators. The circulators of de Santis and Pucci[55,56] show no performance advantages over more conventional microstrip and stripline circulators described in Section 9.5, and the provision of an inhomogeneous magnetic bias increases the complexity and cost.

Mazur[57] proposes an alternative form of the field-displacement circulator using the shorted edge of Araki *et al.*[45] It is shown in Fig. 9.24. It provides normal circulation between ports 1 to 2 and 2 to 3 and isolation between ports 1 and 3. His[57] experimental results showed reasonable performance over the band 4·5 to 6·5 GHz.

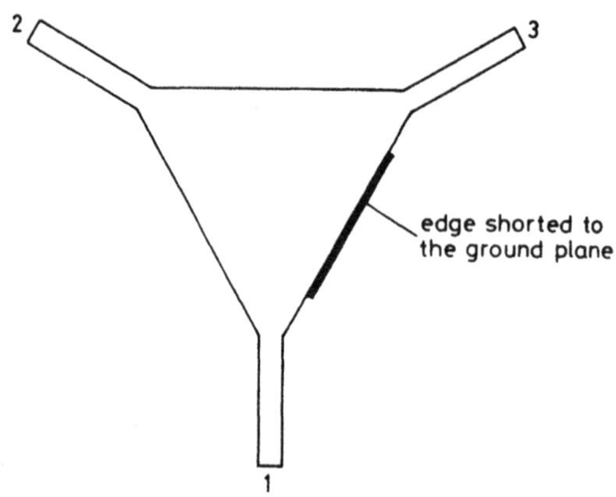

Fig. 9.24 *Alternative geometry for a field-displacement circulator with complete isolation between ports 1 and 3*

9.9 Ferrite resonant cavity

A highly polished sphere of low-loss single-crystal YIG material behaves like a high-Q resonant cavity or circuit. Its resonant frequency is a function of the bias magnetic field, and it can be tuned magnetically over a wide band of frequencies. The resonant frequency of a sphere is given by eqn. 2.14 to be, as shown by Carter[58]

$$\omega_0 = \gamma H_0 \tag{9.12}$$

If the sphere is other than spherical, eqn. 9.12 is modified, as shown by Kittel,[59] by the demagnetisation factors to give the resonant frequency

$$\omega_r^2 = (\omega_0 - N_z\omega_m + N_x\omega_m)(\omega_0 - N_z\omega_m + N_y\omega_m) \tag{9.13}$$

where ω_m is defined in eqn. 2.14 and is a measure of the internal magnetisation in the YIG. Demagnetisation factors are discussed in Sections 6.8 and 12.3. Helszajn[60] gives a full analysis of YIG resonators, and of filters made with them.

High-purity single-crystal YIG is used to make the resonant spheres which are then ground to shape with an exceptionally fine surface finish. The quality of the surface finish determines the microwave *Q*-factor of the ferrite sphere;[61] the ferrimagnetic resonance line width is proportional to the surface roughness. There are also small line-width variations depending on the crystallographic orientation of the sphere in the bias magnetic field. A typical sphere will have a surface finish of about 0·1 μm and will be about 1 mm in diameter, which makes it compatible with microwave stripline and microstrip circuits. However, the YIG resonant sphere may be used with any form of microwave transmission

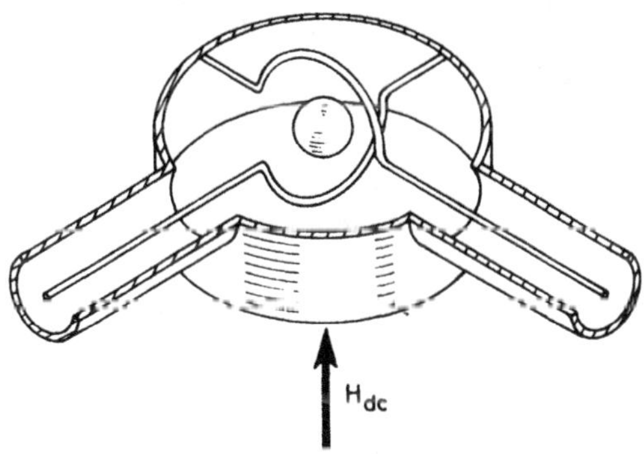

Fig. 9.25 *Ferrite resonant sphere mounted between two coaxial lines*

line and applications with the spheres mounted in waveguide are also given. The sphere is mounted adjacent to a microwave transmission line so that the microwave magnetic field is perpendicular to the magnetic bias field. At frequencies away from resonance, there will only be a small field coupled into the sphere, but, at resonance, more power is coupled into the sphere and large microwave magnetic fields will be supported in the sphere and adjacent to it. In particular, at resonance, a strong magnetic field is generated perpendicular to the bias magnetic field and the original magnetic field. This means that, if the YIG sphere is mounted between two transmission lines at right angles, as shown in Fig. 9.25, a signal is coupled between the input and the output only at the resonant frequency. The device acts as a bandpass filter. It can also be used in conjunction with an active device such as a Gunn diode to provide a magnetically tunable oscillator. The delay line properties of magnetostatic mode propagation in thin-film YIG can also be used to construct oscillators.[62]

Because of its small size, the YIG resonant sphere is ideally suited for use with microstrip and stripline circuits. Fig. 9.26 shows the YIG sphere mounted

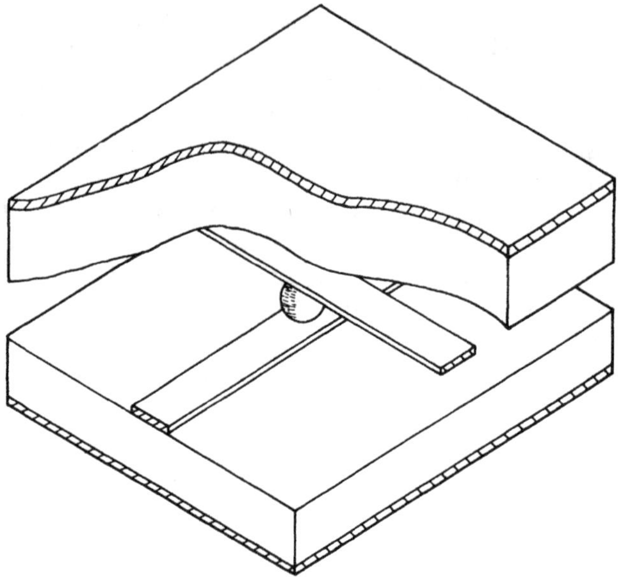

Fig. 9.26 *Ferrite resonant sphere mounted between two strip lines*

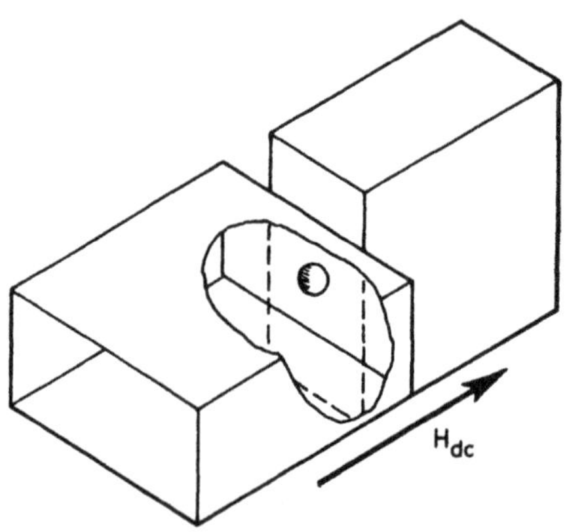

Fig. 9.27 *Ferrite resonant sphere mounted between two waveguides*

between two stripline conductors at right angles to one another. A simple resonant circuit provides a bandpass filter of rather inadequate performance for many applications. Carter[58] discusses the design of such filters in general terms and gives two applications which were realised in waveguide. Better performance filters require a number of resonators suitably coupled together. Matthaei[63] discusses the design of band-stop filters using a number of YIG spheres mounted in waveguide, and Carter[64] the design of multisphere bandpass filters in stripline. Helszajn[60,65] gives useful formulas and a number of design examples. Keane[66] discusses the design of filters in general terms. Details of the design of multiresonator bandpass and band-stop filters are given by Matthaei, Young and Jones,[67] and formulas for the coupled Q of the YIG sphere in a number of different microwave transmission structures are given by Carter[58] and Helszajn.[60]

In order to provide a bandpass performance in waveguide, the YIG sphere needs to be mounted in a coupling hole between two waveguides where the magnetic fields in the waveguides are orthogonal, as shown in Fig. 9.27.

9.10 References

1 GUPTA, K. C., GARG, R. and CHADHA, R.: 'Computer-aided design of microwave circuits', (Artech House, 1981)

2 WHEELER, H. A.: 'Transmission line properties of a stripline between parallel planes', *IEEE Trans.*, 1978, **MTT-26**, pp. 866–876

3 SAAD, T. S.: 'Microwave engineers' handbook. Vol. 1' (Artech House, 1971), pp. 122–127

4 BAHL, I. J. and GARG, R.: 'A designer's guide to stripline circuits', *Microwaves*, 1978, **17**, 1, pp. 90–96

5 EDWARDS, T. C.: 'Foundations for microstrip circuit design' (Wiley, 1981)

6 WHEELER, H. A.: 'Transmission-line properties of a strip on a dielectric sheet on a plane', *IEEE Trans.*, 1977, **MTT-25**, pp. 631–647

7 BAHL, I. J. and TRIVEDI, D. K.: 'A designer's guide to microstrip line', *Microwaves*, 1977, **16**, 5, pp. 174–176, 178, 180, 182

8 HAMMERSTAD, E. and JENSEN, O.: 'Accurate models for microstrip computer-aided design', IEEE MTT-S. Int. Microwave Symp. Digest, 1980, pp. 407–409

9 PRAMANICK, P. and BHARTIA, P.: 'An accurate description of dispersion in microstrip', *Microwave J.*, 1983, **26**, 12, pp. 89–92, 96

10 BHARTIA, P. and PRAMANICK, P.: 'A new microstrip dispersion model', *IEEE Trans.*, 1984, **MTT-32**, pp. 1379–1384

11 MIRSHEKAR-SYAHKAL, D. and DAVIES, J. B.: 'Accurate solution of microstrip and coplanar structures for dispersion and for dielectric and conductor losses', *IEEE Trans.*, 1979, **MTT-27**, pp. 694–699

12 LEWIN, L.: 'A resonance absorption isolator in microstrip for 4 Gc/s', *Proc. IEE*, 1956, **104B**, Suppl. No. 6, pp. 364–365

13 BRODWIN, M. E.: 'Propagation in ferrite-filled microstrip', *IRE Trans.*, 1958, **MTT-6**, pp. 150–155

14 PUCEL, R. A. and MASSE, D. J.: 'Microstrip propagation on magnetic substrates', *IEEE Trans.*, 1972, **MTT-20**, pp. 304–313

15 WHEELER, H. A.: 'Transmission-line properties of parallel strips separated by a dielectric sheet', *IEEE Trans.*, 1965, **MTT-13**, pp. 172–185

16 NISBET, W. T. and HELSZAJN, J.: 'Microstrip synthesized from waveguide equivalent', *Microwave Systems News*, 1981, **11**, 1, pp. 117–121, 136, 137

17 MINOR, J. C. and BOLLE, D. M.: 'Modes in the shielded microstrip on a ferrite substrate', *IEEE Trans.*, 1971, **MTT-19**, pp. 570–577

18 HOFMANN, H.: 'Dispersion of the ferrite-filled microstrip-line', *Arch. Elektronik & Uebertragungstech.*, 1974, **28**, pp. 223–227
19 LERER, A. M., MIKHALEVSKIY, V. S. and SHCHUCHINSKIY, A. G.: 'Electrodynamic theory of microstrip line with a ferrite substrate', *Rad. i Elek.*, 1984, **29**, pp.1039–1048 (English translation in *Rad. Eng. & Elec. Phys.*, **29**, 7, pp. 28–38)
20 YEH, D. T. and BOLLE, D. M.: 'Characteristic impedance and field patterns of the shielded microstrip on a ferrite substrate', *IEEE Trans.*, 1975, **MTT-23**, pp. 585–588
21 JANICZAK, B. and KITLINSKI, M.: 'Analysis of coupled asymmetric microstrip lines on a ferrite substrate', *Electron. Lett.*, 1983, **19**, pp. 779–781
22 LANGE, F. J. K.: 'Analysis of shielded strip- and slot-lines on a ferrite substrate transversely magnetised in the plane of the substrate', *Arch. Elektronik & Uebertragungstech.*, 1982, **36**, pp. 95–100
23 KAL, S., BHATTAHARYA, D. and CHAKRABORTI, N. B.: 'Development of yttrium iron garnet paste and fabrication of non-reciprocal microwave components', *J. Instn. Electronics & Telecom. Engrs. (India)*, 1982, **28**, pp. 71–75
24 ZIENIUTYCZ, W.: 'Modes of propagation in slot line with layered substrate containing magnetised ferrite', *Electron. Lett.*, 1983, **19**, pp. 135–136
25 BOCK, G.: 'New multilayered slot-line structures with very high nonreciprocity', *Electron. Lett.*, 1983, **19**, pp. 966–968
26 ZIENIUTYCZ, W.: 'Nonreciprocal phase shift in finline with layered substrate containing magnetised ferrite', *Electron. Lett.*, 1984, **20**, pp. 183–185
27 HAYASHI, Y. and MITTRA, R.: 'An analytical investigation of finlines with magnetised ferrite substrate', *IEEE Trans.*, 1983, **MTT-31**, pp. 495–498
28 VON AULOCK, W. H. and FAY, C. E.: 'Linear ferrite devices for microwave applications' (Academic Press, 1968), pp. 116–129
29 NISBET, W. T. and HELSZAJN, J.: 'Mode charts for microstrip resonators on dielectric and magnetic substrates using a transverse-resonance method', *IEE J. MOA*, 1979, **3**, pp. 69–77
30 WU, Y. S. and ROSENBAUM, F. J.: 'Wide-band operation of microstrip circulators', *IEEE Trans.*, 1974, **MTT-22**, pp. 849–856
31 KHILLA, A. M.: 'Field theory treatment of microstrip circulators for wide-band operation', *Arch. Elektronik & Uebertragungstech.*, 1981, **35**, pp. 445–451
32 HALL, P. S., PRIOR, C. J. and RAMSDALE, P. A.: 'First-order design method and experimental results for microstrip surface circulators with external matching circuits', *Electron. Lett.*, 1982, **18**, pp. 1043–1044
33 BRAAS, M. and SCHIEBLICH, C.: 'E-type circulator for fin lines', *Electron. Lett.*, 1981, **17**, pp. 701–702
34 GOEBEL, U. and SCHIEBLICH, C.: 'Broad-band finline circulators', IEEE MTT-S Symp. Digest, 1982, pp. 249–251
35 DENISOV, A. A.: 'The switching effect in a five-arm stripline ferrite loaded junction', *Radiotek.*, 1983, **38**, 6, pp. 71–73 (English translation in *Telecom. & Rad. Eng.*, **37/38**, 6, pp. 128–130)
36 DENISOV, A. A.: 'A miniature M-arm circulator based in a symmetrical star configuration of striplines', *Radiotek.*, 1983, **38**, 8, pp. 72–74 (English translation in *Telecom. & Rad. Eng.*, **37/38**, 8, pp. 65–67)
37 JONES, R. R.: 'A slow wave digital ferrite strip transmission line phase shifter', *IEEE Trans.*, 1966, **MTT-14**, pp. 684–688
38 HANSSON, E. R. B., ADITYA, S. and LARSSON, M. S.: 'Planar meanderline ferrite-dielectric phase shifter', *IEEE Trans.*, 1981, **MTT-29**, pp. 209–215
39 HINES, M. E.: 'Reciprocal and nonreciprocal modes of propagation in ferrite stripline and microstrip devices', *IEEE Trans.*, 1971, **MTT-19**, pp. 442–451
40 DE SANTIS, P.: 'Fringing-field effects in edge-guided wave devices', *IEEE Trans.*, 1976, **MTT-24**, pp. 409–415
41 COURTOIS, L., FORTERRE, G. and CHIRON, B.: 'Improvement in broad band ferrite

isolators', Proc. 20th Conf. Magn. & Magn. Mat. 1974, A.I.P. Conf. Proc, No. 24, pp. 501–502

42 FORTERRE, G., CHIRON, B. and COURTOIS, L.: 'A survey of broad band stripline ferrite isolators', *IEEE Trans.*, 1975, **MAG-11**, pp. 1279–1281

43 COURTOIS, L., BERNARD, N., CHIRON, B. and FORTERRE, G. E.: 'A new edge-mode isolator in the very high frequency range', *IEEE Trans.*, 1976, **MTT-24**, pp. 129–135

44 DYDYK, M.: 'Edge guide: one path to wideband isolator design', *Microwaves*, 1977, **16**, 1, pp. 54, 56–58, and 2, pp. 50, 52, 54–56

45 ARAKI, K., KOYAMA, T. and NAITO, Y.: 'A new type of isolator using the edge-guided mode', *IEEE Trans.*, 1975, **MTT-23**, p. 321

46 DE SANTIS, P.: 'Edge-guided modes in ferrite microstrips with curved edges', *App. Phys.*, 1974, **4**, pp. 167–174

47 BOLLE, D. M.: 'The edge guided mode on ferrite loaded stripline', IEEE MTT-S Int. Microwave Symp. Digest, 1976, pp. 257–259

48 BOLLE, D. M.: 'The modal spectrum of ferrite-loaded striplines', IEEE MTT-S Int. Microwave Symp. Digest, 1977, pp. 519–522

49 TALISA, S. H. and BOLLE, D. M.: 'On the modeling of the edge-guided mode stripline isolators', *IEEE Trans.*, 1979, **MTT-27**, pp. 584–591

50 TALISA, S. H. and BOLLE, D. M.: 'A five-region model for the edge-guided mode ferrite-loaded stripline isolator', *IEEE Trans.*, 1979, **MAG-15**, pp. 1001–1003

51 BOLLE, D. M. and TALISA, S. H.: 'The edge-guided mode nonreciprocal phase shifter', *IEEE Trans.*, 1979, **MTT-27**, pp. 878–882

52 LAVROVA, T. I. and NIKOL'SKII, V. V.: 'A method of determining the field of a stripline', *Izv. Vys. Uch. Zav. Rad.*, 1980, **23**, pp. 448–494 (English translation in *Rad. Phys & Quant. Elect.*, **23**, pp. 338–343)

53 HELSZAJN, J. and DOWNS, H.: 'Field-displacement, circular polarisation, scalar permeabilities and differential phase shift in edge mode ferrite devices', IEE Coll. Ferrite materials, devices and techniques, Digest No. 1985/15, 1985, pp. 4.1–4.5

54 GARDIOL, F. E.: 'Anisotropic slabs in rectangular waveguides', *IEEE Trans.*, 1970, **MTT-18**, pp. 461–467

55 DE SANTIS, P. and PUCCI, F.: 'Novel type of MIC symmetrical 3-port circulator', *Electron. Lett.*, 1972, **8**, pp. 12–13

56 DE SANTIS, P. and PUCCI, F.: 'The edge-guided-wave circulator', *IEEE Trans.*, 1975, **MTT-23**, pp. 516–519

57 MAZUR, J.: 'Analysis of quasicyclic ferrite junction by point-matching technique', *Electron. Lett.*, 1983, **19**, pp. 140–142

58 CARTER, P. S.: 'Magnetically-tunable microwave filters using single-crystal yttrium-iron-garnet resonators', *IRE Trans.*, 1961, **MTT-9**, pp. 252–260

59 KITTEL, C.: 'On the theory of ferromagnetic resonance absorbtion', *Phys. Rev.*, 1948, **73**, pp. 155–161

60 HELSZAJN, J.: 'YIG resonators and filters' (Wiley, 1985)

61 LECRAW, R. C., SPENCER, E. G. and PORTER, C. S.: 'Ferromagnetic resonance line width in yttrium iron garnet single crystals', *Phys. Rev.*, 1958, **110**, pp. 1311–1313

62 CARTER, R. L., OWENS, J. M. and DE, D. K.: 'YIG oscillators: Is a planar geometry better?', *IEEE Trans.*, 1984, **MTT-32**, pp. 1671–1674

63 MATTHAEI, G. L.: 'Magnetically tunable band-stop filters', *IEEE Trans.*, 1965, **MTT-13**, pp. 203–212

64 CARTER, P. S.: 'Side-wall-coupled, strip-transmission-line magnetically tunable filters employing ferrimagnetic YIG resonators', *IEEE Trans.*, 1965, **MTT-13**, pp. 306–315

65 HELSZAJN, J.: 'Principles of microwave ferrite engineering' (Wiley, 1969), pp. 214–233

66 KEANE, W. J.: 'Narrow-band YIG filters aid wide-open receivers', *Microwaves*, 1978, **17**, 9, pp. 50–54, 56

67 MATTHAEI, G. L., YOUNG, L. and JONES, E. M. T.: 'Microwave filters, impedance matching networks and coupling structures' (McGraw-Hill, 1964)

Chapter 10

Millimetre Wave Devices

10.1 Introduction

Microwave is usually considered to cover the frequency range 0.3–30 GHz where the wavelength is of the same order of scale as the human body and microwave structures are relatively easy to construct and handle. Millimetre waves cover the higher frequency range 30–300 GHz or up to 1000 GHz. In the millimetre wave frequency range, scaling becomes a problem so that copies of microwave devices become difficult to manufacture with adequate accuracy. Then special techniques suitable only for millimetre wave devices become necessary. This chapter concentrates on those techniques which are peculiar to ferrite devices operating at millimetre wave frequencies. The rest of this section discusses the different transmission structures that have been developed particularly for use as millimetre wave lines. In the next section, we consider the limitations due to the fact that the ferrite material is much less efficient as a gyromagnetic material at these higher frequencies. For free-space propagation, there are a number of atmospheric gas absorption frequencies which limit the range of useful propagation, but there are also windows at 35, 94, 140 and 230 GHz, which is why there are a large number of publications concentrating particularly on these frequencies.

Rectangular waveguide becomes small at these frequencies, thus it is no longer bulky, but there are problems of manufacture of very small sizes at the highest frequencies, for example, the standard waveguide for operation at 300 GHz has a broad dimension of 0·86 mm. Alternatives are being investigated which are larger structures compared to the wavelength. One is the dielectric waveguide. Marcatili[1] and Goell[2] have investigated the dielectric rectangular waveguide for integrated optics but the waveguide is also suitable for the highest millimetre wave frequencies. It consists of a rectangular rod of dielectric material immersed in another dielectric of slightly smaller permittivity, as shown in Fig. 10.1*a*. An alternative is the supported dielectric waveguide shown in Fig. 10.1*b*. A dielectric rod of high permittivity is supported on a sheet of dielectric of low permittivity. Another alternative is the guided surface wave of the dielectric image guide described by Schlisinger and King.[3] It consists of a semicircular or rectangular dielectric strip on a conducting ground plane as shown in Fig. 10.1*c*. The rectangular dielectric image guide has been further investigated by Solbach.[4,5] Zhou and Itoh[6,7] propose a further modification, which radiates less and ought to be useful on bends, of the trapped image guide as shown in Fig. 10.1*d*. Yoneyama and Nishida[8] propose another alternative which consists of the rectangular dielectric waveguide sandwiched between two parallel conducting plates, as shown in Fig. 10.1*e*, which they call the nonradiative dielectric waveguide. They also propose an improved version,[9] as shown in Fig. 10.1*f*, called the insulated nonradiative dielectric waveguide. These dielectric waveguide structures are all compatible with bulk optical style generators and waves can be launched from rectangular metallic waveguide, but they are not as useful with semiconductor circuit devices as microstrip.

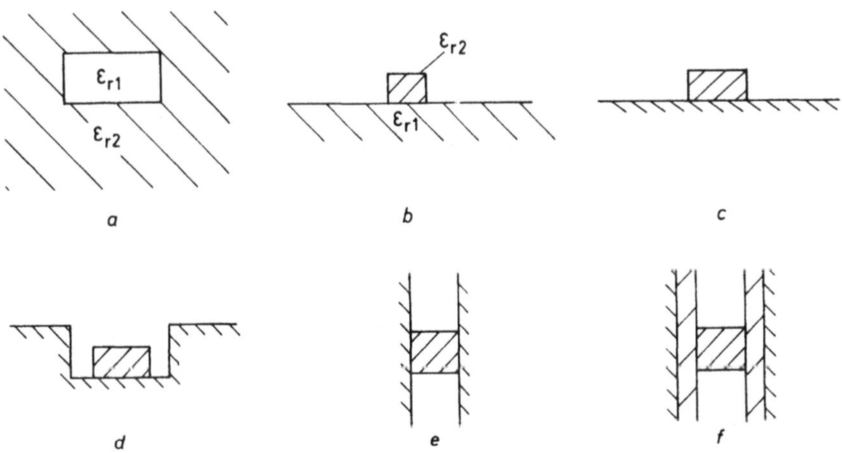

Fig. 10.1 *Different millimetre waveguide structures*
a Rectangular dielectric waveguide
b Supported dielectric waveguide
c Dielectric image guide
d Trapped image guide
e Nonradiative dielectric waveguide
f Insulated nonradiative dielectric waveguide

Semiconductor devices operate satisfactorily into the millimetre wave frequency region so that stripline or microstrip circuits compatible with integrated circuits are used at these frequencies. The dimensions of microstrip become very small and can cause difficulties. An alternative is to use slotline as discussed in Section 9.4. The slotline may be used with an adjacent groundplane, as shown in Fig.9.7*c*, or without, as shown in Fig. 10.2*a*. Samardzija and Itoh[10] have investigated the double-layered slotline particularly for use with millimetre waves. It is shown in Fig. 10.2*b* and makes a low-loss transmission medium when $\varepsilon_{r2} > \varepsilon_{r1}$. This is similar to the ferrite dielectric layered substrate investigated by Zieniutycz[11] and Bock,[12] as shown in Fig. 9.9*a*. Solbach[13] has investigated the performance of finline as a medium for mounting semiconductor devices in millimetre wave systems. His finline is slotline mounted across the centre *E*-plane line of standard rectangular waveguide. Three different forms of finline are shown in Fig. 10.3. In a theoretical investigation, the finline using a

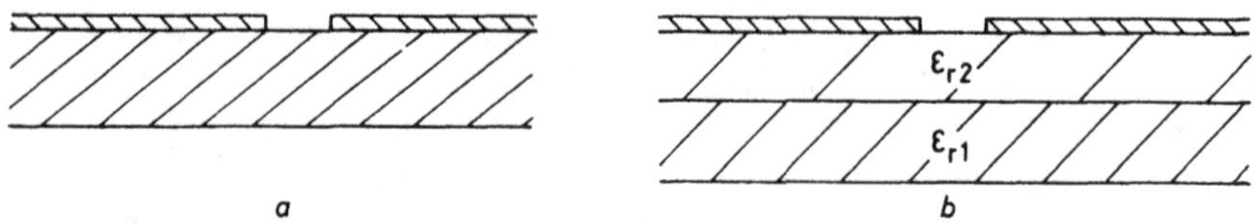

Fig. 10.2 *a* Slotline without a backing plane
b Double-layered slotline $\varepsilon_{r1} > \varepsilon_{r2}$

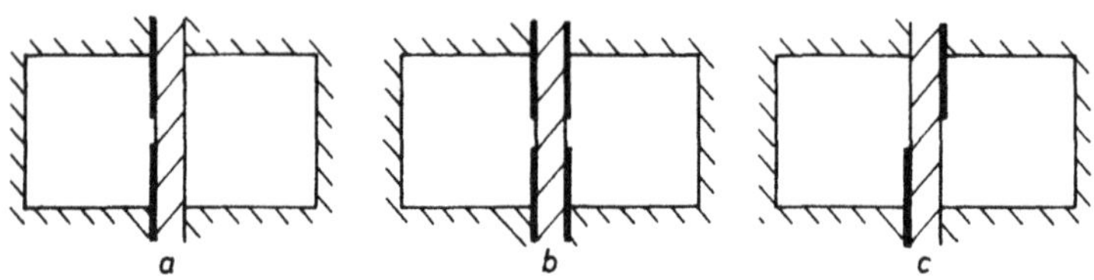

Fig. 10.3 *a* Unilateral finline
b Bilateral finline
c Antipodal finline

dielectric support having a relative permittivity of unity is the same as ridge waveguide having a very narrow ridge. Normally slotline is not shielded and is made with a substrate having a high relative permittivity so as to concentrate most of the microwave power in the substrate. Because it is shielded, finline does not need to use a substrate with a high relative permittivity, and so it is usually made with a substrate having a low relative permittivity. The theoretical propagation constant of a number of different finline structures has been investigated by Mirshekar-Syahkal and Davies[14] and by Vahldieck.[15]

A discussion of the properties of slotline and finline with ferrite substrates is given in Section 9.4. Applications to particular devices for use at millimetre wave frequencies are given in the rest of this chapter.

10.2 Limitations of ferrite materials

Ideally there is no limit to the frequency at which ferrite materials will demonstrate gyromagnetic properties. However, eqn. 2.15 shows certain characteristic effects as the frequency is increased. Away from resonance, at very high frequencies, $\omega \gg \omega_0$, and

$$\mu = \mu_0; \qquad \kappa = \mu_0 \omega_m / \omega \tag{10.1}$$

where $\omega_m = \gamma M_0/\mu_0$. For a material of any given saturation magnetisation, M_0, the cross-diagonal component of the permeability tensor, κ, is inversely proportional to frequency. Therefore, at higher frequencies, κ gets smaller. Thus rotation or differential phase change per wavelength also gets smaller as the frequency is increased. In microstrip circulators, Wu and Rosenbaum[16] have shown that broadband circulation is possible when $\frac{1}{2} < |\kappa/\mu| < 1$. It then becomes necessary to use a ferrite having the highest possible saturation magnetisation. At the present, nickel ferrite or lithium ferrite, both having a saturation magnetisation of 0·50 T, are recommended for use in millimetre wave devices. Using these ferrites, $|\kappa/\mu| = \frac{1}{2}$ at 28 GHz, so that the condition for wideband operation of microstrip circulators cannot be met at millimetre wave frequencies. What is more, it has been shown experimentally that the bandwidth of stripline devices decreases as the operating frequency is increased. In 1975, Forterre, Chiron and Courtois[17] showed that the bandwidth of broadband stripline isolators decreased from $2\frac{1}{4}$ octaves to $1\frac{1}{4}$ octaves when the upper operating frequency was doubled from 10 GHz to 20 GHz.

For higher frequencies above 300 GHz, the use of the nonreciprocal properties of the magnetised plasma in a semiconductor material is proposed. Bolle and Talisa[18] have shown theoretically that there is a field-displacement effect using a high-quality moderately-doped *n*-type GaAs at liquid nitrogen temperature. They investigate a thin semiconductor region backed by a dielectric and, in another paper,[19] a thin semiconductor region on a layered dielectric. There appears to be useful nonreciprocal effects at 300–500 GHz.

10.3 Permanent magnet materials

For ferrite devices operating at resonance, such as resonance isolators or ferrite resonant cavities, the magnetising field required for resonance is proportional to the resonant frequency. For a spherical ferrite body, the resonant frequency is given by eqn. 9.12. At a frequency of 84 GHz, the biasing magnetic field needs to be 3 T, which requires a very bulky electromagnet or a superconducting magnet. An alternative is to use a ferrite material having a large internal magnetisation, or anisotropy field. If the applied magnetic field is in the same direction as the anisotropy field, the resonant frequency is given by

$$\omega_0 = \gamma(H_0 + H_a) \tag{10.2}$$

where H_0 is the applied magnetic field intensity and H_a is the internal anisotropy field. As above, when quoting magnitudes for the magnetic fields, they will be given in tesla, the units of flux density, because the bias magnetic field is always an externally applied magnetic flux. It is also easy to remember the usual value of γ for ferrites as 28 GHz/T. Permanent magnet ferrite materials are discussed in Section 1.7. Barium and strontium hexagonal ferrite materials have large internal magnetic fields and can be used in millimetre wave ferrite devices.

Harrison[20] discusses the different compounds that may be used to provide different internal magnetic fields having a selfresonance at different frequencies. Adjustment of the operating frequency is then possible with a small additional external magnetic field.

Taft, Harrison and Hodges[21] describe a range of millimetre wave resonance isolators using hexagonal ferrites in rectangular waveguide. A thin sheet of ferrite is mounted on a dielectric slab parallel to the narrow wall of the waveguide, as shown in Fig. 10.4. A number of different ferrite materials are used in one isolator to provide broadband performance. An external bias

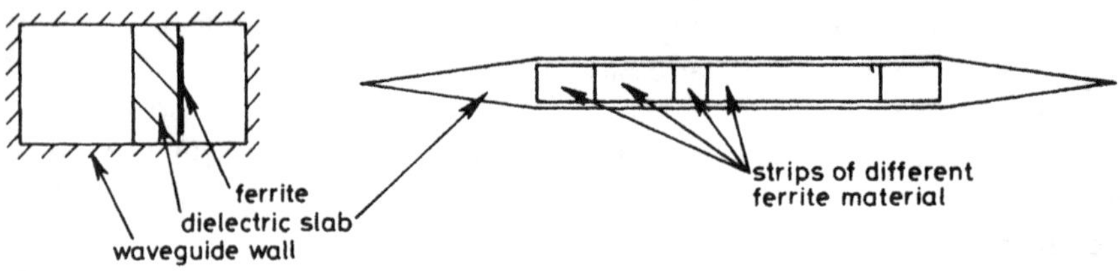

Fig. 10.4 *Wideband resonance isolator*

magnetic flux density of about 0·1 T is provided for final adjustment. A similar system using thin slabs of permanent magnet ferrite to provide resonance isolation in finline is described by Helszajn and Thorpe.[22] Two or three thin slabs of different ferrite material are laid on top of one another, thus providing a more compact device than that shown in Fig. 10.4. These hexaferrite materials can be expected to provide gyromagnetic resonance up to about 140 GHz.

Antiferromagnetic materials have even higher effective internal magnetic fields available for resonance. Heller, Stickler and Thaxter[23] discuss the properties of antiferromagnetic materials and their interaction with an electromagnetic wave at resonance. For the simplest antiferromagnetic material there are two sublattices in the crystal having equal and opposite magnetisations which cancel out and provide no external magnetisation. Using an antiferromagnetic material where the internal magnetic field is strongly aligned to one crystallographic direction, a single crystal may be used to provide controlled interaction with an electromagnetic wave. If an external biasing magnetic field is used, there are two resonant frequencies given by the external field aiding or reducing the internal field. However, the two modes of resonance precess in opposite directions of circular polarisation and may be used to construct nonreciprocal devices. They[23] give some experimental results using Cr_2O_3 at 77 K and 140 GHz in a resonance isolator in a rectangular waveguide configuration.

The ferrite sphere used as a magnetically tunable resonant cavity, as described in Section 9.9, is also limited for use at millimetre wave frequencies by the size of the bias magnetic field for resonance. However, single-crystal barium ferrite hexagonal permanent magnet material may be used to make the ferrite resonant sphere. Reisch, Grant, Lind, Espinosa and Goldberg[24] used spheres made from single crystals of $Ba_3Zn_2Fe_{24}O_{41}$ (ZnZ) for filters in the frequency range 26–40 GHz. Lemke, Hoppe, Tolksdorf and Welz[25] used single crystal hexagonal barium ferrite ($BaFe_{12}O_{19}$, BaM) spheres for filters up to 80 GHz. The material has a low millimetre wave loss and a high H_a. From their measurements, they obtained a value for the gyromagnetic ratio $\gamma = 28$ GHz/T. They used polished spheres of about 0·3 mm diameter which were allowed to rotate freely on the tip of a small quartz tube so that the internal magnetisation could align parallel to the external bias magnetic field. For permanent structures, as opposed to experimental configurations, the ferrite sphere can be held in an aligning magnetic field while the fixing adhesive sets.

10.4 Faraday rotation devices

Faraday rotation devices are described in Sections 5.1 to 5.3 for application at usual microwave frequencies. The theory of longitudinally magnetised ferrite sections of circular shape which are normally used to provide Faraday rotation in devices is given in Chapter 3. Faraday rotation devices at millimetre wave frequencies, made to scale, require ferrite rods of extremely small diameter supported at the centre of a small circular waveguide. One possible alternative is to use ferrite-filled circular waveguide, and a mode chart for such a waveguide is given in Fig. 3.4. That figure shows that there is a range of normalised waveguide radii where the dominant HE_{11}-mode is the only mode of propagation given by

$$0{\cdot}090 < \frac{r}{\lambda_0} < 0{\cdot}135$$

However, the rotation per unit length will vary with frequency. The results in Fig. 3.4 are calculated for ferrite properties appropriate to operation around 10 GHz, so that, at millimetre wave frequencies, κ_r will be quite small and the nonreciprocal phase change will be much smaller than that indicated by Fig. 3.4. Using a larger radius ferrite-filled waveguide in a region where higher-order modes are able to propagate, Fig. 3.4 shows that there is a range of waveguide radius where the differential phase change of the HE_{11}-mode is approximately constant. If the HE_{11}-mode in the ferrite-filled waveguide is generated from the dominant TE_{11}-mode in circular waveguide, it is possible that higher-order mdoes will not be excited and such a design ought to provide a broadband rotator.

Barnes[26] has suggested the use of a longitudinally magnetised ferrite rod waveguide (with the metal waveguide wall at infinity) to provide broadband Faraday rotation devices at millimetre wave frequencies. He gives details of a Faraday rotation isolator operating over the frequency band 50–60 GHz, using a 45° rotator as described in Section 5.2. He uses a common lower-microwave-frequency ferrite and a low magnetising field. If it is required for switching, the absence of conducting waveguide walls makes fast switching easy. For the ferrite acting as a dielectric rod waveguide, if the diameter of the ferrite rod is sufficiently large, the microwave power is concentrated in the ferrite rod and the fields of the HE_{11}-dominant mode in the ferrite approximate to those of a plane wave in an unbounded ferrite medium. Then the rotation per unit length is given by eqn. 2.58, which is seen to be independent of frequency. This is why the devices described by Barnes[26] are intrinsically broadband.

The ferrite rod dielectric waveguide rotator has been further investigated by Williamson and Koh[27] in producing an isolator for operation over the frequency band 26–40 GHz. When the ferrite was surrounded by a low permittivity dielectric, it was found that adding a conductive wall to the outside of the dielectric created a different waveguide system which could support several modes and degrade the device performance. Without that waveguide wall, only the dominant mode appeared to propagate. Good performance was obtained when there was a maximum difference between the permittivity of the ferrite rod and that of the dielectric surrounding medium, so that the ferrite rod was supported in air in their final design. A diagram of their isolator is given in Fig. 10.5. It has an insertion loss of 1·0 dB and an isolation better than 20 dB over the frequency band 26–40 GHz. The input and output polarised absorbers are resistive films along a diameter of the dielectric tapers.

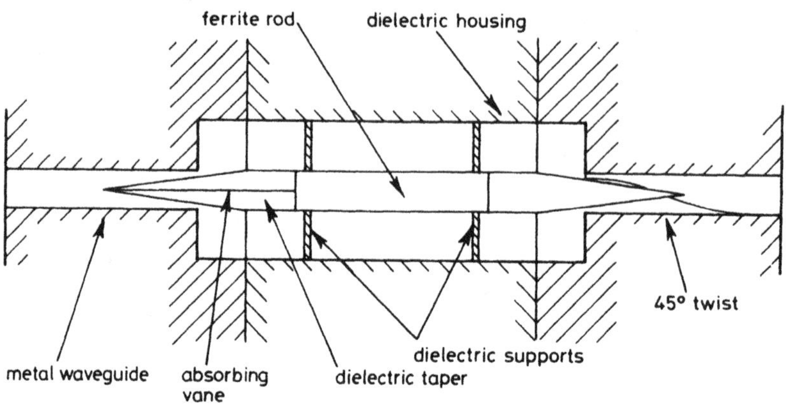

Fig. 10.5 *Wideband rotator isolator*

Boyd[28] has scaled his design[29] of the dual-mode phase changer for use up to 60 GHz. The basic principle of the dual-mode phase changer is given in Sections 5.3 and 5.6. The whole device is constructed in ferrite-filled circular waveguide including the birefringent quarter-wave plates. At these high frequencies, the size of the ferrite becomes very small, but it has still been found possible to provide a transverse four-pole magnetic field on the ferrite rod. The dual-mode phase changer makes use of the nonreciprocal phase change effect of longitudinally magnetised ferrite material on a circularly polarised wave. Boyd[28] does not use the ferrite-rod waveguide described in the previous paragraph to give broadband performance, but uses ferrite-filled circular waveguide with the conducting waveguide wall on the outside of the ferrite rod.

10.5 Transversely magnetised structures

A number of different configurations of transversely magnetised ferrite in various transmission lines have been investigated to provide phase change or isolation. Some will be described in this section. Babbitt and Stern[30] describe a scaled-down version of the rectangular waveguide nonreciprocal phase changer, as described in Section 7.2, for use at 35, 65 and 94 GHz. The best performance is obtained when the centre of the ferrite toroid is filled with a material having a high permittivity. Babbitt and Stern produce the ferrite toroid for their millimetre wave phase changers by arc plasma spray deposition of a ferrite powder on to a dielectric former. A lithium ferrite powder is sprayed on to a lithium titanate dielectric core. The lithium ferrite has a high saturation magnetisation, up to 0.5 T, and lithium titanate has a relatively permittivity of 26. The thermal expansion of the dielectric also matches that of the ferrite, hence avoiding stresses and cracks in the ferrite. The ferrite is fired after spraying. Further details of the fabrication techniques are given in another paper by the same authors.[31]

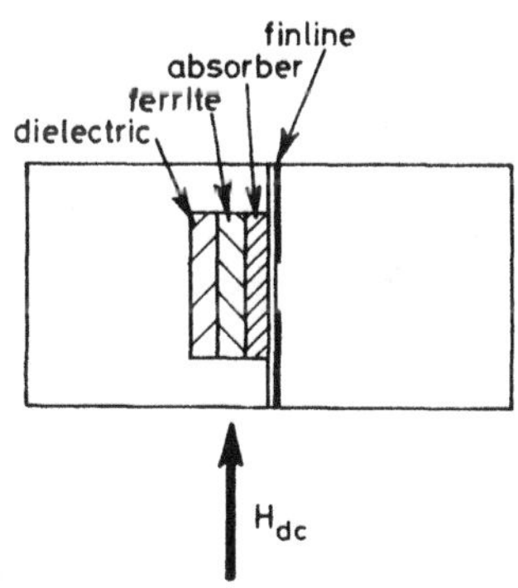

Fig. 10.6 *Field-displacement finline isolator*

A resonant isolator using permanently magnetised ferrite in rectangular waveguide[21] has been described in Section 10.3 and is shown in Fig. 10.4. A field-displacement isolator in ferrite loaded finline is described by Beyer and Solbach[32] and analysed in more detail by Beyer and Wolff.[33] They calculate the field distribution in the finline using the ferrite dielectric substrate shown in Fig. 9.9*a*. There is a maximum field strength in the plane of the metallisation at the slot for one direction of propagation and there is a minimum at the ferrite interface for the opposite direction of propagation. They constructed an isolator as shown in Fig. 10.6 and measured its performance at about 11 GHz. Further investigation of the layered ferrite dielectric slotline as used in this finline isolator has been given by Zieniutycz[11] and Bock.[12,34] A permanent magnet resonance isolator in finline has already been described in Section 10.3.[22] A number of different permanent-magnet-ferrite thin slabs are mounted on a dielectric spacer on a finline conductor similar in geometry to that shown in Fig. 10.6.

Yun and Itoh[35,36] describe an isolator structure in a modified trapped image guide. A ferrite slab backed by a conductor is coupled to an image guide as shown in Fig. 10.7. With a transverse magnetic field, the conductor backed ferrite slab gives a nonreciprocal phase change. The thicknesses of the ferrite and image guide are chosen so that the propagation constants of the modes in the two are nearly equal for propagation in one direction and different for propagation in the opposite direction. The wave propagating on the image guide

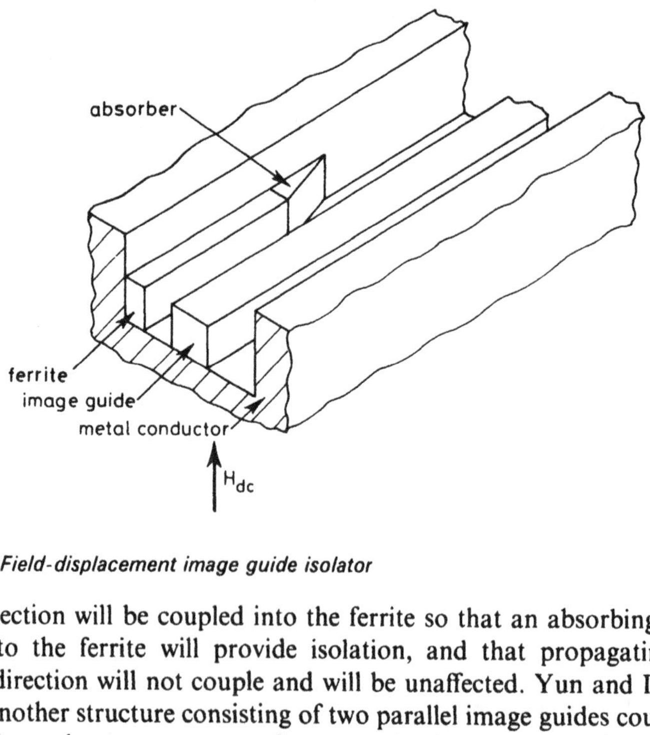

Fig. 10.7 *Field-displacement image guide isolator*

in one direction will be coupled into the ferrite so that an absorbing material adjacent to the ferrite will provide isolation, and that propagating in the opposite direction will not couple and will be unaffected. Yun and Itoh[37] also propose another structure consisting of two parallel image guides coupled by a ferrite slab overlay to construct a four-port circular used as an isolator. They obtain about 10 dB isolation.

Muraguchi, Araki and Naito[38] propose the use of a transversely magnetised ferrite disk as a nonreciprocal travelling-wave resonator. It is critically coupled to a dielectric image guide and provides narrowband resonance isolation. Isolation occurs for opposite directions of propagation at different frequencies corresponding to the resonances of the circularly polarised modes of opposite hand in the disk. At millimetre wave frequencies quite a small ferrite disk is sufficiently large to act as a resonant cavity.

10.6 Waveguide junction circulators

The Y-junction circulator is probably the most used microwave ferrite component. Versions have been made in most of the transmission lines used at millimetre waves. We start by discussing circulators in waveguide, and, in the next section, those in microstrip, coplanar line, finline, image line and supported dielectric waveguide. In some cases these are not new designs but modifications of known microwave designs, making allowance for the small gyromagnetic effect at these high frequencies.

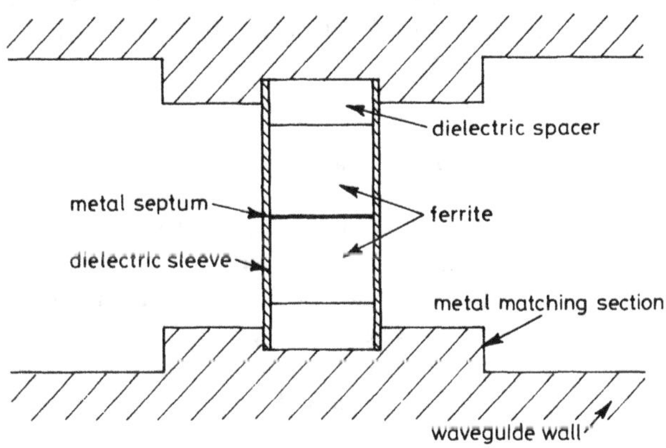

Fig. 10.8 *Waveguide Y-junction circulator, showing mounting of the central ferrite cylinders*

Piotrowski and Raue[39] have devised a design of *H*-plane Y-circulator which uses two cylinders of ferrite at the centre of the junction separated by a conducting septum, as shown in Fig. 10.8. The conducting septum enables half-height ferrite resonators to be used in a standard height waveguide design. In addition, mechanical stability and strength is provided by the dielectric sleeve sitting in a recess in each metallic matching transformer. The absence of any epoxy or glue contributes to a very low-loss circulator. The diameter of the matching transformer is equal to the diameter of the ferrite plus $\lambda_0/2$. The optimum step height of the transformer is obtained experimentally. This design provides a good broadband match which could be scaled to provide a 7 GHz bandwidth of 27–34 GHz or 31–38 GHz. Piotrowski and Schell[40] adapt the design for use at 93–100 GHz. A larger ferrite cylinder is used which will support properly selected higher-order modes, because the ferrite has unsufficient gyromagnetism at these higher frequencies. The overall construction is similar to that shown in Fig. 10.8, except that there is no metallic septum cutting the ferrite cylinder in half, and, in proportion, the ferrite diameter is much larger. The greater volume of ferrite gives rise to a larger insertion loss than that obtained by Piotrowski and Raue[39] but it is still less than 0·5 dB over a 7 GHz band.

The *E*-plane Y-junction circulator in waveguide has been investigated by Solbach.[41,42] It combines simply with finline circuits[13] and has a larger bandwidth than the simple finline circulator described by Braas and Schieblich,[43] shown in Fig. 9.10*a* and discussed in Section 9.5. The *E*-plane circulator is of very simple design, having a single ferrite disk whose diameter fits exactly into the diameter available at a waveguide Y-junction. Compared with Fig. 8.8, there is only one ferrite disk, there are no metal inserts and there is no circular metal region at the junction. The magnetic force due to the bias magnetic field is used to clamp the ferrite disk to the waveguide wall. The centre frequency of the circulator is determined by the dielectric resonance of the ferrite disk which is controlled by the only available variable, the disk thickness. Solbach[41] gives a formula for the circulator centre frequency in terms of the radius r, height h, and relative permittivity ε_r of the ferrite disk

$$f = \frac{c}{2\pi\sqrt{\varepsilon_r}}\sqrt{\left[\left(\frac{3{\cdot}11}{r}\right)^2 + \left(\frac{\pi}{2h}\right)^2\right]} \tag{10.3}$$

where c is the speed of light in vacuum. The radial wavenumber, $3{\cdot}11/r$, was obtained empirically from measured results. The formula in eqn. 10.3 was found to be within 1 GHz for various waveguide sizes and different ferrite materials in the frequency range 30–100 GHz. The useful bandwidth is about 1·5% throughout the frequency range. The bandwidth can be increased by using a larger diameter ferrite disk or by having a triangular shaped dielectric slab on top of the ferrite disk.[42]

Solbach[44] has also investigated the properties of an *E*-plane four-port circulator at about 94 GHz. The basic design is similar in construction to the Y-junction circulator described in the previous paragraph. There is only one ferrite disk whose diameter exactly fits in the aperture at the junction. Tuning is obtained by using a metal plunger at the centre of the junction with variable clearance between the plunger and the surface of the ferrite disk. The metal plunger provides a very simple method of tuning the circulator operating frequency without the necessity for altering the thickness of the ferrite disk or of providing additional dielectric disks.

10.7 Other Y-junction circulators

Circulator designs in microstrip have been realised by scaling and by the use of design criteria suitable for microwave frequency designs. The design theory is given in Section 8.7. Oxley, Scarman and Lowbridge[45] state that 10% bandwidth circulators may be produced at 26–110 GHz. Arain[46] describes a 94 GHz suspended stripline circulator. He uses a higher-order resonance in the ferrite disk so as to use larger diameter ferrite disks at this frequency. The construction of his circulator is similar to that shown in Fig. 8.11. The stripline conductor is fabricated on each side of a thin dielectric substrate of relative permittivity 2·2. A dielectric ring around the outside of the ferrite disk is used for matching. A 1 GHz bandwidth is reported which could be widened by further matching.

In the design of a finline circulator, Goebel and Schieblich[47] have produced a broadband circulator which has already been described in Section 9.5 and is shown in Fig. 9.10*b*. A millimetre wave model gave greater than 20 dB isolation over the band 28–36 GHz.

There has been very little published about circulators for image guide or for dielectric waveguide. Naito, Muraguchi and Tsuji[48] have investigated the effect of adding a transversely magnetised ferrite disk on top of an image guide Y-junction, as shown in Fig. 10.9. Unfortunately, in this form, there is a large radiation from the curved surfaces of the ferrite disk. They[48] propose the use of trapped image guide, as shown in Fig. 10.1*d*, or a wire grid fence parallel to the line of the image guide on each side of the guide and around the open curved surfaces of the ferrite disk. They measured the performance of a device using the wire grid fence at about 60 GHz and showed that circulator action occurs but the insertion loss is about 5 dB.

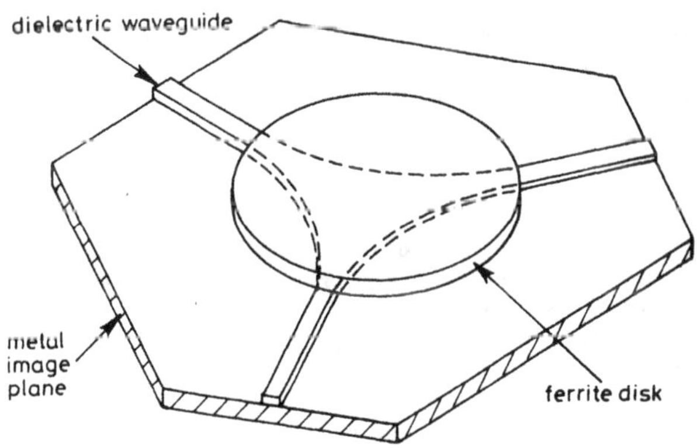

Fig. 10.9 *Y-junction circulator in image guide*

Stern and Babbitt[49] have proposed a circulator using a ferrite element at the centre of a Y-junction in the supported dielectric waveguide shown in Fig. 10.1*b*. The waveguide dielectric is magnesium titanate ($\varepsilon_r = 16$) which is then a good match to the ferrite insert at the centre of the junction. The ferrite element consists of a triangular prism of the same height as the dielectric guide so that there is a simple butt connection between the ferrite and the dielectric waveguide. Satisfactory circulator action was obtained over a 20% bandwidth at 35 GHz.

10.8 References

1 MARCATILI, E. A. J.: 'Dielectric rectangular waveguide and directional coupler for integrated optics', *Bell Syst. Tech. J.*, 1969, **48**, pp. 2071–2102

2 GOELL, J. E.: 'A circular-harmonic computer analysis of rectangular dielectric waveguide', *Bell Syst. Tech. J.*, 1969, **48**, pp. 2133–2160

3 SCHLISINGER, S. P. and KING, D. D.: 'Dielectric image lines', *IRE Trans.*, 1958, **MTT-6**, pp. 291–299

4 SOLBACH, K.: 'The fabrication of dielectric image lines using casting resins and the properties of the lines in the millimeter-wave range', *IEEE Trans.*, 1976, **MTT-24**, pp. 879–881

5 SOLBACH, K.: 'Electric probe measurements on dielectric image lines in the frequency range 26–90 GHz', *IEEE Trans.*, 1978, **MTT-26**, pp. 755–758

6 ZHOU, W. B. and ITOH, T.: 'Analysis of trapped image guides using effective dielectric constants and surface impedances', *IEEE Trans.*, 1982, **MTT-30**, pp. 2163–2166

7 ZHOU, W. B. and ITOH, T.: 'Field distribution in the trapped image guide', *Electromag.*, 1984, **4**, pp. 21–34

8 YONEYAMA, T. and NISHIDA, S.: 'Nonradiative dielectric waveguide for millimetre-wave integrated circuits', *IEEE Trans.*, 1981, **MTT-29**, pp. 1188–1192

9 YONEYAMA, T., FUJITA, S. and NISHIDA, S.: 'Insulated nonradiative dielectric waveguide for millimeter-wave integrated circuits',, *IEEE Trans.*, 1983, **MTT-31**, pp. 1002–1008

10 SAMARDZIJA, N. and ITOH, T.: 'Double-layered slot line for millimeter-wave integrated circuits', *IEEE Trans.*, 1976, **MTT-24**, pp. 827–831

11 ZIENIUTYCZ, W.: 'Modes of propagation in slot line with layered substrate containing magnetised ferrite', *Electron. Lett.*, 1983, **19**, pp. 135–136

12 BOCK, G.: 'New multilayered slot-line structures with very high nonreciprocity', *Electron. Lett.*, 1983, **19**, pp. 966–968

13 SOLBACH, K.: 'The status of printed millimeter-wave E-plane circuits', *IEEE Trans.*, 1983, **MTT-31**, pp. 107–121

14 MIRSHEKAR-SYAHKAL, D. and DAVIES, J. B.: 'An accurate, unified solution to various fin-line structures, of phase constant, characteristic impedance, and attenuation', *IEEE Trans.*, 1982, **MTT-30**, pp. 1854–1861

15 VAHLDIECK, R.: 'Accurate hybrid-mode analysis of various finline configurations including multilayered dielectrics, finite metallization thickness, and substrate holding grooves', *IEEE Trans.*, 1984, **MTT-32**, pp. 1454–1460

16 WU, Y. S. and ROSENBAUM, F. J.: 'Wide-band operation of microstrip circulators', *IEEE Trans.*, 1974, **MTT-22**, pp. 849–856

17 FORTERRE, G., CHIRON, B. and COURTOIS, L.: 'A survey of broad band stripline ferrite isolators', *IEEE Trans.*, 1975, **MAG-11**, pp. 1279–1281

18 BOLLE, D.M. and TALISA, S. H.: 'Fundamental considerations in millimeter and near-millimeter component design employing magnetoplasmons', *IEEE Trans.*, 1981, **MTT-29**, pp. 916–923

19 TALISA, S. H. and BOLLE, D. M.: 'Performance predictions for isolators and differential phase shifters for the near-millimeter wave range', *IEEE Trans.*, 1981, **MTT-19**, pp. 1338–1343

20 HARRISON, G. R.: 'Hexagonal ferrites for millimeter wave applications', *Proc. SPIE Int. Soc. Opt. Eng.*, 1981, **317**, pp. 251–261

21 TAFT, D. R., HARRISON, G. R. and HODGES, L. R.: 'Millimeter resonance isolators utilising hexagonal ferrites', *IEEE Trans.*, 1963, **MTT-11**, pp. 346–350

22 HELSZAJN, J. and THORPE, W.: '18–26 GHz finline resonance isolators using hexagonal ferrites', *Proc. IEE*, 1985, **132H**, pp. 73–76

23 HELLER, G. S., STICKLER, J. J. and THAXTER, J. B.: 'Antiferromagnetic materials for millimeter and submillimeter devices', *J. Appl. Phys. Suppl.*, 1961, **32**, pp. 307S–312S

24 REISCH, F. E., GRANT, R W., LIND, M. D., ESPINOSA, G. P. and GOLDBERG, I. B.:

'Magnetically tuned Zn_2Z filters for the 18–40 GHz frequency range' *IEEE Trans.*, 1975, **MAG-11**, pp. 1256–1258

25 LEMKE, M., HOPPE, W., TOLKSDORF, W. and WELZ, F.: 'Magnetically tunable millimetre-wave filter with single-crystal barium ferrite', *IEE J. MOA*, 1979, **3**, pp. 253–254

26 BARNES, C. E.: 'Broad-band isolators and variable attenuators for millimeter wavelengths', *IRE Trans.*, 1961, **MTT-9**, pp. 519–523

27 WILLIAMSON, I. M. H. and KOH, L.: 'A wideband K-band faraday rotational isolator', IEE Coll. Ferrite materials, devices and techniques, Digest No. 1985/15, 1985, pp. 6.1–6.6

28 BOYD, C. R.: 'A 60 GHz dual-mode ferrite phase shifter', IEEE MTT-S Symp. Digest, 1982, pp. 257–259

29 BOYD, C. R.: 'A dual-mode latching reciprocal ferrite phase shifter', *IEEE Trans.*, 1970, **MTT-18**, pp. 1119–1124

30 BABBITT, R. W. and STERN, R. A.: 'Non-reciprocal ferrite phase shifters for millimeter applications', IEEE MTT-S Int. Microwave Symp. Digest, 1978 pp. 94–96

31 BABBITT, R. W. and STERN, R. A.: 'Fabrication and performance of ferrite phase shifters for millimeter frequencies', *IEEE Trans.*, 1979, **MAG-15**, pp. 1744–1747

32 BEYER, A. and SOLBACH, K.: 'A new fin-line ferrite isolator for integrated millimeter-wave circuits', *IEEE Trans.*, 1981, **MTT-29**, pp. 1344–1348

33 BEYER, A. and WOLFF, I.: 'Power density distribution analysis of ferrite loaded finlines for the development of integrated nonreciprocal millimeter wave elements', IEEE MTT-S Int. Microwave Symp. Digest, 1984, pp. 342–344

34 BOCK, G.: 'Nonreciprocal propagation characteristics of slot lines with dielectric-ferrite double layer substrate', Proc. 13th Euro. Microwave Conf., 1983, pp. 809–813

35 YUN, S. W. and ITOH, T.: 'A novel distributed millimeter-wave isolator', Proc. 13th Euro. Microwave Conf., 1983, pp. 174–178

36 YUN, S. W. and ITOH, T.: 'A distributed millimeter-wave isolator using nonreciprocal coupling structure', *Int. J. Infrared & mm. Waves*, 1984, **5**, pp. 775–792

37 YUN, S. W. and ITOH, T.: 'Nonreciprocal wave propagation in a hollow image guide with a ferrite layer', Proc. 14th Euro. Microwave Conf., 1984, pp. 341–345

38 MURAGUCHI, M., ARAKI, K. and NAITO, Y.: 'A new type of isolator for millimeter-wave integrated circuits using a nonreciprocal travelling-wave resonator', *IEEE Trans.*, 1982, **MTT-30**, pp. 1867–1873

39 PIOTROWSKI, W. S. and RAUE, J. E.: 'Low-loss broad-band EHF circulator', *IEEE Trans.*, 1976, **MTT-24**, pp. 863–866

40 PIOTROWSKI, W. and SCHELL, S.: 'Low loss 92–100 GHz circulators', IEEE MTT-S Int. Microwave Symp. Digest, 1982, pp. 252–254

41 SOLBACH, K.: 'E-plane circulators 30 through 150 GHz for integrated mm-wave circuits', Proc. 13th Euro. Microwave Conf., 1983, pp. 163–167

42 SOLBACH, K.: 'E-plane circulators aid mm-wave design', *Microwaves & RF*, 1983, **22**, 13, pp. 73–78

43 BRAAS, M. and SCHIEBLICH, C.: 'E-type circulator for fin lines', *Electron. Lett.*, 1981, **17**, pp. 701–702

44 SOLBACH, K.: '94 GHz 4-port E-plane junction circulator', *IEEE Trans.*, 1984, **MTT-32**, pp. 722–724

45 OXLEY, T. H., SCARMAN, R. L. and LOWBRIDGE, P. L.: 'Millimeter-wave hybrid-open microstrip techniques', IEEE MTT-S Int. Microwave Symp. Digest, 1982, pp. 215–217

46 ARAIN, M. H.: 'A 94 GHz suspended stripline circulator', IEEE MTT-S Int. Microwave Symp. Digest., 1984, pp. 78–79

47 GOEBEL, U. and SCHIEBLICH, C.: 'Broadband fin-line circulators', IEEE MTT-S Int. Microwave Symp. Digest, 1982, pp. 249–251

48 NAITO, Y., MURAGUCHI, M. and TSUJI, A.: 'A new type circulator for millimeter integrated circuits', IEEE MTT-S Int. Microwave Symp. Digest, 1980, pp. 250–253

49 STERN, R. A. and BABBITT, R. W.: 'Dielectric waveguide circulator', *Int. J. Infrared & mm. Waves*, 1982, **3**, pp. 11–18

Chapter 11

High-power and nonlinear effects

11.1 Spinwaves and magnetostatic waves

In the rest of this book, the precession of the magnetisation in any small region of ferrite is assumed to be uniform, and any variation of the microwave fields is negligible. However, there are other modes of motion of the magnetisation which vary with very short wavelengths within a small region of ferrite. They are called spinwaves, and can be excited in the ferrite when the microwave magnetic field intensity exceeds a certain critical field value. They contribute to the attenuation in the ferrite and to nonlinear effects at high peak power. Above a threshold critical field, spinwaves are excited which increase exponentially and contribute to excess microwave loss owing to the ferrite. The mathematical mechanism for their generation and propagation is beyond the scope of this book but it has been given by Suhl[1] and subsequently by others[2-7] in textbooks. At any one frequency, the critical magnetic field varies as the static magnetic bias field is varied, as shown in Fig. 11.1. The minimum microwave critical field is given by[7]

$$H_c = \frac{2\Delta H_k(\omega/\omega_m)\,[1 - (\omega_0/\omega)]}{1 - (\omega_m/2\omega) + [1 + (\omega_m/2\omega)^2]^{1/2}} \tag{11.1}$$

where ω_0 and ω_m are given by eqn. 2.14 and ΔH_k is the spinwave linewidth. This equation shows that high peak power capability is improved by increasing the spinwave linewidth ΔH_k or by decreasing the saturation magnetisation M_0.

The wavelength of spinwaves is primarily determined by the effective exchange field which arises from the exchange energy, and aligns the electron spins in a magnetic material. Under the influence of an external disturbing magnetic field, the electron spins precess as a single unit and it is this precession which gives ferrites their useful magnetic properties. If the uniformity of the motion is disturbed slightly, as always occurs owing to thermal agitation, strong demagnetising and exchange fields are generated. Under certain conditions, these local disturbances can grow at the expense of the external disturbing magnetic field and will then propagate through the ferrite as spinwaves. They have a short wavelength so that they may be analysed as plane waves even within a small ferrite sample. Spinwaves are excited when the disturbing magnetic field has an amplitude greater than a certain critical value as discussed in the previous paragraph. Then the spinwaves absorb power from the disturbing magnetic field and heat up the ferrite material.

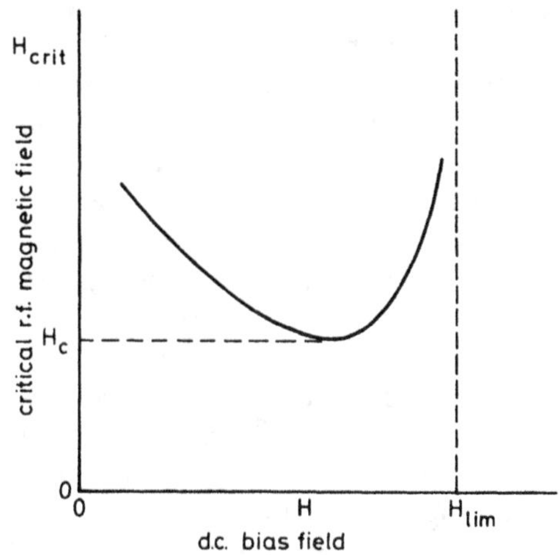

Fig. 11.1 *Variation of critical microwave magnetic field with change of d.c. bias magnetic field*

In a small ferrite sample, there is another class of spinwaves whose wavelength is no longer small compared with the size of the sample and whose field strength varies across the sample. The frequency of oscillation of these modes is dependent on the size and shape of the sample. Their wavelength is still much smaller than the wavelength of any electromagnetic wave. They are governed by the magnetostatic form of Maxwell's equations, where the rate of change with regard to time is small compared with the other field quantities in eqns. 2.29 to 2.32 so that

$$\nabla \cdot \boldsymbol{B} = 0 \tag{11.2}$$

$$\nabla \times \boldsymbol{H} = 0 \tag{11.3}$$

Therefore, they are called *magnetostatic modes* of oscillation in the sample. They were first investigated by Walker[8] and the theory is given in a number of textbooks.[2-4,6]

11.2 High-power devices

There are two effects which degrade the performance of ferrite devices at high powers:

1 excessive heating owing to large microwave mean powers
2 excessive loss and consequent heating in the ferrite owing to the excitation of spinwaves at high peak powers.

Devices designed to overcome the effects of heating at high mean powers are described in this section and devices designed to suppress the excitation of spinwaves at high peak powers are described in Section 11.3. As a ferrite material heats up, its microwave properties deteriorate and thermal runaway can occur. To design for high average powers, it is necessary to have an adequate mechanism for removing heat from the ferrite elements and to use a ferrite material having a low loss at elevated temperatures.

Industrial microwave heating uses high powers, particularly in the UHF industrial bands, 896 and 915 MHz and at 2·45 GHz. There is a requirement for a high-power isolator working at these frequencies in order to protect the oscillator in these systems. Leppin[9] describes a resonance isolator at 915 MHz using four thin ferrite strips in rectangular waveguide. The thickness of the ferrite strips is varied along the length of the isolator so as to equalise the rate of heat generation along the length of the device. One broad face of each ferrite strip is metallised and the strip is soldered to the broad wall of the waveguide. Cooling water is passed through a cavity in the waveguide wall under the ferrite strips between the ferrite and the bias magnets. Good thermal design keeps the temperature difference between the ferrite and the cooling water to about 5 K.

One problem of operation at high temperatures is the differential expansion between the ferrite and the waveguide wall to which it is attached. High-power devices use thin sections of ferrite tightly bonded to water-cooled sections of metal waveguide wall. Buck and King[10] found that using a hard epoxy adhesive was unsatisfactory. Differential expansion caused the adhesive film to fracture, the unsupported section of ferrite then to overheat and crack and vaporise. The use of a flexible silicone adhesive having only about 1/10th of the strength of the epoxy cured the problem.

One approach to high-power circulator design is to use the differential phase changer circulator described in Section 7.3, thus only half the average power is being transmitted through each ferrite device and the unwanted reverse power is absorbed in an external load. The results given by Buck and King[10] were obtained in the differential phase change sections of such a device. Fournet-Fayas, Priou and Forterre[11] describe a 50 kW CW ferrite circulator for use at the industrial heating frequency of 2·45 GHz. It is similar in construction to the differential phase change circulator shown in Fig. 7.11. However, they used only one pair of ferrite slabs in each differential phase change section as shown for the isolator in Fig. 7.16, and each ferrite strip had an alumina dielectric strip fixed adjacent to it. At high power, when the ferrite elements are hot, the differential phase change is less than it is when the ferrite is cold. In this circulator, the differential phase change of each element is adjusted to be 90° when it is at the temperature appropriate to operation at 50 kW. This gave the circulator a minimum insertion loss at high power while still having satisfactory performance at low powers.

The differential phase change circulator discussed in the last paragraph is large, heavy and expensive compared with the Y-junction circulator used as an isolator. Much effort has been expended on the design of waveguide Y-junction circulators for use at about 915 MHz. Okada, Ohwi and Mori[12] describe a Y-junction circulator for 30 kW CW at 915 MHz. They use a thin circular ferrite disk mounted on a circular metal transformer section as shown in Fig. 11.2.

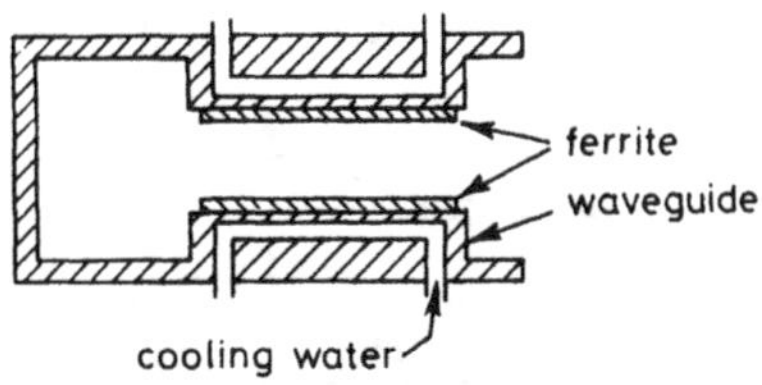

Fig. 11.2 *High-power circulator*

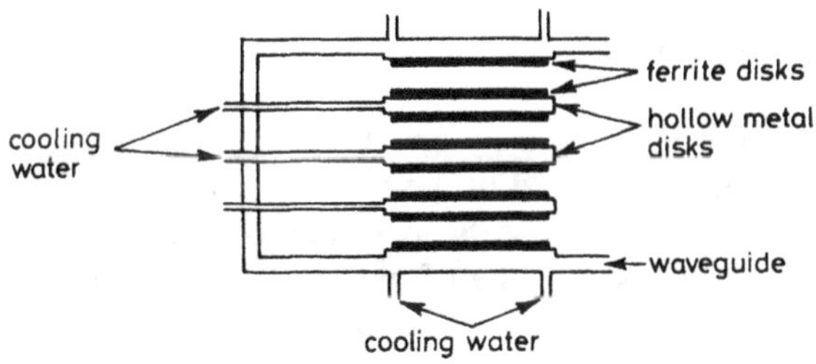

Fig. 11.3 *Very-high-power circulator*

Okada and Ohwi[13] have extended the design to use a number of ferrite disks attached to each side of a number of water cooled metal disks in the junction, as shown in Fig. 11.3. Such a device can handle 100 kW at 915 MHz. One further problem in these high-power low-frequency circulators is the very small thickness-to-diameter ratio of the ferrite disks. They are prone to fracture during handling in manufacture and owing to the thermal stresses when in use at high power. Okada, Ohwi and Sumi[14] have investigated the performance of a high-power circulator similar to that shown in Fig. 11.2 with the ferrite disk in a number of discrete pieces having a distinct gap between adjacent pieces. The performance of the circulator having its ferrite disk in distinct pieces is as good as one using a large complete ferrite disk.

At higher frequencies, the size of the ferrite disk becomes much smaller. One of the problems of a high-power circulator is removing the heat generated in the ferrite. If part of the ferrite is replaced by a lower loss dielectric material, the device will withstand a higher power. Helszajn[15] suggests replacing some of the ferrite cylinder at the centre of the junction with a low-loss high-thermal-conductivity dielectric material. In the low-frequency high-power circulators,[12,13] the ferrite is used in the form of a large diameter thin disk or as a number of thin disks supported on metal disks, as shown in Fig. 11.3. Helszajn[15] uses a smaller diameter disk with additional dielectric loading, as shown in Fig. 11.4. Reducing the volume of ferrite at the junction reduces the effective gyromagnetic effect of the ferrite so that a compromise is necessary between the microwave performance and the power capability. Such a circulator operated at 6·5 kW at 2·35 GHz without cooling. Helszajn and Sharp[16] also describe a similar circulator for operation with 2·0 kW at 8·4 GHz using forced air cooling.

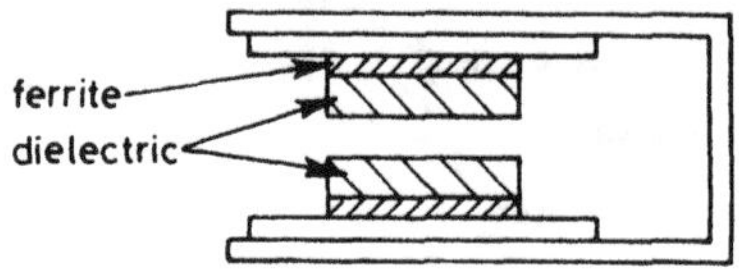

Fig. 11.4 *Centre of a high-power circulator described by Helszajn*[15]

11.3 High-peak-power devices

The onset of high-power nonlinearity and the excitation of spinwaves occurs when the critical r.f. magnetic field in the ferrite is exceeded. There is also the problem that, as a ferrite material heats up, its microwave properties deteriorate and thermal runaway can occur. The r.f. magnetic field strength in the ferrite must be less than the critical field at which spinwaves are excited. Very often the peak power limitation occurs at much lower average powers than otherwise would cause overheating in the ferrite material.

From eqn. 11.1, it can be seen that high-peak-power capability is accomplished by increasing the spinwave linewidth, ΔH_k, and by reducing the saturation magnetisation, M_0, in the material. It may also be accomplished by adjusting the geometry of the device to reduce the maximum intensity of the microwave magnetic field in the ferrite. For the twin-slab latching ferrite phase changer in rectangular waveguide, as shown in Fig. 7.4*a*, Schlomann[17] has analysed its power handling capability in terms of the critical fields in the ferrite. Mizobuchi and Kurebayashi[18] produce an improved design similar to that shown in Fig. 7.6 with better high-power capability. The design reduces the maximum intensity of the microwave field in the ferrite, thus improving the peak-power capability, and, by using a dielectric with high thermal conductivity and improved contact between the ferrite and the waveguide wall, the average power capability is also improved.

The onset of high-power nonlinearity and the excitation of spinwaves does not occur even when the critical r.f. magnetic field in the ferrite is exceeded under certain conditions. The value of this critical magnetic field varies as shown in

Fig. 11.1. The minimum value of the critical magnetic field is given by eqn. 11.1, and this is the criterion to use to ensure that high-peak-power loss never occurs. However, at other values of bias magnetic field, the critical value of the r.f. magnetic field is larger. Above a certain value of bias field, H_{lim}, spinwaves are not excited and there is no high-peak-power loss. Lagrange, Lahmi and Vallatin[19] produced a high-peak-power junction circulator operating with the magnetic bias field above H_{lim}. If the normal resonance absorption curve is measured at high-peak powers, a subsidiary resonance is observed at the bias magnetic field appropriate to H_c in Fig. 11.1, which is due to spinwaves being excited in the ferrite. There is a region of low loss above the limit field, H_{lim}, before the main ferromagnetic resonance is reached, as shown in Fig. 11.5.

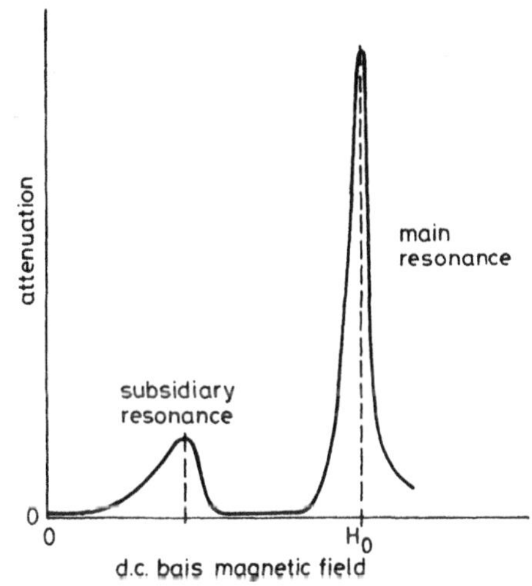

Fig. 11.5 *Attenuation versus d.c. bias magnetic field at high peak power showing the subsidiary and main resonances*

Helszajn and Walker[20] have analysed the conditions for operating in this low-loss region between the subsidiary and main resonances for waveguide high-power differential phase changers in a circulator. Their circulator operated at a peak power level of 1 MW at 9·6 GHz. Helszajn and Powlesland[21] applied the same theory to the design of microstrip Y-junction circulators and were able to show that their devices operated with a low loss up to a peak-power level of greater than 2 MW at 9·375 GHz. Nisbet, Hastings and Helszajn[22] investigate the bandwidth available from such a device. Nicolas[7] gives a simple formula for H_{lim}

$$H_{lim} = \frac{\omega}{2\gamma} + \frac{NM_0}{\mu_0} \tag{11.4}$$

where N is the demagnetisation factor in the direction of application of the magnetic bias field H_{lim}. Demagnetisation factors are discussed in Section 12.3.

11.4 Low-field loss

Low-field loss is not strictly a high-power effect but it will be discussed here because it is an unwanted loss mechanism that can occur in some ferrite materials at low values of bias magnetic field. In practical designs its effect is overcome by using the correct grade of ferrite material for that application. When the ferrite material is unmagnetised or magnetised below saturation, each crystal is split up into a number of domains, each magnetised to saturation but in arbitrary directions. The effective field within each domain is called the anisotropy field, H_a. There will be a resonance owing to this field in the domains. There are two mechanisms for the change of field in a domain: change of direction of field in the domain and domain wall movement. Both these mechanisms contribute to resonance losses in the material. Smit and Wijn[23] show that the frequency for resonance is dependent on the angle between the anisotropy field in the domain and the r.f. magnetic field direction. It varies between

$$\omega_{min} = \gamma H_a \tag{11.5}$$

and

$$\omega_{max} = \gamma(H_a + M_0/\mu_0) \tag{11.6}$$

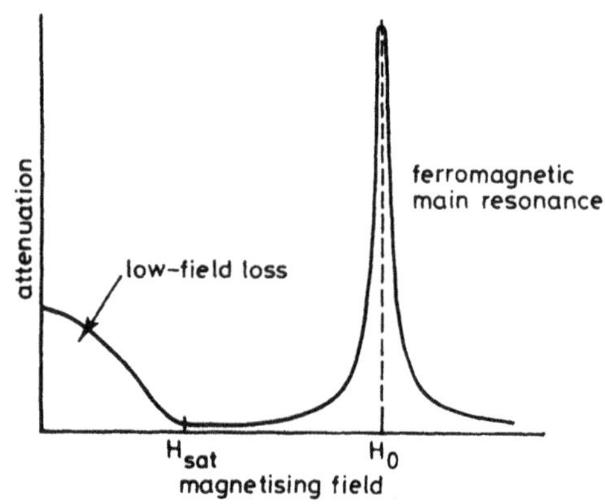

Fig. 11.6. *Attenuation versus d.c. bias magnetic field at low powers showing low-field loss*

where γ is the gyromagnetic ratio and M_0 is the saturation magnetisation in the ferrite. In a polycrystalline material, there is a random distribution of crystallite directions and probably a random distribution of domain orientations so that resonances will occur at any frequency between ω_{max} and ω_{min}. If the material is magnetised to saturation, there is no longer any possibility of change of direction of the anisotropy field or of domain wall motion, thus this resonance does not occur. It only occurs at low fields below saturation. Then the absorption loss of a ferrite material operating between ω_{max} and ω_{min} is as shown in Fig. 11.6.

It becomes necessary to operate any microwave device above saturation or above ω_{max} in order to avoid the effects of low-field loss. A fuller explanation of the derivation of eqns. 11.5 and 11.6 has been given by Soohoo[24] and in a very similar form by Helszajn.[4]

For devices such as isolators operating at ferromagnetic resonance, provided that the resonance condition occurs at a field above that required to saturate the ferrite, low-field loss does not cause problems. However, for control devices operating below saturation or at remanence, a ferrite material must be chosen having no low-field loss at the frequency of operation. For microwave devices, it is possible to operate at frequencies above ω_{max}. For example, a typical ferrite would have H_a of the order of 20 kA/m and M_0 of 0·2 T. From eqn. 11.6, ω_{max} is then 6·2 GHz and that ferrite will not exhibit low-field loss when operating at frequencies above 6·2 GHz. Ferrites having larger saturation magnetisations are suitable for operation at higher frequencies and specially tailored ferrites with low saturation magnetisations are suitable for use at low microwave frequencies. An alternative for low frequencies is to operate above the resonant magnetic field.

Low-field loss will also affect the ferromagnetic resonance condition at low frequencies. Above saturation, there cannot be any domain wall motion or change of direction of anisotropy field due to an r.f. magnetic field. However, as the bias magnetic field drops below that required for saturation, more and more domains will be created and the low-field loss will increase as shown by Fig. 11.6. At low microwave frequencies, ferromagnetic resonance occurs at magnetic fields below that required to saturate the ferrite and the mechanisms of low-field loss increase the linewidth of the ferromagnetic resonance curve. The linewidth increases as the bias magnetic field is reduced in the same way that the low-field loss increases.

11.5 Power limiting

The process of spinwave generation and microwave power absorption can be used to provide microwave power limiting. In Section 11.3 we discussed how to eliminate the conditions for spinwave generation so that devices were able to operate at high peak power. In this section we will be considering how to lower the critical field level so that power limiting can be obtained at low-power levels. Spinwaves are excited when the precession in the ferrite material reaches a particular critical angle. For the subsidiary resonance condition shown in Fig. 11.5, moderately large levels of r.f. magnetic field are required to provide the critical angle for the precession in the material. However, at the main ferromagnetic resonance, the same critical precession angle is produced at a much lower level of applied r.f. magnetic field. Then the spinwave resonance and the ferromagnetic resonance occur at the same magnetic bias condition, which gives it the name *coincidence limiting*. Microwave power limiting is discussed by Lax and Button[25] and by Helszajn[4] and only a summary of the principles will be given here.

One way to obtain low levels of power limiting is to use a ferrite resonant cavity, as described in Section 9.9. A small highly-polished ferrite sphere is mounted between two orthogonal transmission lines similar to that shown in Fig. 9.26. If the ferrite is biased to resonance, spinwave excitation occurs at a critical r.f. magnetic field given approximately by[25,26]

$$h_c = \frac{\mu_0 \Delta H \Delta H_k}{M_0} \tag{11.7}$$

where ΔH is the ferromagnetic resonance linewidth, ΔH_k is the spinwave linewidth and M_0 is the saturation magnetisation in the ferrite. The relationship between the output power and the input power is similar to that shown in Fig. 11.7. The output power is directly related to the angle of the precession in the

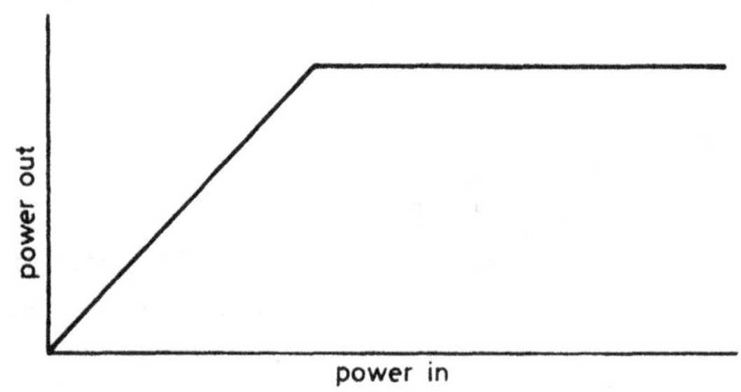

Fig. 11.7 *Typical power limiter performance*

ferrite sphere. At low powers, the precession angle is proportional to the input power so that the input and output powers are the same apart from the insertion loss of the ferrite resonant cavity. When the precession in the ferrite reaches its critical angle, any further increase in input power only serves to increase the power in the spinwaves and the precession in the ferrite remains at the critical angle; the output power remains constant irrespective of the input power. Such a device provides useful limiting to protect sensitive receivers in radar and other systems. Sansalone and Spencer[26] measured a critical power of 67 μW using a single crystal YIG sphere at 3 GHz and room temperature. They showed that the critical power was much lower for the same device at the lower temperatures of 77 and 4·2 K.

Sansalone and Spencer[26] also investigated the effect of limiting pulse power signals. They discovered that there was a leading edge spike at the output of the device at the beginning of each pulse. As the input pulse power increased above the critical value, the peak power of the output leading edge spike increased but the spike was of shorter duration, as shown in Fig. 11.8. The spike duration is of the order of 30 μs. As far as they could measure, the area under the spike curve remained constant. At the beginning of a disturbance there is an exponential increase of the spinwave amplitude. The initial rate of increase depends on the power pumped into the spinwaves so that the build-up time for spinwave resonance becomes shorter as the input power is increased. Then the total energy in the spike remains roughly constant. Helszajn and McStay[27.4] derive the condition for coincidence limiting to occur to be

$$\omega < 2N_t \gamma M_0/\mu_0 \tag{11.8}$$

where N_t is the demagnetisation factor in the plane of the r.f. magnetic field. As the ferrite must also be magnetically saturated, the frequency range of operation is limited to

$$N_t\omega_m < \omega < 2N_t\omega_m \tag{11.9}$$

where ω_m is given by eqn. 2.14.

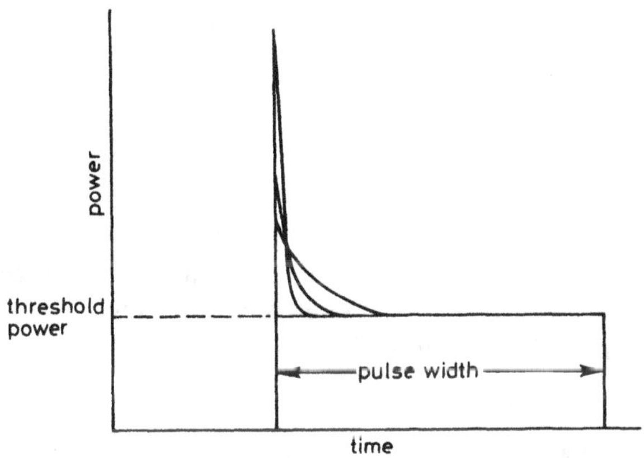

Fig. 11.8 *Spikes in the output from a microwave power limiter when the input microwave power is pulsed. Four output pulses having different input powers superimposed*

If the ferrite is operated well away from resonance in the region where subsidiary spinwave resonances can be excited, subsidiary resonance limiting can occur. A fuller discussion of subsidiary resonance spinwave excitation has already been given in Section 11.3 while discussing peak-power limiting. However, subsidiary resonance limiting can also be used; it gives a limiting characteristic similar to that shown in Fig. 11.7 but at a much higher power level than coincidence limiting, of the order of 100 W. If the ferrite limiter is used in a microwave cavity or other structure to increase the r.f. magnetic field in the ferrite, a subsidiary resonance limiter can be made giving a limited power output of about 1 W.

As the ferrite structure used for coincidence limiting is similar to that described in Section 9.9 to make tunable cavities and tunable filters, it is very easy to produce tunable filter limiters.

11.6 Frequency doubling

The nonlinear effects in a ferrite material, particularly at high power, can be used for harmonic generation of microwaves. The effect was first used for frequency doubling by Melchor, Ayres and Vartanian,[28] and also described by Lax and Button.[25] The second harmonic component occurs most obviously when the ferrite is excited by a linearly polarised microwave magnetic field H_x acting perpendicularly to the bias magnetic field H_{z0}, as shown by Fig. 11.9.

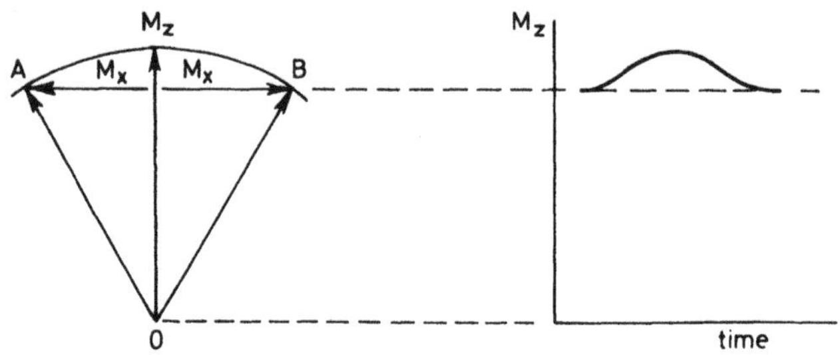

Fig. 11.9 *Frequency doubling in ferrites*

There is a second harmonic component of the microwave field parallel to the d.c. magnetic field. Mathematically, in Section 2.3 all the second-order small terms were neglected. In particular, the more exact expression for the z-directed component of eqn. 2.7 ought to be

$$j\omega M_z = -\gamma H_x M_y + \gamma H_y M_x \tag{11.10}$$

The higher-order terms in the other two components of eqn. 2.7 can still be neglected so that the first two equations of eqn. 2.8 are still valid and may be substituted into eqn. 11.10, giving

$$M_z = \frac{\kappa\gamma}{\omega}(H_x^2 + H_y^2) \tag{11.11}$$

instead of the third equation of eqn. 2.8. Melchor *et al.*[28] show, from a physical consideration similar to that given in Fig. 11.9, that there is no output if the r.f. magnetic fields in the plane perpendicular to the direction of the d.c. magnetic field are circularly polarised. Circular polarisation is given by

$$H_y = \pm jH_x$$

which, when substituted into eqn. 11.11, gives $M_z = 0$. Otherwise there is an output flux density parallel to the d.c. magnetic field which is proportional to the square of the magnetic field intensity in the plane perpendicular to it. Owing to the square law relationship, the alternating component of M_z is at twice the exciting frequency. Using the condition for maximum efficiency of frequency conversion, i.e. a linearly polarised magnetic field, $H_y = 0$, and substituting eqn. 2.9 into eqn. 11.11 gives

$$M_z = -\frac{\gamma^2 M_0}{\gamma^2 H_0^2 - \omega^2} H_x^2 \tag{11.12}$$

Then the output will be maximum when the input is at the ferromagnetic resonance frequency. An approximate value for the amplitude of the magnetic flux density at resonance can be obtained by replacing κ in eqn. (11.11) with its value given by eqn. 2.20, which includes a loss term and allows for the shape of the resonance curve. This is equivalent to replacing γH_0 in eqn. 11.12 by $(\gamma H_0 + j\omega\alpha)$. Then for the first-order approximation when $\omega = \omega_0 = \gamma H_0$

$$M_z = \frac{j\gamma^2 M_0}{2\omega_0^2 \alpha} H_x^2 \tag{11.13}$$

Substituting for α from eqn. 2.27 into eqn. 11.13 gives the result in terms of the resonant linewidth of the ferrite material, ΔH

$$M_z = j\frac{\gamma M_0}{\omega_0 \Delta H} H_x^2 \tag{11.14}$$

Owing to the squaring operation, the amplitude of the doubled frequency component of M_z is

$$M_{z,2\omega} = j\frac{\gamma M_0}{2\omega_0 \Delta H} H_x^2 \tag{11.15}$$

In experiments, Ayres[29] showed that the measured outputs varied in proportion to the material properties according to eqn. 11.15 for a number of different ferrite materials.

As with any frequency doubling system, careful design is required to handle the signals at the two frequencies. For the ferrite frequency doubler, the input microwave magnetic field is perpendicular to the bias magnetic field and the output doubled frequency component is parallel to it. If the ferrite is transversely magnetised in a rectangular waveguide system, the *E*-planes of the input and output waveguides need to be perpendicular, as shown by Fig. 11.10. Using the normal 2:1 ratio rectangular waveguide, the doubled frequency component output will be able to propagate in the TE_{01}-mode in the input waveguide. It is necessary to provide filters in the waveguide so that the $2f$ component cannot propagate to the source and the fundamental frequency component cannot propagate to the output. Melchor *et al.*[28] showed frequency doubling from 9 to 18 GHz with a conversion efficiency of better than −20 dB and as good as −7 dB.

Frequency conversion using ferrites does not appear to be preferable to frequency conversion using diodes but the ferrites can be operated at much higher powers than the diodes. As we have discussed in Sections 11.3 and 11.5, nonlinearities occur at high powers, and, consequently, frequency conversion to higher harmonics does also occur in ferrites. Schied and Weis[30] have investigated frequency doubling, tripling and quadrupling using a rectangular ferrite slab in a rectangular cavity resonator. They obtained a conversion efficiency of 40% with frequency doubling.

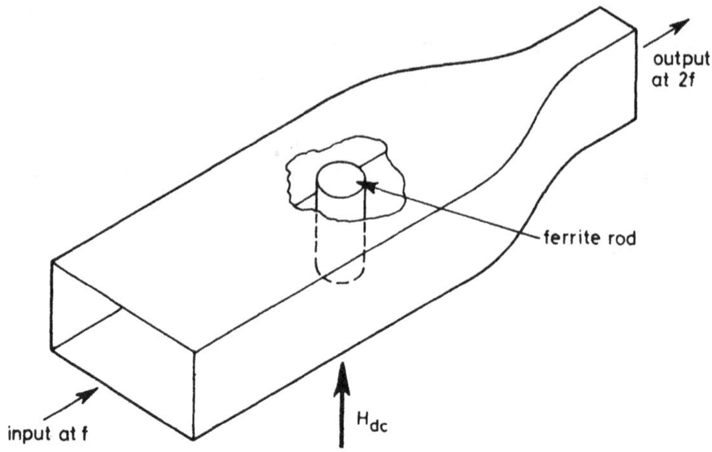

Fig. 11.10 *Ferrite frequency doubler in waveguide described by Melchor et al.*[28]

11.7 Amplification

The nonlinear properties of the ferromagnetism equation discussed in this chapter can also be used to obtain microwave amplification. The mode of operation is that of the three-frequency parametric amplifier. If the ferrite is pumped at a frequency f_p, then a signal at f_s will be amplified provided that the idler or difference frequency f_i, is allowed to exist

$$f_i = f_p - f_s \tag{11.16}$$

The ferromagnetic amplifier was first analysed by Suhl[31] and also discussed in the textbooks by Lax and Button[25] and Helszajn.[4] Suhl[31] identified three different modes in which a ferrite sample can be used as a ferromagnetic amplifier. In each one, the pump power is used to excite the uniform mode of precession in the ferrite at its ferromagnetic resonant frequency. They are:

1 *Electromagnetic operation* (EM): the ferrite sample is mounted in a cavity which is resonant at the signal and idler frequencies. The ferrite sample is resonant to the pump frequency and also provides the coupling between the signals at other frequencies.
2 *Magnetostatic operation* (MS): two magnetostatic modes[8] in the ferrite support the signal and idler frequencies. The pump excites the main ferromagnetic resonance in the ferrite. Nonlinearities in the ferrite response provide the

coupling between the various frequency signals. The magnetostatic modes provide some fields external to the ferrite sample which can be used to couple to the input and output circuits. Pure spinwave modes do not provide this external coupling and cannot be used in this application.

3 *Semistatic operation* (SS): one magnetostatic mode and one electromagnetic cavity mode are used to support the idler and signal frequencies, respectively.

There is one other mode of operation identified by Lax and Button,[25] *the modified semistatic mode of operation* (MSS): the main ferromagnetic resonance is used for the idler frequency and cavity modes are used to support the signal and pump frequencies. This provides an efficient method of coupling pump power in and signal power out of the amplifier.

In the electromagnetic and modified semistatic modes of operation, a microwave cavity or circuit needs to be provided which is resonant at two microwave frequencies. This appears to make those modes utilizing at least one magnetostatic mode in addition to the main ferromagnetic resonance, MS and SS, preferable for ease of design. In the amplifier, if the pump power is increased gradually, a condition is reached at which the amplifier oscillates at both the idler and signal frequencies when there is no loading on the signal circuit. Loading the signal circuit quenches the oscillation and the device amplifies signals at the signal frequency. Increasing the pump power increases the precession angle in the ferrite. At a lower critical power level, amplification occurs. As the pump power is increased further, a higher critical power level is reached at which the precession angle is the critical angle for the generation of spinwaves. Then any further increase in pump power only serves to increase the power going into spinwaves which are dissipated as heat in the ferrite; the precession angle in the ferrite does not increase and there can be no increase in the energy going into the amplification.

11.8 References

1 SUHL, H.: 'The nonlinear behaviour of ferrites at high microwave signal levels', *Proc. IRE*, 1956, **44**, pp. 1270–1284

2 CLARRICOATS, P. J. B.: 'Microwave ferrites' (Chapman & Hall, 1961), pp. 68–86

3 LAX, B. and BUTTON, K. J.: 'Microwave ferrites and ferrimagnetics' (McGraw-Hill, 1962), pp. 170–231

4 HELSZAJN, J.: 'Principles of microwave ferrite engineering' (Wiley, 1969), pp. 30–51 and 66–100

5 HELSZAJN, J.: 'Nonreciprocal microwave junctions and circulators' (Wiley, 1975), pp. 332–335

6 HELSZAJN, J.: 'YIG resonators and filters' (Wiley, 1985), pp. 50–75

7 NICOLAS, J.: 'Microwave ferrites', *in* WOHLFARTH, E. P.(Ed.) 'Ferromagnetic materials. Vol.2' (North-Holland, 1980), pp. 243–296

8 WALKER, L. R.: 'Magnetostatic modes in ferromagnetic resonance', *Phys. Rev.*, 1957, **105**, pp. 390–399

9 LEPPIN, J.: 'Thermal design of ferrite isolators for industrial microwave equipment', *J. Microwave Power*, 1974, **9**, pp. 251–261

10 BUCK, D. C. and KING, M. L.: 'Thermal stress-induced breakdown in an S-band isolator', *IEEE Trans.*, 1978, **MTT-26**, pp. 357–360
11 FOURNET-FAYAS, C., PRIOU, A. C. and FORTERRE, G. E.: 'A 50 kW CW ferrite circulator in S band', *IEEE Trans.*, 1978, **MTT-26**, pp. 360–363
12 OKADA, F., OHWI, K. and MORI, M.: 'The development of a high-power microwave circulator for use in breaking of concrete and rock', *J. Microwave Power*, 1975, **10**, pp. 171–180
13 OKADA, F. and OHWI, K.: 'Design of a high-power CW Y-junction waveguide circulator', *IEEE Trans.*, 1978, **MTT-26**, pp. 364–369
14 OKADA, F., OHWI, K. and SUMI, W.: 'A high-power circulator with disk consisting of sectral ferrite pieces', Ferrites, Proc. 3rd. Int. Conf. on Ferrites. (Japan), 1980, pp. 854–856
15 HELSZAJN, J.: 'High-power waveguide circulators using quarter-wave long composite ferrite/dielectric resonators', *Proc. IEE*, 1981, **128H**, pp. 268–273
16 HELSZAJN, J. and SHARP, J.: '2000 W mean-power X-band waveguide junction circulator using a composite turnstile resonator', *Proc. IEE*, 1983, **130H**, pp. 275–276
17 SCHLOMANN, E.: 'Theoretical analysis of twin-slab phase shifters in rectangular waveguide', *IEEE Trans.*, 1966, **MTT-14**, pp. 15–23
18 MIZOBUCHI, A. and KUREBAYASHI, H.: 'Nonreciprocal remanence ferrite phase shifters using grooved waveguide', *IEEE Trans.*, 1978, **MTT-26**, pp. 1012–1016
19 LAGRANGE, A., LAHMI, H. and VALLATIN, B. R.: 'X-band high-peak-power junction circulator: influence of the static magnetic field', *IEEE Trans.*, 1973, **MAG-9**, pp. 531–534
20 HELSZAJN, J. and WALKER, P. N.: 'Operation of high peak power differential phase shift circulators at direct magnetic fields between subsidiary and main resonances', *IEEE Trans.*, 1978, **MTT-26**, pp. 653–658
21 HELSZAJN, J. and POWLESLAND, M. E.: 'Low-loss high-peak-power microstrip circulators', *IEEE Trans.*, 1981, **MTT-29**, pp. 572–578
22 NISBET, W. T., HASTINGS, W. I. and HELSZAJN, J.: 'Microwave bandwidth of high-power ferrite devices biased between the main and subsidiary resonances', *Proc. IEE*, 1984, **131H**, pp. 276–278
23 SMIT, J. and WIJN, H. P. J.: 'Physical properties of ferrites', *Advances in Electronics and Electron Physics*, 1954, **6**, pp. 69–136
24 SOOHOO, R. F.: 'Theory and application of ferrites' (Prentice-Hall, 1960), pp. 83–88
25 LAX, B. and BUTTON, K. J.: 'Microwave ferrites and ferrimagnetics', (McGraw-Hill, 1962), pp. 637–702
26 SANSALONE, F. J. and SPENCER, E. G.: 'Low-temperature microwave power limiter', *IRE Trans.*, 1961, **MTT-9**, pp. 272–273
27 HELSZAJN, J. and MCSTAY, J.: 'Simplified theory of nonlinear phenomena in ferrimagnetic materials', *Proc. IEE*, 1967, **114**, pp. 1585–1591
28 MELCHOR, J. L., AYRES, W. P. and VARTANIAN, P. H.: 'Microwave frequency doubling from 9 to 18 KMC in ferrites', *Proc. IRE*, 1957, **45**, pp. 643–646
29 AYRES, W. P.: 'Millimeter-wave generation experiment utilizing ferrites', *IRE Trans.*, 1959, **MTT-7**, pp. 62–65
30 SCHIED, E. and WEIS, O.: 'Microwave-harmonic generation in ferrites at high power', *J. Magn. & Magn. Mater.*, 1984, **45**, pp. 377–381
31 SUHL, H.: 'Theory of the ferromagnetic microwave amplifier', *J. Appl. Phys.*, 1957, **28**, pp. 1225–1236

Chapter 12

Perturbation theory and measurements

12.1 Introduction

The microwave properties of ferrite and other dielectric materials are often measured by observing the change in resonant frequency and Q-factor of a microwave resonant cavity when a small sample of ferrite or other material is inserted. If the ferrite sample is small, it will have only a small effect on the undisturbed fields in the cavity. Calculations can be simplified, if it is assumed that the fields external to the ferrite body are unchanged from those of the empty cavity and it is only the fields in the ferrite body which are different. The theory is called *perturbation theory* because it is assumed that the ferrite makes only a small perturbation to the fields existing in the unperturbed activity. Perturbation theory, as it is applied to a resonant cavity, is given in this chapter. Then it is applied to the measurement of ferrite material properties. The particular cavity systems suitable for the measurement of permittivity and tensor permeability are described and theoretical calculations are given.

Perturbation theory may be applied also to calculating the effect of any small insert into a waveguide system. It is applicable when the fields external to the perturbing body are virtually unchanged from those of the empty waveguide. In the early days of the analysis of structures of ferrite in waveguide, before an exact field analysis was available, perturbation theory was applied to the analysis of many common structures of ferrite in waveguide. Such analysis is a major part of textbooks on ferrites, such as Lax and Button[1] or Clarricoats.[2] However, most applications of ferrite in waveguide involve a perturbation which cannot be considered to be small and for which perturbation theory is not accurate, hence the full theoretical analyses which are given in Chapters 3, 4 and 6. This is also the reason why perturbation theory is not given the prominence in this book that it is given in some earlier texts. However, perturbation theory has been used to analyse the effect of a transversely magnetised ferrite slab in rectangular waveguide in Section 6.8, and the development of the formula derived from the application of perturbation theory in a transmission structure is also given in Section 12.7.

Perturbation theory has long been used to estimate the effect of small changes or small inclusions in resonant cavities. One of the earliest treatments was given by Slater,[3] who considered the perturbation due to a small deformation of the boundary surface of the cavity. Casimir[4] has extended the theory to the introduction of a small body into the cavity. Artman and Tannenwald[5] used cavity techniques to measure the microwave properties of small ferrite spheres and applied perturbation techniques to the calculation of the results.

12.2 Perturbation theory for a cavity

A small ferrite body will have only a small effect on the electromagnetic fields in a resonant cavity. Calculations are simplified if it is assumed that the fields external to the ferrite body are unchanged from those of the empty cavity, and it is the fields in the ferrite body alone which are different. It is assumed that the ferrite is sufficiently small to cause only a small perturbation in the resonant frequency of the cavity. The theory was first applied to cavity measurements of microwave ferrite properties by Artman and Tannenwald.[5] The theory was subsequently given in more detail by Waldron[6,7] and in the textbooks by Waldron,[8] Lax and Button[1] and Helszajn.[9] The perturbation theory applied to resonant cavities, given here, is similar to that given by Waldron.[7]

The fields in the cavity and in the ferrite sample are governed by Maxwell's equations, eqns. 2.31 and 2.32

$$\nabla \times \boldsymbol{E} = -\frac{\partial \boldsymbol{B}}{\partial t} \tag{12.1}$$

$$\nabla \times \boldsymbol{H} = \frac{\partial \boldsymbol{D}}{\partial t} \tag{12.2}$$

Measurements are made of the resonant frequency of the cavity when it is empty and the fields in the empty cavity are defined to be

$$\boldsymbol{E} = \boldsymbol{E}_0 \exp \mathrm{j}\omega t \tag{12.3}$$

$$\boldsymbol{H} = \boldsymbol{H}_0 \exp \mathrm{j}\omega t \tag{12.4}$$

With a ferrite sample inserted into the cavity, there will be a small change in the resonant frequency, and then the fields in the cavity and in the sample are defined to be

$$\boldsymbol{E} = (\boldsymbol{E}_0 + \boldsymbol{E}_1) \exp \mathrm{j}(\omega + \delta\omega)t \tag{12.5}$$

$$\boldsymbol{H} = (\boldsymbol{H}_0 + \boldsymbol{H}_1) \exp \mathrm{j}(\omega + \delta\omega)t \tag{12.6}$$

Substituting for both situations from eqns. 12.3 and 12.5 into eqn. 12.1 gives

$$\nabla \times \boldsymbol{E}_0 = -\mathrm{j}\omega \boldsymbol{B}_0 \tag{12.7}$$

and

$$\nabla \times (\boldsymbol{E}_0 + \boldsymbol{E}_1) = -\mathrm{j}(\omega + \delta\omega)(\boldsymbol{B}_0 + \boldsymbol{B}_1) \tag{12.8}$$

Subtracting eqn. 12.7 from eqn. 12.8 gives

$$\nabla \times \boldsymbol{E}_1 = -\mathrm{j}\,[\omega \boldsymbol{B}_1 + \delta\omega(\boldsymbol{B}_0 + \boldsymbol{B}_1)] \tag{12.9}$$

Similarly, substituting from eqns. 12.4 and 12.6 into eqn. 12.2 and taking the difference gives

$$\nabla \times \boldsymbol{H}_1 = \mathrm{j}[\omega \boldsymbol{D}_1 + \delta\omega(\boldsymbol{D}_0 + \boldsymbol{D}_1)] \tag{12.10}$$

Here

$$\boldsymbol{B}_0 = \mu_0 \boldsymbol{H}_0 \quad \text{and} \quad \boldsymbol{D}_0 = \varepsilon_0 \boldsymbol{E}_0 \tag{12.11}$$

and, outside the sample

$$\boldsymbol{B}_1 = \mu_0 \boldsymbol{H}_1 \quad \text{and} \quad \boldsymbol{D}_1 = \varepsilon_0 \boldsymbol{E}_1 \tag{12.12}$$

and, inside the sample

$$\left.\begin{aligned} \boldsymbol{B}_1 &= [\boldsymbol{\mu}(\boldsymbol{H}_0 + \boldsymbol{H}_1) - \mu_0 \boldsymbol{H}_0] \\ \boldsymbol{D}_1 &= [\varepsilon(\boldsymbol{E}_0 + \boldsymbol{E}_1) - \varepsilon_0 \boldsymbol{E}_0] \end{aligned}\right\} \tag{12.13}$$

where $\boldsymbol{\mu}$ and ε are the tensor permeability and permittivity of the ferrite.

Take the scalar product of $\boldsymbol{H}_0$ and eqn. 12.9 and that of $\boldsymbol{E}_0$ and eqn. 2.10 and take the sum, giving

$$\begin{aligned} \boldsymbol{H}_0 \cdot \nabla \times \boldsymbol{E}_1 + \boldsymbol{E}_0 \cdot \nabla \times \boldsymbol{H}_1 = {} & \mathrm{j}\omega\,[\boldsymbol{E}_0 \cdot \boldsymbol{D}_1 - \boldsymbol{H}_0 \cdot \boldsymbol{B}_1] \\ & + \mathrm{j}\delta\omega\,[\boldsymbol{E}_0 \cdot (\boldsymbol{D}_0 + \boldsymbol{D}_1) \\ & - \boldsymbol{H}_0 \cdot (\boldsymbol{B}_0 + \boldsymbol{B}_1)] \end{aligned} \tag{12.14}$$

Take the vector identity

$$\begin{aligned} \nabla \cdot (\boldsymbol{H}_0 \times \boldsymbol{E}_1 + \boldsymbol{E}_0 \times \boldsymbol{H}_1) \equiv {} & \boldsymbol{E}_1 \cdot \nabla \times \boldsymbol{H}_0 - \boldsymbol{H}_0 \cdot \nabla \times \boldsymbol{E}_1 \\ & + \boldsymbol{H}_1 \cdot \nabla \times \boldsymbol{E}_0 - \boldsymbol{E}_0 \cdot \nabla \times \boldsymbol{H}_1 \end{aligned} \tag{12.15}$$

Substituting from Maxwell's equations for $\nabla \times \boldsymbol{H}_0$ and $\nabla \times \boldsymbol{E}_0$ into eqn. 12.15 and rearranging gives

$$\begin{aligned} \boldsymbol{H}_0 \cdot \nabla \times \boldsymbol{E}_1 + \boldsymbol{E}_0 \cdot \nabla \times \boldsymbol{H}_1 = {} & \mathrm{j}\omega(\boldsymbol{E}_1 \cdot \boldsymbol{D}_0 - \boldsymbol{H}_1 \cdot \boldsymbol{B}_0) \\ & - \nabla \cdot (\boldsymbol{H}_0 \times \boldsymbol{E}_1 + \boldsymbol{E}_0 \times \boldsymbol{H}_1) \end{aligned} \tag{12.16}$$

Subtracting eqn. 12.14 from eqn. 12.16 gives

$$\nabla\cdot(\boldsymbol{H}_0\times\boldsymbol{E}_1+\boldsymbol{E}_0\times\boldsymbol{H}_1) = \mathrm{j}\omega(\boldsymbol{E}_1\cdot\boldsymbol{D}_0-\boldsymbol{H}_1\cdot\boldsymbol{B}_0-\boldsymbol{E}_0\cdot\boldsymbol{D}_1+\boldsymbol{H}_0\cdot\boldsymbol{B}_1)-\mathrm{j}\delta\omega[\boldsymbol{E}_0\cdot(\boldsymbol{D}_0+\boldsymbol{D}_1)-\boldsymbol{H}_0\cdot(\boldsymbol{B}_0+\boldsymbol{B}_1)] \tag{12.17}$$

Now integrate eqn. 12.17 over the volume of the cavity. First apply Gauss's theorem to the integral of the divergence on the left-hand side of eqn. 12.17, which gives

$$\int_{\text{volume}}\nabla\cdot(\boldsymbol{H}_0\times\boldsymbol{E}_1+\boldsymbol{E}_0\times\boldsymbol{H}_1)\,\mathrm{d}V = \int_{\text{surface}}(\boldsymbol{H}_0\times\boldsymbol{E}_1+\boldsymbol{E}_0\times\boldsymbol{H}_1)\cdot\mathrm{d}\boldsymbol{A} \tag{12.18}$$

where the integral on the left-hand side is the integral throughout the volume of the cavity and the integral on the right-hand side is an integral over the surface and $\mathrm{d}\boldsymbol{A}$ is a vector perpendicular to the surface proportional to a small element of the surface area. For a microwave resonant cavity, it may be assumed that the surface of the cavity is perfectly conducting and the component of the electric field vector parallel to that surface is zero. Then the product of the terms inside the bracket on the right-hand side of eqn. 12.18 is a vector parallel to the surface and the integral is zero everywhere on the surface. Substituting from eqn. 12.11 and 12.12 shows that the terms in $\mathrm{j}\omega$ in eqn. 12.17 are zero everywhere inside the cavity outside the sample. Then applying the integral to eqn. 12.17 gives

$$\mathrm{j}\omega\int_{\text{sample}}(\boldsymbol{E}_1\cdot\boldsymbol{D}_0-\boldsymbol{H}_1\cdot\boldsymbol{B}_0-\boldsymbol{E}_0\cdot\boldsymbol{D}_1+\boldsymbol{H}_0\cdot\boldsymbol{B}_1)\,\mathrm{d}V = \mathrm{j}\delta\omega\int_{\text{cavity}}[\boldsymbol{E}_0\cdot(\boldsymbol{D}_0+\boldsymbol{D}_1)-\boldsymbol{H}_0\cdot(\boldsymbol{B}_0+\boldsymbol{B}_1)]\,\mathrm{d}V \tag{12.19}$$

So far, apart from assuming that the cavity has perfectly conducting walls, eqn. 12.19 is exact. Now assume that the effect of the ferrite sample is small on the resonant frequency of the cavity. Then we may assume that $\delta\omega \ll \omega$ and that $\boldsymbol{B}_1$ and $\boldsymbol{D}_1$ are small compared with $\boldsymbol{B}_0$ and $\boldsymbol{D}_0$, and eqn. 12.19 is simplified to give

$$\frac{\delta\omega}{\omega} = \frac{\int_{\text{sample}}(\boldsymbol{E}_1\cdot\boldsymbol{D}_0-\boldsymbol{H}_1\cdot\boldsymbol{B}_0-\boldsymbol{E}_0\cdot\boldsymbol{D}_1+\boldsymbol{H}_0\cdot\boldsymbol{B}_1)\,\mathrm{d}V}{\int_{\text{cavity}}(\boldsymbol{E}_0\cdot\boldsymbol{D}_0-\boldsymbol{H}_0\cdot\boldsymbol{B}_0)\,\mathrm{d}V} \tag{12.20}$$

In the region of the sample itself, $\boldsymbol{B}_1$ and $\boldsymbol{D}_1$ are probably of the same order as $\boldsymbol{B}_0$ and $\boldsymbol{D}_0$, but their contribution to the total integral in the denominator of eqn. 12.20 is small because the sample is small. Introducing the ferrite material constants from eqn. 12.12 into the numerator of eqn. 12.20 gives

$$\left.\begin{aligned} -\boldsymbol{E}_0\cdot\boldsymbol{D}_1 + \boldsymbol{E}_1\cdot\boldsymbol{D}_0 &= -\varepsilon\boldsymbol{E}_0\cdot\boldsymbol{E}_0 - \varepsilon\boldsymbol{E}_0\cdot\boldsymbol{E}_1 \\ &\quad + \varepsilon_0\boldsymbol{E}_0\cdot\boldsymbol{E}_0 + \varepsilon_0\boldsymbol{E}_1\cdot\boldsymbol{E}_0 \\ &= -(\varepsilon-\varepsilon_0)\boldsymbol{E}_0\cdot(\boldsymbol{E}_0+\boldsymbol{E}_1) \\ &= -\chi_e^e\boldsymbol{E}_0\cdot\boldsymbol{E}_0 \\ \boldsymbol{H}_0\cdot\boldsymbol{B}_1 - \boldsymbol{H}_1\cdot\boldsymbol{B}_0 &= (\boldsymbol{\mu}\cdot\boldsymbol{H}_0)\cdot\boldsymbol{H}_0 + (\boldsymbol{\mu}\cdot\boldsymbol{H}_0)\cdot\boldsymbol{H}_1 \\ &\quad - \mu_0\boldsymbol{H}_0\cdot\boldsymbol{H}_0 - \mu_0\boldsymbol{H}_1\cdot\boldsymbol{H}_0 \\ &= [(\boldsymbol{\mu}-\mu_0)\cdot\boldsymbol{H}_0]\cdot(\boldsymbol{H}_0+\boldsymbol{H}_1) \\ &= (\boldsymbol{\chi}^e\cdot\boldsymbol{H}_0)\cdot\boldsymbol{H}_0 \end{aligned}\right\} \qquad (12.21)$$

where $\boldsymbol{\chi}^e$ and χ_e^e are the external forms of the tensor susceptibility and the dielectric susceptibility, respectively, i.e. they relate the flux densities *inside* the sample to the field intensities *outside* the sample assuming that the sample is so small that any variation of the external field quantities across the region of the sample is negligible. Then eqn. 12.20 becomes

$$\frac{\delta\omega}{\omega} = \frac{\int_{\text{sample}} [(\boldsymbol{\chi}^e\cdot\boldsymbol{H})\cdot\boldsymbol{H} - \chi_e^e\boldsymbol{E}\cdot\boldsymbol{E}]\,dV}{\int_{\text{cavity}} (\varepsilon_0\boldsymbol{E}\cdot\boldsymbol{E} - \mu_0\boldsymbol{H}\cdot\boldsymbol{H})\,dV} \qquad (12.22)$$

where the subscript 0 has been dropped in eqn. 12.22 because it now applies to all the field quantities. The field quantities in eqn. 12.22, as defined by eqns. 12.3 to 12.6, still include phase information. Inside any cavity at resonance, as shown by eqns. 12.42 to 12.44 and 12.54 to 12.58, the electric and magnetic fields are out of phase and the minus signs in eqn. 12.22 become plus signs when electric and magnetic field amplitudes are used to evaluate the equation. Other authors, such as Lax and Button[1] or Helszajn,[9] have plus signs and complex conjugate quantities in their versions of eqn. 12.22 where the factors j^2 have been incorporated into their equations.

In eqn. 12.21 we have defined the flux densities inside the sample in terms of the fields external to the sample. This means that the susceptibilities used in eqns. 12.21 and 12.22 are *external* susceptibilities and are functions of the shape of the sample as well as of material properties. It may be observed that the various quantities in eqn. 12.22 are a measure of the electric and magnetic power in the cavity or in the sample. If the sample is placed at a position of zero electric field in the cavity so as to measure the magnetic properties of the material, then eqn. 12.22 may be simplified to be

$$\frac{\delta\omega}{\omega} = \frac{-\int_{\text{sample}} (\boldsymbol{\chi}^e\cdot\boldsymbol{H})\cdot\boldsymbol{H}\,dV}{2\int_{\text{cavity}} \mu_0\boldsymbol{H}\cdot\boldsymbol{H}\,dV} \qquad (12.23)$$

because normally at resonance the electric and magnetic field energies are equal. Similarly, if the sample is placed at a position of zero magnetic field so as to measure the electric properties of the material, eqn. 12.22 becomes

$$\frac{\delta\omega}{\omega} = \frac{-\int_{\text{sample}} \chi_e^e \boldsymbol{E} \cdot \boldsymbol{E}\, \mathrm{d}V}{2 \int_{\text{cavity}} \varepsilon_0 \boldsymbol{E} \cdot \boldsymbol{E}\, \mathrm{d}V} \tag{12.24}$$

These formulas use the values of the fields existing in the empty cavity, thus the approximation is involved in the assumption that the sample does not appreciably alter the fields near to it. The assumption is more likely to be true if the sample has an ellipsoidal shape. As we shall see in the next section, ellipsoidal shaped samples also enable an exact relationship to be obtained between the external susceptibility and the true material properties.

12.3 Sample shape

When a small ferrite sample is inserted into a resonant cavity, the electric and magnetic fields external to the sample will not necessarily be the same as the electric and magnetic fields inside the sample. Calculation is necessary to determine the relationship between them so as to establish the relationship between the external and intrinsic susceptibilities of the sample. A derivation similar to the one given in this section has already been given in Section 6.8 in connection with the application of perturbation theory to the calculation of propagation constants in ferrite loaded waveguide. The effect of the shape of the ferrite sample on the external susceptibilities can be given in terms of the demagnetisation factors due to the shape of the sample. Static demagnetisation factors can be obtained for ellipsoidal shaped samples and Waldron[10] has shown that the same static demagnetisation factors apply to electromagnetic fields in both isotropic and anisotropic media. In particular, they are applicable to the tensor permeability of ferrites.

Let a ferrite body of arbitrary shape be magnetised in the z-direction so that its intrinsic susceptibility is given by

$$\boldsymbol{\chi} = \begin{vmatrix} \chi & \mathrm{j}\kappa & 0 \\ -\mathrm{j}\kappa & \chi & 0 \\ 0 & 0 & 0 \end{vmatrix} \tag{12.25}$$

The internal magnetic field has components, H_{ix}, H_{iy} and H_{iz}, which give rise to the magnetisation, M_x, M_y and M_z where

$$\boldsymbol{M} = \boldsymbol{\chi} \boldsymbol{H}_i \tag{12.26}$$

In the absence of the sample in the cavity, there will be the external magnetic field, H_{0x}, H_{0y} and H_{0z} which is related to the magnetisation in the sample by the external susceptibility

$$\boldsymbol{M} = \boldsymbol{\chi}^{e} \boldsymbol{H}_0 \tag{12.27}$$

where the external susceptibility is defined by

$$\boldsymbol{\chi}^{e} = \begin{vmatrix} \chi^{e}_{xx} & \chi^{e}_{xy} & 0 \\ \chi^{e}_{yx} & \chi^{e}_{yy} & 0 \\ 0 & 0 & 0 \end{vmatrix} \tag{12.28}$$

Introducing the demagnetisation vector $\boldsymbol{N}$, the internal magnetic field is given by

$$\boldsymbol{H}_{i} = \boldsymbol{H}_0 - \frac{\boldsymbol{N} \cdot \boldsymbol{M}}{\mu_0} = \boldsymbol{H}_0 - \frac{\boldsymbol{N} \cdot (\boldsymbol{\chi} \boldsymbol{H}_{i})}{\mu_0} \tag{12.29}$$

whence

$$H_{ix} = H_{0x} - \frac{N_x}{\mu_0} (\chi H_{ix} + j\kappa H_{iy}) \tag{12.30}$$

$$H_{iy} = H_{0y} - \frac{N_y}{\mu_0} (\chi H_{iy} - j\kappa H_{ix}) \tag{12.31}$$

$$H_{iz} = H_{0z} \tag{12.32}$$

Rearranging these equations gives expressions for the internal field components

$$H_{ix} = \frac{(\mu_0 + N_y \chi) H_{0x} - jN_x \kappa H_{0y}}{\mathrm{D}} \tag{12.33}$$

$$H_{iy} = \frac{(\mu_0 + N_x \chi) H_{0y} + jN_y \kappa H_{0x}}{\mathrm{D}} \tag{12.34}$$

$$H_{iz} = H_{0z} \tag{12.35}$$

where

$$D = \mu_0 + (N_x + N_y)\chi + N_x N_y (\chi^2 - \kappa^2)/\mu_0$$

Then the components of the external susceptibility are given by

$$\chi^{e}_{xx} = \frac{\mu_0 \chi + N_y (\chi^2 - \kappa^2)}{D} \tag{12.36}$$

$$\chi^{e}_{yy} = \frac{\mu_0 \chi + N_x (\chi^2 - \kappa^2)}{D} \tag{12.37}$$

$$\chi^{e}_{xy} = -\chi^{e}_{yx} = \frac{j\kappa \mu_0}{D} \tag{12.38}$$

These results have been given by Steinert,[11] and, in a different form, including loss, by Lax and Button.[12] Similarly, for the electric fields inside the sample

$$\left.\begin{aligned} E_{ix} &= E_{0x} - \frac{N_x}{\varepsilon_0}(\varepsilon - \varepsilon_0)\, E_{ix} \\ E_{iy} &= E_{0y} - \frac{N_y}{\varepsilon_0}(\varepsilon - \varepsilon_0)\, E_{iy} \\ E_{iz} &= E_{0z} - \frac{N_z}{\varepsilon_0}(\varepsilon - \varepsilon_0)\, E_{iz} \end{aligned}\right\} \tag{12.39}$$

Then the elements of the external electric susceptibility, in terms of the intrinsic electric susceptibility, χ_e, are

$$\left.\begin{aligned} \chi^e_{exx} &= \frac{\chi_e}{1 + N_x\chi_e/\varepsilon_0} \\ \chi^e_{eyy} &= \frac{\chi_e}{1 + N_y\chi_e/\varepsilon_0} \\ \chi^e_{ezz} &= \frac{\chi_e}{1 + N_z\chi_e/\varepsilon_0} \end{aligned}\right\} \tag{12.40}$$

If the sample is of ellipsoidal shape, the demagnetisation factors can be calculated exactly. Calculation of demagnetisation factors for other than the simplest of shapes is very complicated. Analytical expressions can be obtained for ellipsoidal shaped samples but they involve elliptic integrals. Some results have been given by Stoner[13] but a much more comprehensive set of results have been given by Osborn.[14] He gives tables of demagnetisation factors and curves plotted from those results, spanning the full range of ellipsoid sizes. Similar curves are also given by Lax and Button.[12] The results are given here in Fig. 12.1 in a form similar to that used by Stoner,[13] but plotted from the tabulated results given by Osborn.[14] The axes of the ellipsoid are a, b and c and the demagnetisation factor appropriate to the direction of the a-dimension is N_a and similarly for N_b and N_c. The curves give values for the the demagnetisation factor as the ratio c/a varies for specified values of b/a. The curves for all three demagnetisation factors are combined on one graph because they do not overlap. The inset diagram shows which demagnetisation factor is read from which part of the graph. Separation is determined by the curve appropriate to a prolate spheroid ($b = c$) and that appropriate to an oblate spheroid ($a = b$). To use the curves, a needs to be the largest dimension of the ellipsoid and the condition $a \geqslant b \geqslant c > 0$ needs to be applied. From the graph, it is seen that no curves exist for the condition $(c/a) > (b/a)$. The graph contains redundant information because

$$N_a + N_b + N_c = 1 \tag{12.41}$$

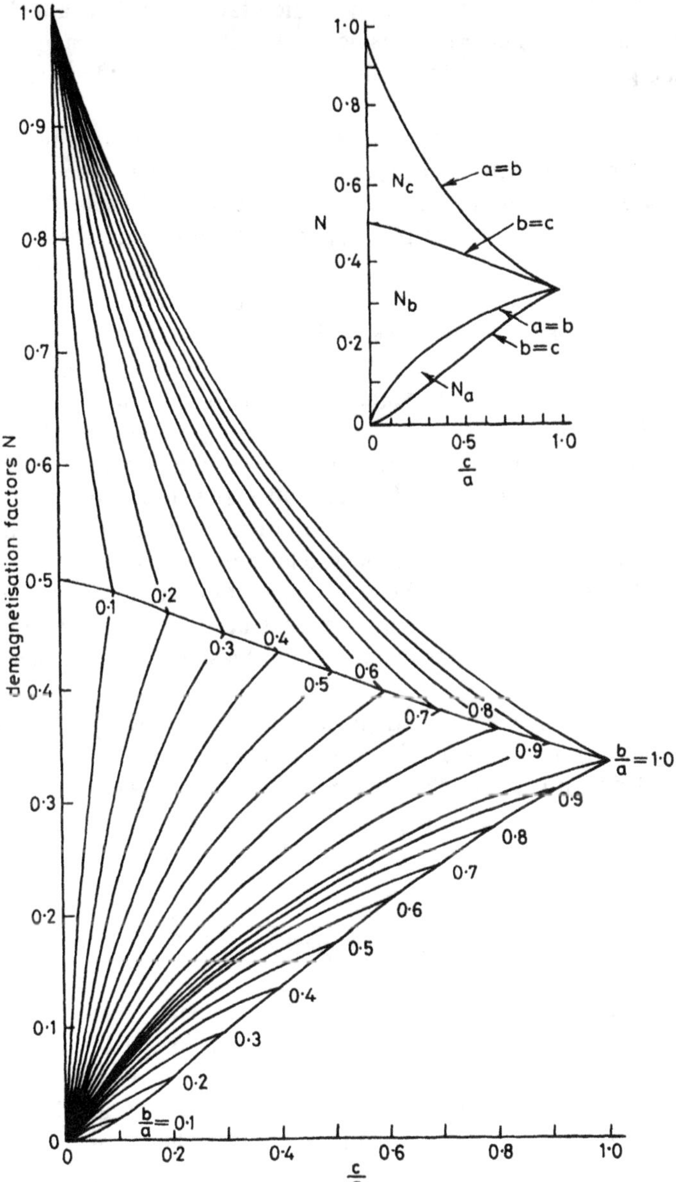

Fig. 12.1 *Demagnetisation factors for a general ellipsoid of axes a, b and c where a ⩾ b ⩾ c. The curves are for N_i versus c/a with b/a as a parameter. The inset diagram shows the part of the curve to use for the different demagnetisation factors. They are plotted from calculated results published by Osborn*[14]

It is still useful to have all three demagnetisation factors easily available, since, for ferrite bodies with an external biasing magnetic field, only demagnetisation factors in the plane perpendicular to the biasing magnetic field are required as shown by eqns. 12.33 to 12.35.

The use of intrinsic and external susceptibilities only becomes necessary in conjunction with perturbation theory. In the exact electromagnetic theory of Chapters 3, 4 and 6, the boundary conditions are applied explicitly and only intrinsic values of microwave susceptibility occur. In situations where the ferrite can be considered to be a small sample perturbing the fields in a waveguide or cavity, the shape of the sample has to be taken into account. In many measurement systems, spherical or ellipsoidal samples can be used and the exact demagnetising factors specified. If the sample is not ellipsoidal, for example, a rectangular shape or a disk, only approximate demagnetisation fators can be used. Brug and Wolf[15] discuss the errors introduced by using demagnetisation factors for non-ellipsoidal samples. They show that for a magnetic disk, the effective demagnetisation factor is dependent on the radial position in the disk and the magnetic equation of state, and is a function of imperfections in the material such as pores and cracks. If there is a linear relationship between the microwave fields, $\boldsymbol{B}$ and $\boldsymbol{H}$, as often occurs, the errors in using effective demagnetisation factors may be quite small. However, care does need to be exercised in using demagnetisation factors to calculate the static biasing magnetic field in the ferrite, H_0.

The electric field also experiences depolarisation effects owing to the shape of the sample, and the same demagnetisation factors enter into the expressions for the external electrical susceptibility in eqn. 12.40. Thus, even though they are called demagnetisation factors, they also enter into expressions describing the relationship between the internal and external electric field strengths.

12.4 Cavity measurement of dielectric properties

In order to apply perturbation theory to the measurement of microwave ferrite properties, a suitable sample of the ferrite material is inserted into a resonant cavity and the change in resonant frequency and Q-factor of the loaded cavity is noted. In order to aid the analysis of the results, the ferrite sample is inserted at a point of zero magnetic field in order to measure its dielectric properties and at a point of zero electric field in order to measure its magnetic properties. Different authors recommend different systems of mounting the ferrite in various cavities. Artman and Tannenwald[5] mount their samples on the end wall of either a cylindrical cavity operating in the TE_{11n}-mode or a rectangular cavity operating in the TE_{10n}-mode. Lax and Button[1] and Helszajn[9] describe the use of the same cavities. However, it is undesirable to have the ferrite sample near to a wall of the cavity, since image charges are induced in the conducting walls and modify the field patterns in the sample. Artman and Tannenwald[5] recognise the problem and mount their samples a small distance from the cavity wall. They have investigated the minimum distance necessary to eliminate any wall proximity errors. However, the procedure to be described here largely follows that described by Spencer, LeCraw and Reggia.[16] They mount their samples at the centre of the cavity so that there can be no possibility of images affecting the fields in the sample.

In order to measure the electrical properties, the ferrite sample is mounted at the centre of a TE_{103}-mode rectangular cavity. From the field pattern, shown in Fig. 12.2, it is seen that, at the centre of the cavity, the magnetic field is zero and

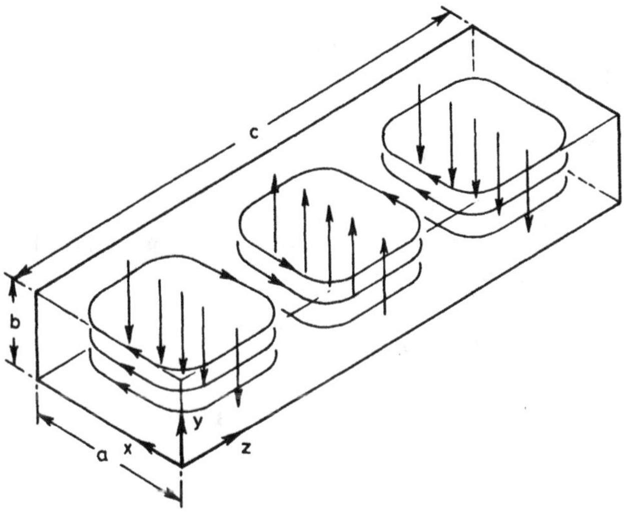

Fig. 12.2 *Fields of the TE_{103}-mode in a rectangular cavity*

the electric field is a maximum. Then eqn. 12.24 can be used to find the change in resonant frequency of the cavity when the sample is inserted. The fields in the cavity are[17]

$$E_y = -\mathrm{j}\frac{\omega\mu a}{\pi} H_0 \sin\frac{\pi x}{a} \sin\frac{3\pi z}{c} \tag{12.42}$$

$$H_x = \frac{3a}{c} H_0 \sin\frac{\pi x}{a} \cos\frac{3\pi z}{c} \tag{12.43}$$

$$H_z = H_0 \cos\frac{\pi x}{a} \sin\frac{3\pi z}{c} \tag{12.44}$$

Substituting these values into eqn. 12.24, the denominator is

$$2\int_{\text{cavity}} \varepsilon_0 E_y E_y \, \mathrm{d}V = 2\varepsilon_0 \left(\frac{\omega\mu_0 a}{\pi} H_0\right)^2 \frac{a}{2}\frac{c}{2} b = \frac{\varepsilon_0}{2}\left(\frac{\omega\mu_0 a}{\pi} H_0\right)^2 v_c \tag{12.45}$$

The sample is at the position of maximum value of E_y which is assumed to be constant across the volume of the sample so that the numerator is given by

$$\left(\frac{\omega\mu_0 a}{\pi} H_0\right)^2 \chi_e^e v_s \qquad (12.46)$$

where v_c and v_s are the volumes of the cavity and the sample, respectively. Then, substituting these expressions in to eqn. 12.24 gives

$$\frac{\delta\omega}{\omega} = \frac{2\chi_e^e v_s}{\varepsilon_0 v_c} \qquad (12.47)$$

For a spherical sample, where $N_y = \frac{1}{3}$ in eqn. 12.40, the permittivity is given by the expression

$$\frac{\delta\omega}{\omega} = -\frac{6(\varepsilon - \varepsilon_0)}{(\varepsilon + 2\varepsilon_0)} \frac{v_s}{v_c} \qquad (12.48)$$

If the loss in the dielectric material is represented by a complex permittivity, eqn. 12.48 is complex and there will be a complex component to the frequency. The complex frequency can be written[5]

$$\omega' - j\omega'' = \omega + \delta\omega + j\Delta\left(\frac{\omega}{2Q}\right) \qquad (12.49)$$

Then taking the real and imaginary parts of eqn. 12.48 gives

$$\frac{\delta\omega'}{\omega} = -\frac{6(\varepsilon' - \varepsilon_0)}{(\varepsilon' - 2\varepsilon_0)} \frac{v_s}{v_c} \qquad (12.50)$$

and

$$\Delta\left(\frac{1}{2Q}\right) = \frac{18\varepsilon_0\varepsilon''}{(\varepsilon' + 2\varepsilon_0)^2} \frac{v_s}{v_c} \qquad (12.51)$$

where $\Delta(1/Q)$ is the change in the unloaded Q of the cavity when the sample is inserted

$$\Delta\left(\frac{1}{Q}\right) = \frac{1}{Q_s} - \frac{1}{Q_e} \qquad (12.52)$$

where Q_e is the Q-factor of the empty cavity and Q_s is that of the cavity with the ferrite sample.

12.5 Cavity measurement of magnetic properties

If the ferrite sample is mounted at the centre of a circular cavity which will support circularly polarised modes, then the tensor permeability becomes a simple scalar effective permeability given by eqn. 2.44. For the field in the sample to be circularly polarised, the sample must be circularly symmetric in the transverse plane and the demagnetising factor in that plane is N_t. Suitable substitution of a relationship between $H_{ix} \pm jH_{iy}$ and $H_{0x} \pm jH_{0y}$ from eqns. 12.29 to 12.38 shows that, for circularly polarised field components

$$\chi^e = \frac{(\chi \pm \kappa)}{1 + N_t(\chi \pm \kappa)/\mu_0} \tag{12.53}$$

A suitable cavity for making measurements is the TE_{11n}-mode cylindrical cavity used as a transmission cavity in the system shown in Fig. 12.3. The rectangular to circular transformer is followed by a polarised absorber arranged to absorb

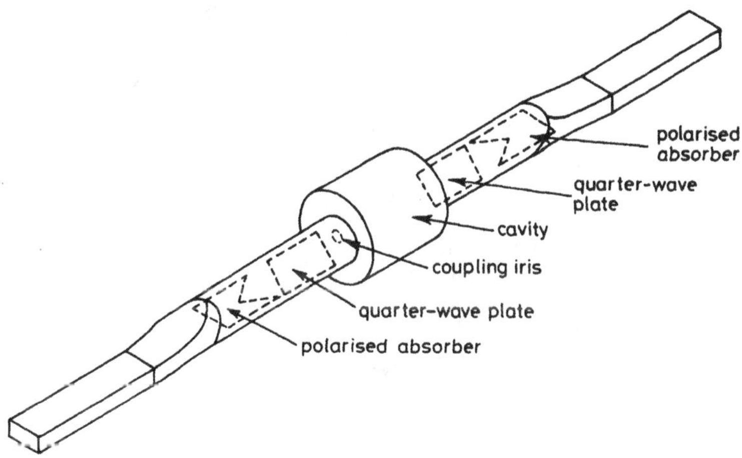

Fig. 12.3 *Waveguide system for applying circular polarisation to a cylindrical cavity*

any crosspolarised wave in the circular waveguide. This is followed by a quarter-wave plate circular polariser and the coupling iris to the cavity. Much of the incident power will be reflected at the iris with the opposite hand of circular polarisation; it will be converted into the perpendicular mode by the quarter-wave plate and absorbed in the polarised absorber. The power coupled through the cavity will be proportional to the response of the cavity and the output waveguide system is the mirror of the input.

We shall now look at the fields of the TE_{11n}-mode in a cylindrical cavity. The fields of the TE_{112}-mode are shown in Fig. 12.4. It is seen that, if the ferrite sample is positioned at the centre of the cavity, it is at a point of zero electric field and maximum magnetic field. The fields in the cavity for the TE_{11n}-mode are then given by[17]

$$E_r = \frac{j\omega\mu}{k_c^2} H_0 \frac{J_1(k_c r)}{r} \sin\theta \sin\frac{n\pi z}{c} \quad (12.54)$$

$$E_\theta = \frac{j\omega\mu}{k_c} H_0 J_1'(k_c r) \cos\theta \sin\frac{n\pi z}{c} \quad (12.55)$$

$$H_r = -\frac{n\pi}{ck_c} H_0\, J_1'(k_c r) \cos\theta \cos\frac{n\pi z}{c} \quad (12.56)$$

$$H_\theta = -\frac{n\pi}{ck_c^2} H_0 \frac{J_1(k_c r)}{r} \sin\theta \cos\frac{n\pi z}{c} \quad (12.57)$$

$$H_z = H_0 J_1(k_c r) \cos\theta \sin\frac{n\pi z}{c} \quad (12.58)$$

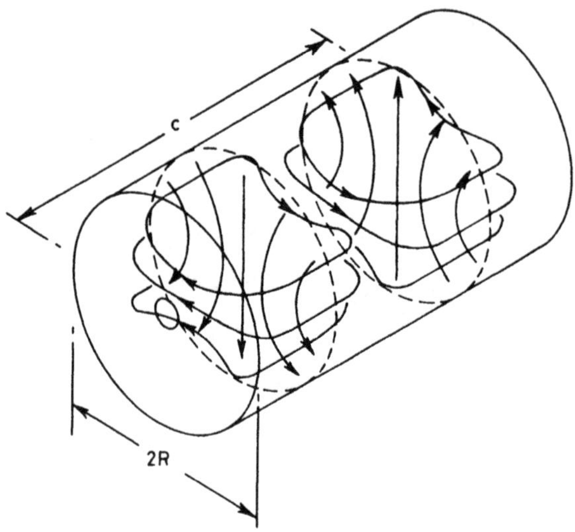

Fig. 12.4 *Fields of the TE_{112}-mode in a cylindrical cavity*

where k_c is given by the boundary condition, $k_c R = 1{\cdot}841$. The change in resonant frequency on inserting the sample is found by substituting the field values into eqn. 12.23, whose denominator becomes

$$2\mu_0 \left(\int_{\text{cavity}} H_r^2 \, dV + \int_{\text{cavity}} H_\theta^2 \, dV + \int_{\text{cavity}} H_z^2 \, dV \right) = \mu_0 \pi c H_0^2$$
$$\times \left[\frac{n^2\pi^2}{c^2 k_c^2} \int_0^R r[J_1'(k_c r)]^2 \, dr + \frac{n^2\pi^2}{c^2 k_c^4} \int_0^R \frac{r[J_1(k_c r)]^2}{r^2} dr + \int_0^R r[J_1(k_c r)]^2 dr \right] \quad (12.59)$$

Each of these integrals of Bessel functions are standard forms and their integral is

$$\mu_0 \frac{\pi c H_0^2}{2}\left[\frac{n^2\pi^2}{c^2 k_c^4}\{(k_c^2R^2 + 1)J_0^2(k_cR) + (k_c^2R^2 - 1)J_1^2(k_cR) - J_0^2(k_cR) - J_1^2(k_cR)\} + \frac{k_c^2R^2}{k_c^2}\{J_1^2(k_cR) - J_0(k_cR)J_2(k_cR)\}\right] \tag{12.60}$$

Inserting numbers from the boundary condition, $k_cR = 1{\cdot}841$, gives

$$\text{denominator} = \mu_0 \frac{\pi c H_0^2}{2k_c^4}\left[\frac{n^2\pi^2}{c^2}(0{\cdot}809) + \frac{1}{R^2}(2{\cdot}741)\right] \tag{12.61}$$

The numerator in eqn. 12.23 is easily evaluated because the sample is situated at the centre of the cavity where the magnetic field is assumed to be uniform and is given by the maximum value of H_r. Therefore

$$\text{numerator} = \chi^e (H_{r\max})^2 v_s = \chi^e \frac{n^2\pi^2}{c^2 k_c^2} H_0^2 [J_1'(0)]^2 v_s \tag{12.62}$$

Then the change in resonant frequency is given by

$$\frac{\delta\omega}{\omega} = -\frac{\chi^e}{\mu_0}\frac{6{\cdot}102\, v_s}{(2{\cdot}913 + c^2/n^2R^2)v_c} \tag{12.63}$$

Spencer *et al.*[16] used a square section (i.e. $c = 2R$) cavity, excited in the TE_{112}-mode as shown in Fig. 12.4, with a spherical sample. Using the demagnetisation factor $N_t = \frac{1}{3}$ in eqn. 12.53 and the dimensions of the cavity in eqn. 12.63, gives

$$\frac{\delta\omega_\pm}{\omega} = -3\frac{(\mu \pm \kappa - \mu_0)}{(\mu \pm \kappa + 2\mu_0)}\frac{1{\cdot}56\, v_s}{v_c} \tag{12.64}$$

Similar to eqns. 12.50 and 12.51, the real and imaginary components of the effective permeability are given by

$$\frac{\delta\omega'_\pm}{\omega} = -\frac{3(\mu' \pm \kappa' - \mu_0)}{(\mu' \pm \kappa' + 2\mu_0)}\frac{1{\cdot}56\, v_s}{v_c} \tag{12.65}$$

$$\Delta\left(\frac{1}{2Q_\pm}\right) = \frac{9\mu_0(\mu'' \pm \kappa'')}{(\mu' \pm \kappa' + 2\mu_0)^2}\frac{1{\cdot}56\, v_s}{v_c} \tag{12.66}$$

If measurements are made using both hands of circular polarisation, values for the complex components of the tensor permeability may be obtained. The bias magnetic field is longitudinal, along the direction of propagation through the cavity. It is most easily provided by a coil round the outside of the cavity and a steel yoke having hollow pole pieces to concentrate the field in the cavity. The pole pieces are hollow so that the input and output waveguides can go through the centre of the pole pieces. Changing the hand of circular polarisation in the cavity requires the use of switched circular polarisers or the physical dismantling of the waveguide system to make the change. However, reversing the direction of the biasing magnetic field has the same effect as changing the hand of circular polarisation and is much easier to achieve. A quicker method of obtaining the real components of the tensor permeability is to excite the cavity with a linearly polarised wave. The ferrite then separates the linearly polarised mode into two circularly polarised modes of opposite hand and each will resonate at a different frequency, giving a double humped frequency response for the cavity.

12.6 Exact methods

One of the problems associated with the use of cavity perturbation theory in the measurement of ferrite properties is that, when the sample is sufficiently small for the theory to be accurate, the frequency change may be very small and difficult to measure accurately. Conversely, using a bigger sample so that the frequency change is larger, may mean that the approximations associated with the use of perturbation theory are no longer negligible. The only way to overcome these difficulties is to use exact theory to calculate the resonant frequency of the cavity. For an arbitrary shaped sample this becomes very difficult, but for a symmetrical shape, such as a rod or circular disk in a cylindrical cavity or a longitudinal slab or rectangular disk in a rectangular

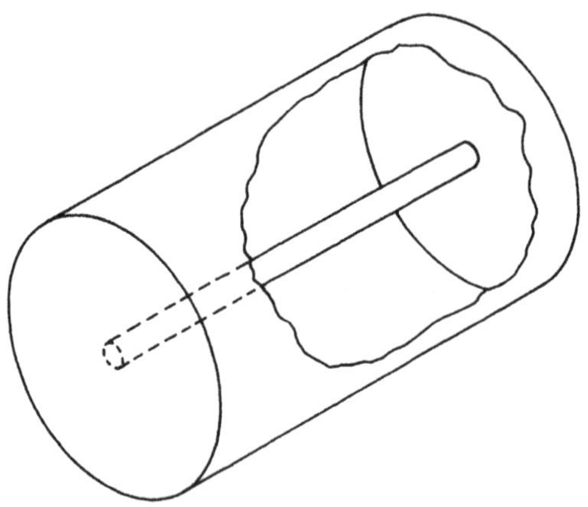

Fig. 12.5 *Rod at the centre of a cylindrical cavity*

cavity, the exact theory is available. The sample needs to be either the same length as the length of the cavity or the same cross-section as the cavity.

Consider a ferrite rod in the centre of a cylindrical cavity, as shown in Fig. 12.5. It is magnetised along its length and may be analysed by the theory given in Section 3.5 as a length of ferrite loaded circular waveguide in the same way that the empty cavity is analysed as a length of empty circular waveguide. For the particular geometry used, it will be necessary to draw up the relationships between resonant frequency and the ferrite properties. In this case it is not easy to separate the effects of the permittivity and permeability of the ferrite. An alternative uses a ferrite disk completely filling one end of a cylindrical cavity as shown in Fig. 12.6. This is analysed as a short length of completely filled

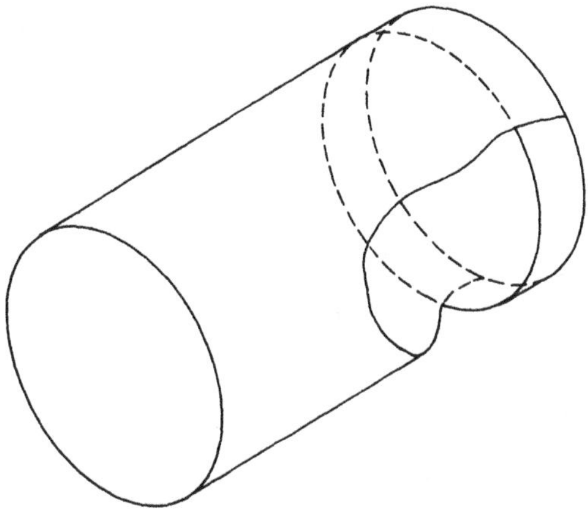

Fig. 12.6 *Disk at one end of a cylindrical cavity*

circular waveguide by the theory given in Section 3.4, coupled to a length of empty circular waveguide. Because the theory used is exact, no demagnetisation factors enter into the expressions used for the r.f. properties of the ferrite. Ito, Ohkawa and Yamamoto[18] measure the Faraday rotation due to a longitudinally magnetised thin ferrite rod in circular waveguide, and give formulas for obtaining the microwave ferrite parameters from the measurements.

A transversely magnetised ferrite slab may be mounted in a rectangular waveguide cavity, similar to that shown in Fig. 6.2. It is analysed in Section 6.3 and the propagation constant is given by eqn. 6.29. Again, for measurement purposes, a family of theoretical results needs to be produced relating the resonant frequency to the ferrite properties. For the theory to be exact, the ferrite has to fill the cross-section of the cavity in one plane. Srivastava, Patni and Hanumantha Rao[19] describe a technique using a thin ferrite slab completely filling one end of a rectangular cavity operating in the TE_{104}-mode.

12.7 Perturbation theory for waveguide

Perturbation theory can also be applied to the conditions of a small sample disturbing the propagation conditions in a waveguide or transmission line. It is assumed that there is a ferrite body in the waveguide which causes only a small perturbation to the field conditions and the propagation constant in the empty waveguide, such as the transversely magnetised ferrite slab shown in Fig. 12.7, which might be used to construct a resonance isolator. It is assumed that the ferrite loaded waveguide is uniform in the direction of propagation so that all

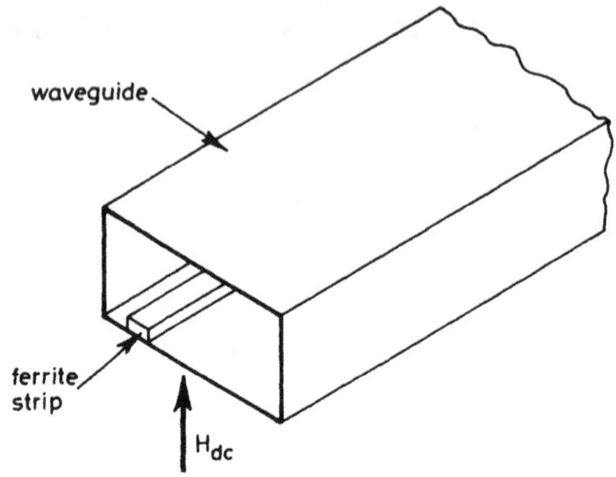

Fig. 12.7 *Transversely magnetised ferrite slab in rectangular waveguide*

the field quantities are characterised by an exp $-\gamma z$ field strength variation. Then the field quantities in the ferrite loaded waveguide are

$$\boldsymbol{E} \exp -\gamma z$$

$$\boldsymbol{H} \exp -\gamma z$$

$$\boldsymbol{D} \exp -\gamma z$$

$$\boldsymbol{B} \exp -\gamma z$$

and in the equivalent empty waveguide

$$\boldsymbol{E}_0 \exp -\gamma_0 z$$

$$\boldsymbol{H}_0 \exp -\gamma_0 z$$

$$\boldsymbol{D}_0 \exp -\gamma_0 z$$

$$\boldsymbol{B}_0 \exp -\gamma_0 z$$

Each of these field quantities, then, depends only upon the transverse co-ordinates (x, y). Using the vector identity

$$\nabla \times a\boldsymbol{A} = a\nabla \times \boldsymbol{A} + \nabla a \times \boldsymbol{A}$$

the curl expressions in Maxwell's equations (eqns 12.1 and 12.2) become

$$\nabla \times \boldsymbol{H} \exp(-\gamma z) = \nabla \exp(-\gamma z) \times \boldsymbol{H} + \exp(-\gamma z)\nabla \times \boldsymbol{H}$$
$$= [-\gamma(\boldsymbol{z} \times \boldsymbol{H}) + \nabla \times \boldsymbol{H}] \exp(-\gamma z) \quad (12.67)$$

where $\boldsymbol{z}$ is the unit vector in the z-direction.

Then Maxwell's equations representing the fields in the perturbed waveguide are

$$\nabla \times \boldsymbol{H} - \gamma(\boldsymbol{z} \times \boldsymbol{H}) = j\omega\varepsilon\boldsymbol{E} \tag{12.68}$$

$$\nabla \times \boldsymbol{E} - \gamma(\boldsymbol{z} \times \boldsymbol{E}) = -j\omega\boldsymbol{\mu}\boldsymbol{H} \tag{12.69}$$

Taking the complex conjugate for all the quantities in the equivalent equations for the unperturbed waveguide gives

$$\nabla \times \boldsymbol{H}_0^* + \gamma_0^*(\boldsymbol{z} \times \boldsymbol{H}_0^*) = -j\omega\varepsilon_0\boldsymbol{E}_0^* \tag{12.70}$$

$$\nabla \times \boldsymbol{E}_0^* + \gamma_0^*(\boldsymbol{z} \times \boldsymbol{E}_0^*) = j\omega\mu_0\boldsymbol{H}_0^* \tag{12.71}$$

Take the scalar product of eqn. 12.68 with $\boldsymbol{E}_0^*$ and eqn. 12.69 with $-\boldsymbol{H}_0^*$ and similarly the product of eqn. 12.70 with $\boldsymbol{E}$ and eqn. 12.71 with $-\boldsymbol{H}$ and take the sum, which gives

$$\begin{aligned}\boldsymbol{E}_0^*\cdot(\nabla \times \boldsymbol{H}) - \boldsymbol{H}_0^*\cdot(\nabla \times \boldsymbol{E}) + \boldsymbol{E}\cdot(\nabla \times \boldsymbol{H}_0^*) - \boldsymbol{H}\cdot(\nabla \times \boldsymbol{E}_0^*)&\\ + \gamma\,[\boldsymbol{H}_0^*\cdot(\boldsymbol{z} \times \boldsymbol{E}) - \boldsymbol{E}_0^*\cdot(\boldsymbol{z} \times \boldsymbol{H})]&\\ + \gamma_0^*[\boldsymbol{E}\cdot(\boldsymbol{z} \times \boldsymbol{H}_0^*) - \boldsymbol{H}\cdot(\boldsymbol{z} \times \boldsymbol{E}_0^*)]&\\ = j\omega\varepsilon\boldsymbol{E}_0^*\cdot\boldsymbol{E} + j\omega\boldsymbol{\mu}\cdot\boldsymbol{H}_0^*\cdot\boldsymbol{H} - j\omega\varepsilon_0\boldsymbol{E}\cdot\boldsymbol{E}_0^* - j\omega\mu_0\boldsymbol{H}\cdot\boldsymbol{H}_0^*&\end{aligned}$$

Simplifying and using the vector identity in eqn. 12.15 gives

$$\begin{aligned}\nabla\cdot(\boldsymbol{H} \times \boldsymbol{E}_0^* + \boldsymbol{H}_0^* \times \boldsymbol{E}) + (\gamma - \gamma_0^*)\,[\boldsymbol{z}\cdot(\boldsymbol{E} \times \boldsymbol{H}_0^* + \boldsymbol{E}_0^* \times \boldsymbol{H})] &=\\ = j\omega[(\varepsilon - \varepsilon_0)\boldsymbol{E}\cdot\boldsymbol{E}_0^* + (\boldsymbol{\mu} - \mu_0)\cdot\boldsymbol{H}\cdot\boldsymbol{H}_0^*]&\end{aligned} \tag{12.72}$$

As in the cavity perturbation theory, we integrate both sides of this equation over a volume which consists of a length of the ferrite loaded waveguide bounded by two transverse sections of the waveguide. Using Gauss's theorem, as given in eqn. 12.18, the integral of the first expression in eqn. 12.72 becomes a surface integral. As discussed with the cavity, this integral is zero over the perfectly conducting walls of the waveguide. At the ends of the waveguide section, the contributions to the integral from each end are equal and opposite and will cancel out, since we are integrating field quantities which are defined to be invariant along the waveguide. Again, similarly to the perturbation theory of a cavity, the expression on the right-hand side of eqn. 12.72 is zero except in the region of the ferrite sample. Because the cross-section of the ferrite loaded waveguide is uniform, the volume integral is equivalent to an integral across the cross-section multiplied by a constant. Then the expression for the complex propagation constant becomes

$$\gamma + \gamma_0^* = \frac{j\omega \displaystyle\int_{\text{sample}} [(\boldsymbol{\chi}\cdot\boldsymbol{H})\cdot\boldsymbol{H}_0^* + \chi_e\boldsymbol{E}\cdot\boldsymbol{E}_0^*]\,dS}{\displaystyle\int_{\text{waveguide}} \boldsymbol{z}\cdot(\boldsymbol{E} \times \boldsymbol{H}_0^* + \boldsymbol{E}_0^* \times \boldsymbol{H})\,dS} \tag{12.73}$$

where it is assumed that the perturbation consists of inserting a perturbing body into an empty waveguide. If the perturbation is of a different nature, then the susceptibilities in eqn. 12.73 need to be replaced by a change in permeability and permittivity, or by the change in the magnetic and electric flux densities due to the perturbation in the region of the perturbing material. In eqn. 12.73 the integrals are across the cross-section of the ferrite sample and across the cross-section of the waveguide.

The condition for perturbation theory to be valid is that the ferrite causes only a small change in the propagating conditions in the equivalent empty waveguide, but, for most applications, a ferrite device is only useful if it causes appreciable change so that perturbation theory will not apply. In this book, it has been used for the theory of cavity measurements, and, in Section 6.8, to obtain an expression for the propagating conditions for a transversely magnetised ferrite slab in rectangular waveguide, such as would be used for the design of a high-power phase changer or a resonant isolator. However, the accuracy of an analysis involving perturbation theory for a ferrite in its resonance absorption condition is likely to be small, because resonance isolation by resonance absorption is not a small perturbation.

12.8 References

1 LAX, B. and BUTTON, K. J.: 'Microwave ferrites and ferrimagnetics' (McGraw-Hill, 1962), pp. 323–353

2 CLARRICOATS, P. J. B.: 'Microwave ferrites' (Chapman & Hall, 1961)

3 SLATER, J. C.: 'Microwave electronics' (Van Nostrand, 1950), p. 80

4 CASIMIR, H. B. G.: 'On the theory of electromagnetic waves in resonant cavities', *Philips Res. Rep.*, 1951, **6**, pp. 162–182

5 ARTMAN, J. O. and TANNENWALD, P. E.: 'Measurement of susceptibility tensor in ferrites', *J. Appl. Phys.*, 1955, **26**, pp. 1124–1132

6 WALDRON, R. A.: 'Theory of the measurement of the elements of the permeability tensor of a ferrite by means of a resonant cavity', *Proc. IEE*, 1957, **104B**, Suppl. No. 6, pp. 307–315

7 WALDRON, R. A.: 'Perturbation theory of resonant cavities', *Proc. IEE*, 1960, **107C**, pp. 272–274

8 WALDRON, R. A.: 'Ferrites' (Van Nostrand, 1961), pp. 98–117

9 HELSZAJN, J.: 'Principles of microwave ferrite engineering' (Wiley, 1969), pp. 52–57

10 WALDRON, R. A.: 'Electromagnetic fields in ferrite ellipsoids', *Brit. J. Appl. Phys.*, 1959, **10**, pp. 20–22

11 STEINERT, L. A.: 'Geometrical anisotropy of magnetic materials in waveguides and cavities', *J. Appl. Phys.*, 1959, **30**, p. 1109

12 LAX, B. and BUTTON, K. J.: 'Microwave ferrites and ferrimagnetics' (McGraw-Hill, 1962), pp. 157–162

13 STONER, E. C.: 'The demagnetizing factors of the general ellipsoid', *Phil. Mag.*, 1945, **36**, pp. 803–821

14 OSBORN, J. A.: 'Demagnetizing factors of the general ellipsoid', *Phys. Rev.*, 1945, **67**, pp. 351–357

15 BRUG, J. A. and WOLF, W. P.: 'Demagnetising fields in magnetic measurements', *J. Appl. Phys.*, 1985, **57**, pp. 4685–4701

16 SPENCER, E. G., LeCRAW, R. C. and REGGIA, F.: 'Measurement of microwave dielectric constants and tensor permeabilities of ferrite spheres', *Proc. IRE*, 1956, **44**, pp. 790–800
17 See, for example, BADEN FULLER, A. J.: 'Microwaves' 2nd edn. (Pergamon, 1979), pp. 88 and 122
18 ITO, A., OHKAWA, S. and YAMAMOTO, H.: 'On a measurement method of the tensor permeability of ferrite using Faraday rotation', *IEEE Trans.*, 1984, **IM-33**, pp. 26–31
19 SRIVASTAVA, C. M., PATNI, M. J. and HANUMANTHA RAO, N. S.: 'An exact solution technique for effective linewidth measurement', Ferrites, Proc. 3rd. Int. Conf. on Ferrites. Japan, 1980, pp. 824–826

Appendixes

Appendix 1: Physical constants

Velocity of light	$c = 2{\cdot}99792 \times 10^{8}$ m/s
Charge of the electron	$e = 1{\cdot}60219 \times 10^{-19}$ C
Mass of the electron	$m = 9{\cdot}10956 \times 10^{-31}$ kg
Planck's constant	$h = 6{\cdot}6262 \times 10^{-34}$ J s
Bohr magneton	$\beta = 9{\cdot}2741 \times 10^{-24}$ A m^2
Permeability constant	$\mu_0 = 4\pi \times 10^{-7} = 1{\cdot}256637 \times 10^{-6}$ H/m
Permittivity constant	$\varepsilon_0 = \dfrac{1}{c^2\mu_0} = 8{\cdot}854188 \times 10^{-12}$ F/m
Euler's constant	$\gamma = 0{\cdot}577215665$

Appendix 2: SI and Gaussian units

The SI system of units, i.e. rationalised practical MKS units, are used in this book. However, the Gaussian or c.g.s. emu units are often used for magnetic quantities, thus a note about the SI units are included here. The defining SI units are the metre, kilogram, second and ampere. The commonest derived mechanical units are the newton, watt and joule for force, power and energy, respectively. The electromagnetic field quantities and their units are

E = electric field intensity, V/m

D = electric flux density, C/m^2

J = current density, A/m^2

H = magnetic field intensity, A/m

B = magnetic flux density, Wb/m^2 = T

M = magnetisation, which is specified in units of magnetic flux density, T, in this book, for reasons given in Section 1.3, although SI recommend that magnetisation is specified in units of magnetic field intensity

Commonly magnetic quantities are specified in Gaussian units, H in oersteds and magnetisation as $4\pi M$ in the same units as B in gauss. Then the appropriate conversion factors are

$$B(\mathrm{G}) \times 10^{-4} = B(\mathrm{T})$$

$$4\pi M(\mathrm{G}) \times 10^{-4} = M(\mathrm{T})$$

$$H(\mathrm{Oe}) \times 79{\cdot}5 = H(\mathrm{A/m})$$

In the Gaussian system of units, $\mu_0 = 1$; in a vacuum, B and H are numerically identical, which can cause a lot of loose thinking about magnetic units.

Appendix 3: Notation

A	arbitrary constant; a parameter
$\mathbf{A}$	arbitrary vector; area
a	constant; broad dimension of rectangular waveguide
$\boldsymbol{a}$	input wave amplitude
B	arbitrary constant
B_1	Bolle function
$\mathbf{B}$, B	magnetic flux density
b	constant; narrow dimension of rectangular waveguide
$\boldsymbol{b}$	output wave amplitude
C	arbitrary constant
c	constant; length of a waveguide resonant cavity
c	speed of light
c	*as subscript*: cavity; critical; cut-off
D	arbitrary constant
$\mathbf{D}$	electric flux density
d	thickness
d	differential operator
E	arbitrary constant
$\mathbf{E}$, E	electric field intensity
e	charge of the electron
e	*as subscript*: effective; electric; empty
e	*as superscript*: effective; external
F	function
f	*as subscript*: ferrite
g	Landé spectroscopic splitting factor
H_0	static magnetic field intensity
H_1	Heller function

$\boldsymbol{H}$, H magnetic field intensity

h height

h Planck's constant

I electric current

$\boldsymbol{I}$ unit matrix

i *as subscript*: internal

$\boldsymbol{J}$ angular momentum

J_n Bessel function of the first kind

j $\sqrt{-1}$

K arbitrary constant

k wavenumber

L arbitrary constant

l length

M magnetic moment

M_0 static magnetisation

$\boldsymbol{M}$, M magnetisation

m constant

$\boldsymbol{m}$ magnetic moment

m mass of the electron

N demagnetisation factor

n integer – order of the Bessel functions

P_1 modified Bolle function

p constant

Q Q-factor of resonator

Q_1 modified Heller function

r a radial distance; radius; radial component in cylindrical co-ordinate system

r *as subscript*: relative

S element of scattering matrix

$\boldsymbol{S}$ scattering matrix

s eigenvalue

s *as subscript*: sample

T transfer matrix

T *as superscript*: transpose

t time; elements in the transfer matrix

t *as subscript*: transverse

$\boldsymbol{u}$ eigenvector

V volume

v voltage; volume

x a dimension in the rectangular co-ordinate system

Y_n Bessel function of the second kind

y a dimension in the rectangular co-ordinate system

Z line impedance

Z_0 wave impedance

z	a dimension in the rectangular co-ordinate system and in the cylindrical co-ordinate system
$\mathbf{z}$	unit vector in the z-direction
α	damping constant
β	Bohr magneton; propagation constant
γ	gyromagnetic ratio; Euler's constant
δ	small increment
ε	permittivity
ε_0	permittivity constant, permittivity of free space
η	impedance of free space
θ	angle; angular component in cylindrical co-ordinate system; rotation
κ	cross-diagonal component of the tensor permeability
λ	damping constant; wavelength
λ_0	characteristic wavelength
μ	permeability; diagonal component of the tensor permeability
μ_0	permeability constant, permeability of free space
$\boldsymbol{\mu}$	permeability tensor
ϕ	an angle; phase change
χ	susceptibility; diagonal component of the tensor susceptibility
χ_e	electric susceptibility
$\boldsymbol{\chi}$	susceptibility tensor
ψ	magnetic flux; rotation per unit length
ω	angular velocity; angular frequency of an alternating quantity
ω_0	resonant angular frequency
Δ	small increment
ΔH_k	spinwave linewidth
∇	differential operator
∂	partial differential operator
$\parallel$	*as superscript*: parallel
$\perp$	*as superscript*: perpendicular

Name Index

Subject Index

www.ingramcontent.com/pod-product-compliance
Lightning Source LLC
LaVergne TN
LVHW011753200926
848484LV00003B/687
9780863410642